RESTORING CREATION

Nature and Environment in the Middle Ages

ISSN: 2399-3804 (Print)
ISSN: 2399-3812 (Online)

Questions of nature, the environment and sustainability are increasingly important areas of scholarly enquiry in various fields. This exciting new series aims to provide a forum for new work throughout the medieval period broadly defined (c.400–1500), covering literature, history, archaeology and other allied disciplines in the humanities. Topics may range from studies of landscape to interaction with humans, from representations of 'nature' in art to ecology, ecotheory, ecofeminism and ecocriticism; monographs and collections of essays are equally welcome.

Proposals or enquiries may be sent directly to the series editor or to the publisher at the addresses given below.

Dr Michael D.J. Bintley, Department of English and Humanities, School of Arts, Birkbeck, University of London, 43 Gordon Square, London, WC1H 0PD

Boydell & Brewer, PO Box 9, Woodbridge, Suffolk, IP12 3DF

Previously published:

1: *The Natural World in the Exeter Book Riddles*, Corinne Dale
2: *Birds in Medieval English Poetry: Metaphors, Realities, Transformations*, Michael J. Warren

RESTORING CREATION

The Natural World in the Anglo-Saxon Saints' Lives of Cuthbert and Guthlac

Britton Elliott Brooks

D. S. BREWER

First published 2019
D. S. Brewer, Cambridge

ISBN 978 1 84384 530 0

D. S. Brewer is an imprint of Boydell & Brewer Ltd
PO Box 9, Woodbridge, Suffolk IP12 3DF, UK
and of Boydell & Brewer Inc.
668 Mt Hope Avenue, Rochester, NY 14620–2731, USA
website: www.boydellandbrewer.com

A catalogue record for this book is available
from the British Library

This publication is printed on acid-free paper

Printed and bound in Great Britain by
TJ International Ltd, Padstow, Cornwall

MIX
Paper from
responsible sources
FSC® C013056
FSC www.fsc.org

CONTENTS

ACKNOWLEDGEMENTS

As with all great adventures, this book could not have been accomplished without the aid of many people. First, I would like to thank my mother, from whom I first learnt of the British Isles and the stories that arose in their fields and mountains, and my father, whose adaptability in the face of change nurtured the flexibility of mind necessary for such an endeavour. I thank too my gracious guides in the shaping and completion of this book, including Malcolm Godden, Annie Sutherland, Winfried Rudolf, Mark Atherton, Daniel Anlezark, and Andy Orchard. The greatest debt rests with Francis Leneghan, whose direction, support, and encouragement saw this project to its conclusion. I would also like to thank Caroline Batten for her thorough and expert proofreading, and Caroline Palmer for her constant support. Every word on these pages was enriched by my fellow Oxonian wanderers of the medieval world, including Stefany Wragg, Hannah Bailey, Michael Hart, Helen Appleton, and Daniel Thomas. I am thankful for the continued support of my brother Cody Brooks, my companion in adventures of mind and foot. I am also indebted to Max Baker-Hytch, Richard Park, Daniel Orton, Kazutomo Karasawa, Koichi Kano, and Hironori Suzuki. Finally, I dedicate this work to my wife, Michelle Brooks, without whom none of this would have been possible, and our son and future scholar, Aidan Takeru Brooks.

ABBREVIATIONS

ASE	*Anglo-Saxon England*
Bede, *HE*	*Bede's Ecclesiastical History of the English People*, ed. Bertram Colgrave and Roger A. B. Mynors, Oxford Medieval Texts (Oxford: Clarendon Press, 1969).
BT	*An Anglo-Saxon Dictionary Based on the Manuscript Collections of the Late Joseph Bosworth*, ed. by T. Northcote Toller (Oxford: Clarendon Press, 1898); *Supplement* ed. by T. Northcote Toller (Oxford: Clarendon Press, 1921); *Revised and Enlarged Addenda*, ed. by Alistair Campbell (Oxford: Oxford University Press, 1972)
CCSL	Corpus Christianorum Series Latina, 201 vols (Turnhout: Brepols, 1953–2014). References are given to volume and page number(s)
CSEL	Corpus Scriptorum Ecclesiasticorum Latinorum, 101 vols (Vienna: Austrian Academy of Sciences, 1866–2015). References are given to volume and page number(s)
DMLBS	*Dictionary of Medieval Latin from British Sources Online* (Brepols) http://clt.brepolis.net/dmlbs/Default.aspx
DOE	*Dictionary of Old English: A to I online*, ed. by Angus F. Cameron, Ashley Crandell Amos, and Antonette diPaolo Healey (pub. online, 2007) http://tapor.library.utoronto.ca/doe/dict/index.html

Fontes	*Fontes Anglo-Saxonici Project*, ed., *Fontes Anglo-Saxonici: World Wide Web Register*, http://fontes.english.ox.ac.uk/
JEGP	*Journal of English and Germanic Philology*
MÆ	*Medium Ævum*
MGH	Monumenta Germaniae Historica
OE Corpus	*Dictionary of Old English Web Corpus*, compiled by Antonette diPaolo Healey with John Price Wilkin and Xin Xiang (Toronto: Dictionary of Old English Project, 2009) http://tapor.library.utoronto.ca/doe/
OEPG	*Das angelsächsische Prosa-Leben des hl. Guthlac*, ed. Paul Gonser, Anglistische Forschungen 27 (Heidelberg: Carl Winters Universitätsbuchhandlung, 1909)
PL	*Patrologia Latina*, ed. Jacques-Paul Migne, 221 vols (Paris: J-P. Garnier Frères, 1841–55). References are given to volume and column number(s)
RES	*Review of English Studies*
VCA	Anonymous, *Vita Sancti Cuthberti*, in *Two Lives of Saint Cuthbert: a Life by an Anonymous Monk of Lindisfarne and Bede's Prose Life*, ed. Bertram Colgrave (Cambridge: Cambridge University Press, 1940)
VCM	Bede, *Vita Sancti Cuthberti* (Metrical), in *Bedas metrische Vita Sancti Cuthberti*, ed. Werner Jaager, Palaestra 198 (Leipzig: Mayer and Müller, 1935)
VCP	Bede, *Vita Sancti Cuthberti* (Prose), in *Two Lives of Saint Cuthbert: a Life by an Anonymous Monk of Lindisfarne and Bede's Prose Life*, ed. Bertram Colgrave (Cambridge: Cambridge University Press, 1940)
VSG	*Felix's Life of Saint Guthlac*, ed. Bertram Colgrave (Cambridge: Cambridge University Press, 1956).

Introduction

EVERY HOUR CHIMES with a new example of ecological crisis: the warming oceans, the loss of biodiversity, and the rise of anti-environmental public policies. In response to our contemporary moment, the Humanities have begun to engage in earnest with questions of ecology. This present study seeks to bring medieval literature into dialogue with these issues, analysing medieval constructions and interpretations of the non-human world as expressed in literature, by considering them in their historical context. This approach highlights how medieval peoples actively reflected upon their own engagement with the non-human world, structured in great part by their theology and philosophy, and articulated them through the artistry of their literature.

Restoring Creation: the Natural World in the Anglo-Saxon Saints' Lives of Cuthbert and Guthlac[1] engages with the growing interest throughout medieval scholarship in the environmental humanities, evidenced by the number of monographs published in the past few years on such topics, including *Water in Medieval Literature* by Albrecht Classen; *Anglo-Saxon Literary Landscapes: Ecotheory and the Environmental Imagination*, by Heidi Estes; and *Inhabited Spaces: Anglo-Saxon Constructions of Place*, by Nicole G. Discenza.[2] This engagement by medieval scholars is heartening, as the majority of the studies in the environmental humanities, in Estes's words, 'dismiss or ignore the medieval, or misrepresent it in discussions of the modern'.[3] This is most evident in the wider, and erroneous, conceptions of the negative role of the natural world in medieval literature, particularly in relation to ecocritical scholarship. For example, Timothy Morton, one of the leading ecocritical theorists, describes the natural world in medieval texts in negative

1 I use capitalised 'Creation' to refer to the physical world created during the Genesis narrative throughout this study.

2 A. Classen, *Water in Medieval Literature: An Ecocritical Reading* (Lanham, MD, 2017); H. Estes, *Anglo-Saxon Literary Landscapes: Ecotheory and the Environmental Imagination* (Amsterdam, 2017); N. G. Discenza, *Inhabited Spaces: Anglo-Saxon Constructions of Place* (Toronto, 2017).

3 Estes, *Anglo-Saxon Literary Landscapes*, p. 10.

and dismissive terms: 'Nature, practically a synonym for evil in the Middle Ages, was considered the basis of social good by the Romantic Period.'[4] This oversimplification of medieval literature is unfortunate, to say the least, and fundamentally distorting, both for the project of literary analysis as a whole, and also for broader discussions in the environmental humanities. The negative vision of the relationship between medieval people and the natural world also appears in more popular arenas, as evidenced by the scholar and writer Alexandra Harris. In her 2015 book *Weatherland: Writers and Artists under English Skies*, she characterises the Anglo-Saxon engagement with nature as follows: 'The impulse of this culture is to favour the controlled, man-made, and essentially social space of the hall – lit by fire and candle [...] The outdoor winter world is dreadful by contrast.'[5] The non-human world was for Anglo-Saxons, according to Morton and Harris, evil, antagonistic, and often utilised as a negative template by which the anthropocentric positive could be defined. Connected but more nuanced views are also found in the work of medieval scholars, such as Jennifer Neville, who categorises the representation of the natural world in Old English poetry as fundamentally unconcerned with the natural world itself, and primarily anthropocentric:

> What emerges is that the representation of the 'natural world' is never an end in itself and is always ancillary to other issues [...] the state of humanity and its position in the universe, the establishment and maintenance of society, the power of extraordinary individuals, the proximity of the deity to creation and the ability of writing to control and limit information.[6]

More recent medieval scholarship has called these views into question. Estes notes not only the multitude of relationships Anglo-Saxons conceived of themselves, 'the land and its nonhuman creatures', but also highlights how even the *Anglo-Saxon Chronicle*, a text anthropocentric in its very structure, 'contains

4 T. Morton, *Ecology without Nature: Rethinking Environmental Aesthetics* (Cambridge, MA, 2007), p. 15.

5 A. Harris, *Weatherland: Writers and Artists under English Skies* (London, 2015), p. 28.

6 J. Neville, *Representations of the Natural World in Old English Poetry* (Cambridge, 1999), p. 18.

numerous references to natural phenomena separate from the mention of human endeavors, suggesting that they are of intrinsic interest'.[7] Corinne Dale has provided a recent and convincing corrective to such anthropocentric readings of Anglo-Saxon literary depictions of the non-human in her 2017 monograph, *The Natural World in the Exeter Book Riddles*, where she carefully argues that 'there is a programme of resistance to anthropocentrism at work in the riddle collection, whereby the riddles challenge human-centered ways of depicting the created world'.[8] This present study continues in this vein, and argues that early medieval constructions of the natural world were neither as negative nor as monolithic as is often argued, and that instead they reveal a sophisticated and considered engagement with the non-human world.

Where this study differs from the approaches taken by scholars like Estes is in its central argument that the relationship between humanity and the non-human world in Anglo-Saxon texts was defined in great part by contemporary theological and philosophical views. I therefore focus less on employing modern ecocritical theory, as Estes does in her book, and more on the manner in which environmental concerns would have been perceived from multiple medieval perspectives. This is an important distinction because often, in our pursuit of understanding the texts we love, we substitute our own views and theoretical perspectives for those expressed in those texts. Discenza begins her book *Inhabited Spaces* by highlighting this very danger:

> As anyone who studies the past knows, it is all too easy to import modern modes of thought into earlier eras. We tend to assume, often unconsciously, that people think as we do. While we can never entirely leave ourselves behind, focused study can help us to identify our preconceptions and distinguish others', so that we recognize where they share our ideas and where they differ.[9]

7 Estes, *Anglo-Saxon Literary Landscapes*, p. 13, notes the inclusion of 'astronomical phenomena such as comets (678, 892, 905, 995), lunar and solar eclipses (744, 773, 806, 809, 904), and, perhaps, the aurora borealis (926, 979)', where the descriptions are not explicitly tied to human concerns.

8 C. Dale, *The Natural World in the Exeter Book Riddles* (Cambridge, 2017), p. 2.

9 Discenza, *Inhabited Spaces*, p. 3.

Restoring Creation: the Natural World in the Anglo-Saxon Saints' Lives of Cuthbert and Guthlac seeks to understand how the Anglo-Saxons themselves conceptualised their relationship with the natural world, and then how those conceptions shaped their literary productions. This study also differs in its subject matter: hagiography. While at first Saints' *Lives* might seem an odd choice, as they are some of the most conventional of all medieval literary genres,[10] they are also some of the most connected to the physical landscape in which the Anglo-Saxons lived. As will be discussed below, early Anglo-Saxon hagiography was concerned with elevating particular saints from very specific locations in the English landscape, most often for the purpose of establishing and expanding their cult. These cult sites commonly became important religious centres, and as such the depictions of the natural world, whether it be elemental features like rivers and dales, mountains and forests, or animate creatures like birds, otters, and seals, are deeply connected to the physical reality which the Anglo-Saxons would have known. This inherent connectedness to the lived experience of these early medieval people allows a unique window into the ways they perceived their relationship with the non-human world, and how they chose to depict that relationship in their literary endeavours.

The Restoration of Creation: Received Exegesis

At the heart of this study of the early Anglo-Saxon *vitae* of saints Cuthbert and Guthlac is a specific exegetical interpretation of the Fall and its effects upon humanity and Creation. This interpretation argues that the indifferent Creation of the Fall can be restored to prelapsarian harmony with humanity by way of sanctity. Such a connection between medieval depictions of the natural world and Creation, Eden, and the Fall, has been noted by various scholars,

10 See M. Lapidge, 'The Saintly Life in Anglo-Saxon England', in *The Cambridge Companion to Old English Literature,* 2nd edn, ed. M. Godden and M. Lapidge (Cambridge, 2013), pp. 251–72, at 262, who notes how it 'was the overall intention of any hagiographer to demonstrate that his saintly subject belonged indisputably to the universal community of saints, and this entailed modelling each *vita* closely on those of earlier authors'.

including Catherine Clarke in her book *Literary Landscapes and the Idea of England, 700–1400*, particularly in connecting the classical topos of the *locus amoenus* with the biblical vision of Eden.[11] Most recently, and in direct connection to the *vitae* of Cuthbert and Guthlac which are the focus of this study, Sally Shockro notes how '[t]he affinity between man and nature was ruptured by Adam's fall, but it was not eternally broken [...] Through piety and purity, Cuthbert and Guthlac have repaired the breach, in themselves, between man and nature, and therefore are able to take their rightful place within the natural world'.[12] Yet Shockro provides little historical, theological, and exegetical context, and subsumes this into her primary argument highlighting the agency and mutuality of animals in these *vitae*. The restoration of Creation has its roots, however, in biblical and patristic eschatology.[13] Acts 3:21 describes the second coming of Jesus Christ as bringing about a restoration of all things: 'Quem oportet quidem caelum suscipere usque in tempora *restitutionis omnium*' (Whom heaven indeed must receive, until the times of the *restitution of all things*').[14] This was interpreted by Paul in Rom. 8:20–21, and by a number of the Church Fathers, as a reference to Creation (sometimes translated as 'creature' here) being restored to its prelapsarian state:[15]

11 See C. A. M. Clarke, *Literary Landscapes and the Idea of England, 700–1400* (Cambridge, 2006), esp. pp. 7–36.

12 S. Shockro, 'Saints and Holy Beasts: Pious Animals in Early-Medieval Insular Saints' *Vitae*', in *Animal Languages in the Middle Ages: Representations of Interspecies Communication*, ed. A. Langdon (Basingstoke, 2018), pp. 51–68, at 65.

13 It should be noted that Shockro, 'Saints and Holy Beasts', concludes her article with an eschatological turn, how 'these stories perhaps present a more complicated statement on the advancement of Christian history than a simple desire to return to a time before the Fall [...] To a pious contemporary reader this might be a heartening sign of the imminent completion of Christian history, in which creation is once again unified under the Creator', pp. 65–66.

14 All translations in this thesis are my own, in consultation with relevant editions, unless otherwise indicated. When I have used translations from standard editions, I have silently emended some features for the sake of consistency, such as capitalising 'Creation' throughout (as explained above), as well as orthographic features, e.g. today for to-day. All translations of the Bible are from the *Holy Bible Douay-Rheims Version, with Challoner Revisions 1749–52* (Baltimore, 1899), likewise silently emended for archaic features, e.g. walks for walketh.

15 Most modern translations render *creaturae* as 'creation' rather than 'creature' (*Douay-Rheims*), in great part due to the exegesis of the Church Fathers, which

> Nam exspectatio creaturae revelationem filiorum Dei exspectat. Vanitati enim creatura subjecta est non volens, sed propter eum, qui subjecit eam in spe: Quia et ipsa creatura liberabitur a servitute corruptionis in libertatem gloriae filiorum Dei.
>
> (For the expectation of the Creation waits for the revelation of the sons of God. For the Creation was made subject to vanity, not willingly, but by reason of him that made it subject, in hope: because the Creation also itself shall be delivered from the servitude of corruption, into the liberty of the glory of the children of God.)

While several patristic texts interpret these verses in a similar manner, the most relevant here, given its influence and availability in Anglo-Saxon England, is St Ambrose's *Exameron*:[16] 'quod Apostolicae quoque liceat astruere auctoritatis exemplo. scriptum est enim [Rom. 8:20] [...] liberabitur autem et ipsa creatura a seruitute corruptionis, cum gratia diuinae remunerationis adfulserit'[17] ('For this belief one may find authority also in the words of the Apostle. It is written [...] however, even Creation itself will also be delivered from its slavery to corruption when the grace of divine reward has shown forth'). Jerome likewise interprets this restoration (*restitutio*) as an eschatological vision of the return to an Edenic natural world: 'Quando autem filii Dei adsumpti fuerint, et ipsa creatura de hoc seruitio liberatur'[18] ('When, moreover, the sons of God shall have attained glory, Creation itself will also be delivered from this slavery'). This restoration of Creation, however, is not only relegated to an eschatological future, but is also glimpsed and made manifest in

I discuss below. I have therefore emended above. See, for example, the *Holy Bible: New Revised Standard Version Catholic Edition* (Oxford, 1999); the *NKJV Study Bible* (Nashville, 2016); the *New English Bible* (Oxford, 1970); and *The Holy Bible: New International Version* (London, 2011).

16 Ambrose's *Exameron* is extant in four MSS, though most date from the tenth century or later; see *Anglo-Saxon Manuscripts: a Bibliographical Handlist of Manuscripts and Manuscript Fragments Written or Owned in England up to 1100*, ed. H. Gneuss and M. Lapidge (Toronto, 2014), no. 20, 61.5, 194, and 778 (all references from this text will use the MS numbers allocated by the editors). The text was clearly known earlier, however, as evidenced by Bede's seventeen uses in his *In Genesim*; see M. Lapidge, *The Anglo-Saxon Library* (Oxford, 2006), p. 194.

17 Ambrose, *Exameron* in *Sancti Ambrosii Opera pars prima, qua continentur libri*, ed. C. Schenkl, CSEL 32.1 (Vienna, 1896), p. 20.

18 Jerome, *Tractatus sive Homiliae in Psalmos, in Marci evangelium, aliaque Varia Argumenta*, ed. G. Morin, CCSL 78 (Turnhout, 1958), p. 344.

the present via the power of saints. The specific exegesis that supports this reading, and which is most relevant for the Anglo-Saxon authors analysed in this study, is that of St Augustine of Hippo, later adapted by the Venerable Bede. The influence of Augustine's and Bede's shared exegesis is evident in the *vitae* of Cuthbert and Guthlac that are the focus of this study, primarily by the way the Anglo-Saxon authors incorporated, often nearly verbatim, sections of the exegesis in all but one of the Latin texts; in the single outlier (the *Anoymous Vita S. Cuthberti*) the underlying influence of the Augustinian interpretation can be discerned from its structure and content. In the case of the OE texts, the exegesis is evident in both the vernacular translation of the Latin sections, as well as in its influence over structure and content.

Patristic Exegesis and Anglo-Saxon Literature

The use of patristic exegesis in interpretations of Anglo-Saxon literature, particularly texts written in Old English, remains contentious. The debate has its roots in Bernard Huppé's 1959 *Doctrine and Poetry: Augustine's Influence on Old English Poetry,* which employed Augustinian exegesis to interpret, among other texts, *Cædmon's Hymn* and *Genesis A*.[19] Huppé viewed Augustine's *De doctrina Christiana* as formulating a 'Christian theory of literature' which provided 'the basic program for a Christian culture', and influenced 'the early practice of poetry in the vernacular, specifically Old English'.[20] The force of Huppé's conviction, however, often led to an indiscriminate application of patristic exegesis, resulting in a number of distortions. Morton Bloomfield, for example, highlights how Huppé's suggestion that the parallel phrases '*Meotodes meahte, his modgeþanc,* and *weorc Wuldorfæder*' in *Cædmon's Hymn* 'may suggest the Trinity' relies on Huppé translating *modgeþanc* as 'wisdom', which Bloomfield argues can only be translated thus 'by a great

[19] B. F. Huppé, *Doctrine and Poetry: Augustine's Influence on Old English Poetry* (New York, 1959).

[20] Huppé, *Doctrine and Poetry,* p. v.

freedom'.[21] Bloomfield and later scholars did not, however, reject the notion of using patristic exegesis as an interpretative tool, but argued for a more measured approach. Judith Garde and Bernard Muir, for example, advocate an approach based on 'the necessary concurrence of *appropriate* patristic analogues and close examination of a text' rather than Huppé's 'indiscriminate imposition', primarily via the liturgy, which they argue is the most profitable avenue for patristic influence on OE poetry.[22] More recently critics such as Larry McKill and Nina Boyd have further argued against Huppé's open application of patristic exegesis.[23] McKill, for example, highlights how Huppé's analysis of *Genesis A* distorts 'the poem's own explicit theme' by its 'exegetical imposition'.[24] What most recent critics agree on is the measured and appropriate application of patristic exegesis, whether via the liturgy, or demonstrable contemporary knowledge of such exegesis.[25] This study will follow in a similar measured vein, and will argue the application of the specific Augustinian and Bedan exegesis is justified for the following reasons: first, the majority of the texts analysed in this thesis are in Anglo-Latin, where the influence of patristic exegesis is much easier to identify; second, as mentioned above, there is a section of the exegesis incorporated nearly verbatim in all the Latin texts except for the *VCA*; third, the OE texts are necessarily influenced by the exegesis as they are translations and adaptations of the Anglo-Latin *vitae*.

21 M. W. Bloomfield, 'Patristics and Old English Literature: Notes on Some Poems', *Comparative Literature* 14 (1962), 36–43, at 41.

22 J. N. Garde and B. J. Muir, 'Patristic Influence and the Poetic Intention in Old English Religious Verse', *Literature and Theology* 2 (1988), 49–68, at 49, 58.

23 L. N. McKill, 'Patterns of the Fall: Adam and Eve in the Old English *Genesis A*', *Florilegium* 14 (1995–96), 25–41; N. Boyd, 'Doctrine and Criticism: A Revaluation of *Genesis A*', *Neuphilologische Mitteilungen* 83 (1982), 230–38. See also C. D. Wright, '*Genesis A ad litteram*', in *Old English Literature and the Old Testament*, ed. M. Fox and M. Sharma (Toronto, 2012), pp. 121–71, for a reading which emphasises the literal, rather than allegorical, meaning of the text.

24 McKill, 'Patterns of the Fall', 26.

25 Garde and Muir, 'Patristic Influence', 58; McKill, 'Patterns of the Fall', 26–27.

Saint Augustine of Hippo's Exegesis

Augustinian exegesis is often difficult to delineate, as his ideas evolved throughout his forty-year exegetical career.[26] This included five attempts at explicating the Genesis narrative: *De Genesi contra Manichaeos* (388–90); *De Genesi ad litteram imperfectus liber* (c.393); the last three books of the *Confessiones* (397–400); *De Genesi ad litteram* (404–15); and book eleven of *De civitate Dei* (417–18).[27] The result of this development in Augustine's thinking about Creation is helpfully summarised by Karla Pollmann, who argues that Augustine has two distinct, though related, positions concerning the effects of the Fall on Creation in Genesis: an early pessimistic view, which gives way to a more optimistic outlook as his thinking matured.[28] In Augustine's early *De Genesi contra Manichaeos*, he perceives the relationship between Creation and humanity through Classical and Hebraic notions of a pessimistic cosmos that starts with a Golden Age of Paradise that is distorted, and which subsequently declines. Augustine's earlier vision understands Creation itself to be transformed by the actions of humans, and that any inimical elements within it are a consequence of those actions.[29] In addition to these classical and Jewish influences, Augustine was also influenced in his early perception by Christian exegetes like Basil the Great and St Ambrose who posit that, though Paradise was itself cursed in the Fall, Creation is now used by God for the development of humanity towards redemption.[30] It is this vision of Creation which shapes Augustine's *De Genesi contra Manichaeos*, evidenced by his interpretation of the critical verse Gen. 3:18, 'spinas et tribulos germinabit tibi et comedes herbas terrae' ('[t]horns and thistles shall it bring

[26] Numerous scholars have noted the evolution of Augustine's ideas. For example, H. Reuling, *After Eden: Church Fathers and Rabbis on Genesis 3:16–21* (Leiden, 2006), p. 159.

[27] For a summary see St Augustine, *On Genesis: Two Books on Genesis against the Manichees and On the Literal Interpretation of Genesis: An Unfinished Book*, trans. R. J. Teske (Washington, DC, 1990), p. 3.

[28] K. Pollmann, 'Human Sin and Natural Environment: Augustine's Two Positions on Genesis 3:18', *Augustinian Studies* 41 (2010), 69–85, at 70.

[29] Pollmann, 'Two Positions', 71.

[30] Pollmann, 'Two Positions', 71.

forth to you: and you shall eat the herbs of the earth'),[31] as portraying a harmonious Creation that only becomes inimical to humanity after the Fall: 'Ergo dicendum est, quod per peccatum hominis terra maledicta sit, ut spinas pareret' ('Therefore it should be said, that through the sin of man the earth was cursed, so as to bring forth thorns').[32] As noted by Hanneke Reuling, Augustine likewise follows the exegeses of Basil and Ambrose in interpreting this punishment as functioning to push humanity towards redemption:[33]

> sed ut peccati humani crimen semper hominibus ante oculos poneret, quo admonerentur aliquando averti a peccatis et ad dei praecepta converti.[34]
>
> (But so that it should always place before the eyes of men the judgment of human sin, whereby they might from time to time be admonished by it to turn away from their sins and back to God's commandments.)

Augustine is not completely convinced by his own exegesis, however, and is incapable of reconciling the prelapsarian Paradise described in Gen. 1:24–25, where God declares that all the beasts (including reptiles) he made are good, with the existence of seemingly useless creatures, including flies and worms.[35] Augustine resolves this paradox by joining these hardships with the exegesis above, where the creatures can function in the same manner as the thorns in Gen. 3: to help push humanity towards redemption.

In Augustine's later work, *De Genesi ad litteram*, the uncertainty expressed in his earlier commentary is replaced by a fully formed and consistent exegesis, and his central interpretation of the Fall has shifted: where before Creation was cursed by the Fall, Augustine here argues that Creation is wholly good, suffering no transformational effects. In his commentary on the same verse, Gen. 3:18,

31 All biblical quotations taken from *Biblia Sacra iuxta Vulgatam versionem*, ed. R. Weber and R. Gryson, 5th edn (Stuttgart, 2007).

32 Augustine, *De Genesi contra Manichaeos*, ed. D. Weber, CSEL 91 (Vienna, 1998), p. 85; see also Pollmann, 'Two Positions', 72–73.

33 See Reuling, *After Eden*, p. 178.

34 Augustine, *De Genesi contra Manichaeos*, p. 85.

35 Augustine, *De Genesi contra Manichaeos*, pp. 92–93; see Pollmann, 'Two Positions', 75.

Augustine reinterprets the emergence of thorns to represent not transformation on the part of Creation, but instead an ontological separation based on the fallen state of humanity:

> *Spinas et tribulos pariet tibi,* ut haec etiam antea terra pariens non tamen homini pareret ad laborem, sed cuiusque modi animalibus conuenientem cibum [...] tunc autem coeperit ista homini parere ad aerumnosum negotium, cum post peccatum coepit in terra laborare.[36]
>
> (*It will bring forth thorns and spiny plants for you,* this may also [be interpreted] that the earth brought them forth before, not to submit man to work, but as fitting food for animals of this kind [...] then, however, that began to subject man to wretched work, when after sin he began to labour on the earth.)

Reuling notes how Augustine's interpretation centres on the pronoun *tibi,* where the thorns and thistles were always present in Creation, but have now 'changed their purpose [...] before sin the thorns and thistles grew for other reasons, and after sin, they were produced *for Adam*'.[37] Charles Mathewes likewise argues that Augustine here sees Creation as inherently good: 'for reasons both Scriptural and metaphysical, Augustine holds that it is the quality of orientation to that world that is the source of our malady'.[38] In Augustine's mature exegesis it is our relational orientation that has been distorted by the Fall, not Creation itself. The harmonious or antagonistic relationship between Creation and humanity depends on, in Pollmann's words, 'the measure of agreement ("congruentia") with deficient (i.e., post-lapsarian) human nature'.[39] The focus of the exegesis is entirely anthropocentric, and Creation retains its status of good regardless of what our perception of it may be. Augustine has re-sanctified Creation and allowed the problem of natural evil to be seen much as the problem of moral evil: a non-entity based purely upon humanity's fallen state of being in relation to God

36 Augustine, *Sancti Aureli Augustini De Genesi ad litteram libri duodecim,* ed. J. Zycha, CSEL 28.1 (Vienna, 1895), p. 84.

37 Reuling, *After Eden,* p. 196.

38 C. Mathewes, 'A Worldly Augustinianism: Augustine's Sacramental Vision of Creation', *Augustinian Studies* 41 (2010), 333–48, at 341.

39 Pollmann, 'Two Positions', 83; See also, Augustine, *De moribus Ecclesiae Catholicae et De moribus Manichaeorum libri duo,* ed. J. B. Bauer, CSEL 90 (Vienna, 1992), Bk II.17, p. 102.

and Creation.[40] This disharmony with nature is therefore dependent upon the level of moral and spiritual holiness of the individual human and, as Pollmann argues, the inimical relational position between humanity and Creation can be sometimes reversed.[41]

The Venerable Bede's Exegesis

Bede's *In Genesim* is an early text in his career, composed sometime between 717 and 725, which places it after the *VCM* (705–16), and either concurrent with or slightly before his *VCP* (c.721).[42] In the text Bede's interpretation of the Fall in the pivotal Gen. 3:18 verse draws heavily on Augustine's *De Genesi contra Manichaeos*:[43]

> Per peccatum enim hominis terra maledicta est, ut spinas pareret, non ut ipsa poenas sentiret quae sine sensu est, sed ut peccati humani crimen semper hominibus ante oculos poneret, quo admonerentur aliquando auerti a peccatis et ad Dei precepta conuerti.[44]
>
> (For through the sin of man the earth was cursed, so that it brings forth thorns, not so that the earth itself, which is without sense, would feel the punishments, but so that it should always place before the eyes of men the judgment of human sin, whereby they might from

[40] This is a reference to Augustine's argument for evil as a non-entity. For example, D. X. Burt, 'Courageous Optimism: Augustine on the Good of Creation', *Augustinian Studies* 21 (1990), 55–66, at 56, summarises the argument: 'Augustine finally came to understand that evil is not a substance as the Manichaeans claimed. It is not a "thing" at all. Quite the opposite, it is a "no-thing", an absence of being. Disease in animals is an absence of health. Vice in humans is an absence of virtue'. Also, see C. Mathewes, *Evil and the Augustinian Tradition* (Cambridge, 2001), esp. p. 75, for a more in-depth exploration of the topic.

[41] Pollmann, 'Two Positions', 83.

[42] For the date of *In Genesim*, see Bede, *On Genesis*, ed. C. B. Kendall (Liverpool, 2008), pp. 45–53; for the *VCM* see M. Lapidge, *Anglo-Latin Literature, 600–899* (London, 1996), p. 34; and for the *VCP*, see *Two Lives of Saint Cuthbert: a Life by an Anonymous Monk of Lindisfarne and Bede's Prose Life*, ed. and trans. B. Colgrave (Cambridge, 1940), p. 16.

[43] See Lapidge, *Anglo-Saxon Library*, p. 282, for evidence of the circulation of *De Genesi contra Manichaeos* in England.

[44] Bede, *In Genesim, Bedae Venerabilis opera. Pars 2, Opera exegetica*, ed. C. W. Jones, CCSL 118A (Turnhout, 1967), p. 68. All references are from this edition by page number. Italics indicate text derived from Augustine.

> time to time be admonished by it to turn away from their sins and back to God's commandments.)

Bede likewise includes the section in which Augustine highlights the pedagogical function of fallen Creation which, via a metaphor of unfruitful trees, points to the allegorical orthodoxy to be learnt from the book of Creation:

> Nam et herbae uenenosae ad poenam uel ad exercitationem mortalium creatae sunt. [...] Per infructuosas quoque arbores insultatur hominibus, ut intellegant quam sit erubescendum sine fructu bonorum operum esse in agro Dei.[45]
>
> (For poisonous plants were created for the punishment and discipline of mortals. [...] Likewise men are insulted by unfruitful trees, so that they might know how shameful it is to be without the fruit of good works in the field of God.)

Bede diverges, however, from Augustine's initial exegesis while commenting on Gen. 1:29–30, by arguing explicitly that the animal kingdom was equally harmonious before the Fall:

> ne ipsae aues raptu infirmorum alitum uiuebant, nec lupus insidias explorabat ouilia circum [...] uniuersa concorditer herbis uirentibus ac fructibus uescebantur arborum.[46]
>
> (Birds themselves did not live by stealing the food of weaker animals, nor did the wolf seek an ambush around the sheepfold [...] everything in harmony fed on the green plants and the fruits of the trees.)

Bede's *In Genesim,* therefore, presents Creation as transformed by the sin of Adam into its present state of predation and decay. He agrees with Augustine that the function of such facets of Creation is to push humanity towards redemption. Bede also believes, however, that this postlapsarian relationship between humanity and Creation is not fixed, as it is fundamentally tied to Adam and Eve's fall from perfect holiness. Thus, Bede argues, the relationship can be restored to a prelapsarian state by way of saints, whose sanctity is great enough that the relational effects of the Fall are taken away: 'Denique testimonium primae creationis legimus uiris

[45] Bede, *In Genesim,* p. 68.

[46] Bede, *In Genesim,* pp. 29–30.

sanctis atque humiliter Deo seruientibus et aues obsequium praebuisse' ('Finally, as evidence of the first creation, we read that birds have rendered obedience to saints humbly serving God').[47] The obedience of the saints results in the restoration of humanity's place in the divine hierarchy, regaining the dominion that God originally gave it in Gen. 1:28.[48] This is exegetically connected to the common interpretation of Jesus as the New Adam, a role which the saints typologically take on here.[49] That Bede utilises the image of a bird and saint in his exegesis here speaks to the influence of texts like the *VCA* in his interpretative development, and also highlights the importance of Cuthbert in Bede's thought. Likewise, Bede's emphasis of this otherwise minor detail highlights the importance of this Augustinian exegesis in his thought, and given Bede's influence over Anglo-Saxon hagiography in general, as well as the specific saints explored here, their shared interpretation has far-reaching implications. These two exegetes therefore share an understanding of three key points that are central to the depiction of the relationship between Creation and the saints Cuthbert and Guthlac in their early Anglo-Saxon *vitae*: first, that the unpleasant and indifferent portions of Creation function in the postlapsarian world to urge humanity, the saints included, towards greater holiness and eventual redemption; second, that the Fall produced a relational breach between humanity and Creation (whether actual or ontological); third, that the effects of the Fall can be temporarily removed by restoring a portion of Creation into its prelapsarian state by means of sanctity.[50] This shared exegetical position creates

47 Bede, *In Genesim*, p. 29.

48 Augustine in particular was focused on the sense of order in the cosmos, though Bede likewise perceived of the universe in hierarchical terms. See C. Harrison, *Beauty and Revelation in the Thought of Saint Augustine* (Oxford, 1992), p. 132, for a discussion of this order in relation to Creation and beauty.

49 The typological interpretation of Christ as the New Adam was an exegetical commonplace. See P. J. Stapleton, '*Kontrastimitation* and Typology in Alcuin's York Poem', *Viator* 43 (2012), 67–78, at 76, for a discussion of its relevance in an Anglo-Saxon context. For St Paul's Christological interpretation of Adam in Scripture, see 1 Cor. 15: 21–22, 47–48, and Rom. 5.

50 This study argues that the restoration accomplished by Cuthbert and Guthlac is temporary, and includes only specific portions of Creation. For the sake of brevity I will employ 'restoration of Creation' with this specific meaning throughout. For a discussion of the relationship between sanctity and the

a place for saints to work miracles, and is at the heart of each of the texts analysed in this book. In the *VCA* the Augustinian/Bedan exegesis of the Fall is implicit, while in the remaining texts (*VCM, VCP, VSG, OEPG,* and *Guthlac A*) it is physically present in part or whole.

The focus of this study is the early Anglo-Saxon *vitae* of Cuthbert and Guthlac in relation to the restoration of Creation, and in accordance with this focus I have imposed certain limitations. I have chosen these two eremitic saints because of their direct and transformative interaction with Creation, and have excluded other, non-eremitic, *vitae*, such as the *Vita S. Wilfridi*. As a structural frame, I have excluded all the later manifestations (post-950) of both saints, in text or material culture: for example, works such as the *Guthlac Roll* and the Old English poem *Durham*. I have also excluded Ælfric's adaptation of Cuthbertine material in his *Catholic Homilies* for the following reasons: first, Ælfric's *Homily II.X* was likely composed around 995,[51] at least half a century later than the latest text analysed in the thesis, the *Old English Prose Life of Guthlac* (*c.* 920); second, the scholarship on Ælfrician hagiography is comparatively extensive, and this study seeks to redress that imbalance by focusing on the pre-Ælfrician corpus of Anglo-Saxon *vitae*; third, *Homily II.X* removes not only most of the landscape detail from the Cuthbertine tradition, but also removes the framing exegesis concerning the restoration of Creation discussed above, a point I address in the conclusion. Finally, I have not included the Old English poem *Guthlac B* in my analysis because, first, this study focuses on the saints' living relationship with Creation, and the content of *Guthlac B*, the saint's death and ascension, places it outside the scope of this study; and second, the only scene of Guthlac's interaction with Creation in *Guthlac B* is that of the birds who honour the saint by singing after he feeds them.[52] This scene is used in the text as an

sanctification of the landscape in later Anglo-Saxon literature, see H. Powell, 'Sanctifying Landscapes: Topographical References in Anglo-Latin Saints' Lives', Unpub. D.Phil. (Oxford University, 2007).

51 *Ælfric's Catholic Homilies: the Second Series*, ed. M. Godden (London, 1979), p. xciii.

52 *The Guthlac Poems of the Exeter Book*, ed. J. Roberts (Oxford, 1979), ll. 916–923b.

allegorical parallel for Guthlac's support of pilgrims with spiritual sustenance, and is not primarily concerned with the relationship between the saint and physical Creation.

This book is organised in the chronological sequence in which the texts were composed, as each of the texts, after the first, draws on its predecessors. Chapter 1 considers how the Anonymous *Vita Sancti Cuthberti* explores Cuthbert's ability to restore Creation by means of obedience, focusing on the use of the Latin word *praecipio*. The *VCA* delineates how Cuthbert's monastic obedience contributes to the restoration of prelapsarian moments, and how the imitative order of a monastery is a part of how the world can be transformed. The author grounds this presentation of sanctity's transformative effect on Creation firmly in the world of eighth-century Northumbria through his highly physical depiction of the landscape, as well as his textual identification of potential sites for lay pilgrimage in the immediate landscape of Farne and Lindisfarne.

Chapter 2 analyses Bede's metrical *Vita Sancti Cuthberti*, arguing that in this overlooked early text, Bede is already fashioning Cuthbert into a saint of universal relevance through his depiction of Cuthbert as an idealised Gregorian monk-pastor; this is in contrast with most previous studies that have located this transformation in Bede's later prose *Vita Sancti Cuthberti*. Bede achieves this transformation with a novel development in hagiography, where he focuses on the distinctly monastic obedience of Cuthbert to the Divine Office, and the role of that obedience in the restoration of Creation. I argue that the *VCM*, as the poetic first half of an *opus geminatum*, both in form and content, was not only a ruminative and poetic exercise for the young Bede himself, but also functioned as such for its intended readers. It is in the *VCM* that Augustine's exegetical framework of the Fall enters Anglo-Saxon hagiography.

Chapter 3 shows how Bede's second account of Cuthbert's life, the prose *Vita Sancti Cuthberti*, further transforms Cuthbert into an idealised monk-pastor, harmoniously balancing the active and contemplative lives. Here Bede alters the very nature of Cuthbert's sanctity from the static predestination of the *VCA* and *VCM*, into a model of growth towards spiritual maturity based on the Evagrian *Vita Sancti Antonii*. The restoration miracles in the *VCP* are made to function in two primary ways: first, to provide an impetus for Cuthbert's maturation; and second, to provide evidence of that saintly progression. Together with these changes Bede further develops the exegetical framework of the restoration of Creation by borrowing directly from Augustine in order to explain the miracles.

Chapter 4 transitions to St Guthlac, a warrior who renounced his aristocratic heritage to pursue the ascetic life on Crowland, an island amidst the East Anglian fens. This chapter focuses on the first *vita* of St Guthlac, Felix's *Vita Sancti Guthlaci,* which details the life and death of the early eighth-century saint. Uniquely in early Anglo-Saxon hagiography, Felix's *Vita* was commissioned not by an ecclesiastical establishment, but instead at the request of King Ælfwald of the East Angles. I argue here that Felix presents a distinctively physical depiction of the landscape, as compared to the general depictions of his hagiographical sources (the *Vita Antonii* for example), employing structures of thought that echo representations of land and space seen in contemporary boundary clauses. Such a depiction connects to the likelihood that both Felix and his patron would have had first-hand knowledge of the East Anglian fens. I explore how Felix builds on Evagrian and Bedan models in order to display Guthlac's progression towards sanctity as centrally connected to this physical landscape. Felix takes the restoration of Creation further than his models and in the portrayal of Guthlac's death and ascension the entire island of Crowland and the surrounding fens are filled with ambrosial smells and heavenly light.

Chapter 5 discusses two of the vernacular *vitae* of Guthlac: the *Old English Prose Life of Guthlac* and *Guthlac A*. First, it shows how the *OEPG* adapts Felix's text to create a physical landscape of Anglo-Saxon England with deeper focus and wider appeal by utilising a lexis of landscape shared with contemporary boundary clauses. The result is a markedly physical vernacular landscape which emphasises to a greater degree than Felix the connection between Guthlac's spiritual progression and the delineated landscape. Second, it argues that *Guthlac A* not only exaggerates the role of the landscape to the extent that the central conflict lies in the competition between Guthlac and the demons for Crowland, but also that the poem more explicitly connects Guthlac with the doctrine of replacement, which joins his arrival to the eremitic space even more with the Edenic paradise. The arrival of Guthlac is also imagined in terms connected to the arrival of Adam to Eden, where both are the rightful guardians of their respective landscapes; *Guthlac A* thereby depicts the saint finding a uniquely intense joy in his restored relationship with animate and elemental Creation.

The journey this book makes reveals the ways Anglo-Saxons actively considered humanity's relationship with the non-human world, and represented it in their literary endeavours: representations shaped by Augustinian and Bedan exegesis, but which were altered, developed, added to, and subtracted from, creating distinct and potent images of Cuthbert and Guthlac interacting with a Creation experientially familiar to many who heard or read these *vitae*. There is a solidity to such an analysis that seeks for inclusion in our discourses about the ways humans have interacted with the non-human throughout history. It reminds us that people, even within a coherent cultural tradition, are never monolithic, but are syncretic, developmental, reflective, and that we should approach the medieval period with care and discernment, as well as a healthy humility of perspective. *Restoring Creation: the Natural World in the Anglo-Saxon Saints' Lives of Cuthbert and Guthlac*, I hope, reminds us as well that our interactions with the non-human are equally syncretic, shaped by a variety of cultural, philosophical, theological, linguistic, literary, and personal factors, and that the stories we tell about such interactions are born from this rich soil.

1

Monastic Obedience and Prelapsarian Cosmography: The Anonymous *Vita Sancti Cuthberti*

THE ANONYMOUS *Vita Sancti Cuthberti* (*VCA*)[1] has typically been interpreted as the ground from which Bede built his two lives of Cuthbert, rather than as a text with its own internal coherence, structure, and themes.[2] Cuthbert's interactions with Creation in the *VCA* have likewise been read in relation to Bede's adaptation of them in his *vitae*. When the *VCA* is examined on its own, however, in the light of its theological context, which informs the thematic and structural choices in the text, these miracles are better understood as functioning within a postlapsarian world delineated by Augustinian/Bedan exegesis. The *VCA* explores the nature of Cuthbert's ability to restore Creation by means of obedience, focusing on the fundamental moment of disobedience at the Fall; by Cuthbert's perfect saintly obedience humanity's rightful place in the divine hierarchy is restored, and for a moment the world returns to its prelapsarian state. Creation in these miracles is not merely a set of

1 *Two Lives of Saint Cuthbert: a Life by an Anonymous Monk of Lindisfarne and Bede's Prose Life*, ed. B. Colgrave (Cambridge, 1940).

2 For example, C. Newlands, 'Bede and Images of Saint Cuthbert', *Traditio* 52 (1997), 73–109; S. Shockro, 'Bede and the Rewriting of Sanctity', *Haskins Society Journal* 21 (2009), 1–19; P. Cavill, 'Some Dynamics of Story-Telling: Animals in the Early *Lives* of St Cuthbert', *Nottingham Medieval Studies* 43 (1999), 1–20; M. Clayton, 'Hermits and the Contemplative Life in Anglo-Saxon England', in *Holy Men and Holy Women: Old English Prose Saints' Lives and Their Contexts*, ed. P. E. Szarmach (Albany, NY, 1996), pp. 147–75, esp. 155–56; S. Coates, 'The Bishop as Pastor and Solitary: Bede and the Spiritual Authority of the Monk-Bishop', *Journal of Ecclesiastical History* 47 (1996), 601–19, at 613.

signifiers pointing to divine truth, nor is it a simple point of contrast for humanity. The *VCA* instead reveals how monastic obedience can participate in the restoration of prelapsarian divine cosmography, how the imitative nature of order in the monastery is not purely an earthly construct, but part of how the world is transformed. The *VCA* grounds this depiction of sanctity's transformative ability in the world of early eighth-century Northumbria through physical descriptions of the landscape, as well as by the textual creation of potential sites for a specific kind of pilgrimage. The result is a set of miracles firmly fixed in the traversable and familiar landscape, which points to the ability of the obedient to transform the daily hardships of life, and thereby participate in the miraculous.

The *VCA* was composed at some time between the translation of Cuthbert's uncorrupt corpse in 698 and the death of King Aldfrith in 705.[3] This was a period of novel stability for the monastery at Lindisfarne after considerable turmoil, including the excommunication of Bishop Wilfrid and his followers after the council at Austerfield in 703.[4] In light of this stability, it is unsurprising that Lindisfarne was keen to promote its newly translated saint and expand its influence. As has been noted by a number of scholars, the structure of monastic life at Lindisfarne was fundamentally synchronistic, involving elements from Irish, Gaulish, and Roman forms of monasticism.[5] The ordering of a monastery like Lindisfarne would have involved some form of *regula mixta,* drawn from the authoritative sources available.[6] As will be discussed in more

3 See C. Stancliffe, 'Cuthbert and the Polarity between Pastor and Solitary', in *St. Cuthbert, His Cult and His Community to A.D. 1200*, ed. G. Bonner, D. Rollason, and C. Stancliffe (Woodbridge, 1989), pp. 21–44, at 24; A. Thacker, 'Lindisfarne and the Origins of the Cult of Saint Cuthbert', ibid., pp. 103–22, at 115; A. Thacker, 'The Making of a Local Saint', in *Local Saints and Local Churches in the Early Medieval West*, ed. A. Thacker and R. Sharpe (Oxford, 2002), pp. 45–74.

4 Thacker, 'Lindisfarne and the Origins', p. 117.

5 For example, S. Foot, *Monastic Life in Anglo-Saxon England* (Cambridge, 2006), p. 56; A. Thacker, 'The Social and Continental Background to Early Anglo-Saxon Hagiography', Unpub. D.Phil. (Oxford, 1976), pp. 93–96.

6 R. Jayatilaka, 'The *Regula Sancti Benedicti* in Late Anglo-Saxon England: The Manuscripts and Their Readers', Unpub. D.Phil. (Oxford University, 1996), pp. 30–39.

detail later, the *VCA* states that Cuthbert himself compiled just such a mixed rule upon his arrival at Lindisfarne.[7] This type of organisation was in part representative of early insular monasticism, as Sarah Foot has demonstrated, whereby the leaders of communities would craft a 'synthetic model of holy living for the guidance of their own communities',[8] drawn from a common stock of authorities, including the rules of Augustine, Basil, Benedict, and Cassian.[9] For Lindisfarne, texts by the Irish St Columbanus would also have been influential,[10] and thus the milieu in which Cuthbert's *vita* was produced was structured in part on texts like the twin *regulae* written by Columbanus: the *Regula monachorum* and the *Regula coenobialis*. The *Regula monachorum* is notable for what Jane Barbara Stevenson calls its 'extraordinary integrity and rigour',[11] and its focus on the role of obedience,[12] for example insisting that the monks should obey their superiors even to death.[13] While the *VCA* is not primarily a cenobitic text, it shares this focus on the role of obedience and may, I would suggest, have been composed with a monastic audience in mind.

7 See below, pp. 42–3.

8 Foot, *Monastic Life*, p. 52.

9 Foot, *Monastic Life*, p. 53.

10 Thacker, 'The Social and Continental Background', notes the influence of Gallic monasteries at Lindisfarne, p. 94.

11 J. B. Stevenson, 'The Monastic Rules of Columbanus', in *Columbanus: Studies on the Latin Writings*, ed. M. Lapidge (Woodbridge, 1997), pp. 203–16, at 216.

12 Stevenson, 'Rule of Columbanus', p. 207: 'The most common problems which the community faced, if this *regula* is anything to go by, were related to anger and failure of absolute obedience. Monks seem to have found it far more difficult to subdue their will absolutely to that of the abbot than to undertake a life of poverty, chastity and physical hardship.'

13 *Sancti Columbani opera, Scriptores Latini Hiberniae II*, ed. and trans. G. S. M. Walker (Dublin, 1957), p. 124. It should be noted that a focus on obedience is not restricted to the *Regula* of Columbanus, as it is a fundamental part of monastic practice. What is unique, as suggested above, is the level to which this becomes the central focus of Columbanus' *Regula*.

Physical and Spiritual Miracles

Scholars have often interpreted Creation's role in hagiography in terms of its allegorical function, treating every appearance of animal or natural force as little more than either a signifier for divine truth, or a mirror for human society.[14] Susan Crane, however, has rebutted this interpretative paradigm, arguing instead that the Creation miracles are focused on the respective saints' interactions with postlapsarian reality.[15] Likewise, Shockro, referencing both Cuthbert and Guthlac, argues that animals are 'colleagues in God's creation' that 'are able to participate in productive relationships with saints'.[16] This study builds on such approaches, and argues that the *VCA* focuses on the physical landscape in which Cuthbert lived, with a predilection towards an almost proto-historical compiling of place names and relevant physical descriptions.[17] The author of the *VCA* has provided his own answer to these charges with an interpretation taken from, as I have argued elsewhere, Jerome's *Commentary on Matthew*.[18] The *VCA* utilises the material from the commentary to make a distinction between physical and spiritual miracles in the vision given to Cuthbert of Bishop Aidan's soul ascending to heaven. The vision itself has hagiographical precedent, employing the familiar visionary trope as well as the biblical source of Jacob and his ladder.[19] The *VCA* alters the depiction of this hagiographical commonplace by delineating how the opening of heaven is

14 See, for example, D. Alexander, *Saints and Animals in the Middle Ages* (Woodbridge, 2008), pp. 60–61, where he discusses this in relation to Adomnán's *Vita Columbae*, and pp. 148–49 in relation to Cuthbert; see also Neville, *Representations*.

15 See S. Crane, *Animal Encounters: Contacts and Concepts in Medieval Britain* (Philadelphia, 2012).

16 Shockro, 'Saints and Holy Beasts', p. 51.

17 This can be seen in his consistent use of place names and the naturalistic detail of Cuthbert's journeys. For example, see *VCA* I.5, in which Cuthbert is described guarding sheep near the river Leader, or *VCA* II.5, in which the river Teviot is named in the eagle-and-fish miracle.

18 For the full version of this argument, see my article, 'A New Source for the Anonymous *Vita S. Cuthberti*', *Notes and Queries* 62 (2015), 356–58.

19 The narrative of Jacob's Ladder occurs in Gen. 28:11–12; though patristic exegesis interpreted the image of angels ascending and descending in John 1:51 as related, and therefore applied Christological typology to both.

specifically a blessing of holy perception, rather than a description of the physical world:

Jerome:	*VCA*:
Aperiuntur autem *caeli non reseratione elementorum, sed spiritalibus oculis* quibus et Hiezechiel in principio uoluminis sui apertos eos esse commemorate.[20]	hoc est *coelo aperto non reseratione elementorum, sed spiritalibus oculis* intuens.[21]
(*The heavens are opened,* however, *not by an unfastening of the elements, but* [they are opened] *to the spiritual eyes;* just as Ezekiel at the beginning of his book records that they were opened.)	(For through *the opened heaven – not by a parting asunder of the natural elements but by* the sight *of his spiritual eyes* –)

The inclusion of this interpretative line has no precedent in hagiographical history, and its application here can be read as a deliberate alteration in the *VCA* to delineate interaction between types of miracles and Creation. Whereas Bede later allegorises this moment,[22] the *VCA* explicitly demonstrates that certain miracles do not directly affect Creation, while others do. The subsequent miracles, therefore, involving the restoration of Creation, belong to the category of physical miracle, a transformation of a physical portion of the postlapsarian world.

The same lexical care is taken by the author in a parallel episode to the vision of Aidan's soul, *VCA* IV.10, the only other vision Cuthbert is given of heaven, where he sees the soul of a brother being led to heaven in the hands of angels.[23] While sharing little in terms of

20 *Sancti Hieronymi Presbyteri. Commentariorum in Mathaeum libri iv*, ed. D. Hurst and M. Adrian, CCSL 77 (Turnhout, 1969), p. 19. Shared lines are in italics.

21 *VCA*, pp. 68–69. Shared lines are in italics.

22 See *VCP*, p. 166.

23 *VCA*, p. 126. The image of the choir of saints set amid the angels in heaven has a lexical parallel in Cassiodorus' *Expositions on the Psalms*. Compare *VCA*, 'et in *choro* angelorum *sanctorum martyrum* collocatam',

phrasing with the vision of Aidan, and including no biblical framing, the first sentence of this chapter makes the same distinction: 'Fidelissima abbatissa Aelfleda de sancto episcopo aliud *scientie spiritalis miraculum* mihi reuelauit' ('The most faithful abbess Ælfflæd related to me another *miracle of spiritual knowledge* concerning the holy bishop').[24] This is the only time the *VCA* describes something specifically as a spiritual miracle, and it aligns precisely with the vision of Aidan and the description of Cuthbert's spiritual eyes. The *VCA* thereby distinguishes between miracles that are primarily physical and those that are of a different order, primarily spiritual, which do not affect Creation.

The contrast between these spiritual miracles and those that directly affect the created world is displayed in the most unadorned of all the restoration miracles in the *VCA*, where Cuthbert saves his foster-mother's house from a fire in the village of Hruringham (II.7). While the miracle has its hagiographical precedent in Sulpicius Severus' *Vita Martini*, the *VCA*'s presentation of Cuthbert's power over wind bears little lexical similarity with this influential *vita*, and functions distinctly. In the *Vita Martini*, St Martin lights a fire to destroy an important pagan shrine, and it is only when the fire threatens an adjoining house that Martin climbs up in front of the blaze and miraculously deflects it.[25] Sulpicius interprets this as the fire acting on the orders of Martin: 'Ita vertute Martini ibi tantum ignis est operatus, *ubi iussus est*' ('Thus, by the virtue of Martin, the fire only acted where *it was ordered to do so*').[26] This episode is distinct from Cuthbert's fire miracle in two fundamental ways: first, the fire is the direct result of Martin's actions, whereas the fire Cuthbert

with *Expositio Psalmorum*, Library of Latin Texts (based on CCSL 97–98, ed. M. Adriaen), online (Turnhout, 2010), Ps. 118, linea 449, http://clt.brepolis.net/llta/pages/Toc.aspx?ctx=1108218. All references taken from this edition: 'Ecce de illo *choro sanctorum* et *martyrum* uerba sonuerunt'.

24 *VCA*, pp. 126–27. I have emended Colgrave's spelling of the Abbess' name from 'Aelfflaed' to 'Ælfflæd' to reflect Old English orthographic convention.

25 *Vita di Martino, Vita di Ilarione, in memoria di Paola*, ed. C. Mohrmann, A. A. R. Bastiaensen, J. W. Smit, L. Canali, and C. Moreschini' (Milan, 1983), Ch. XIV.2, p. 36. All references to this text will be from this edition.

26 *Vita Martini*, Ch. XIV.2, p. 36.

contends with is given no specific cause; second, the purpose of the episode in the *Vita Martini* is to relate how Martin's eradication of pagan idolatry resulted in the conversion of the people who used to worship at the shrines, whereas in Cuthbert's case paganism is nowhere present. This episode in the *VCA*, therefore, bears only a passing resemblance to its nearest hagiographical precedent.

What is most apparent when the *VCA*'s fire episode is contrasted with that of the *Vita Martini* is the *VCA*'s naturalistic depiction of both the conflagration and Cuthbert's miraculous intervention. The *VCA* gives no explanation for the fire, except the statement that during Cuthbert's stay a house started burning on the eastern edge of the village, and a fierce wind began to blow from the same direction.[27] There is no cause suggested for the fire, and the wind is not given any diabolic connotation. At best the *VCA*'s narrative resembles an event which must have occurred in villages built primarily out of wood and thatch; it is therefore focused on the physical postlapsarian world, rather than the demonic forces central to the narrative of the *Vita Martini*. Cuthbert's miracle is situated within this mundane moment, and the title's framing points to this emphasis: 'De eo quod nutricis suae habitacula ab urente flamma *imperio* suo custodiuit' ('How by his *command* he preserved the dwellings of his nurse from a blazing fire').[28] The miracle is here enacted by the saint's command ('imperio'). The potency of Cuthbert's command, however, is contingent upon his sanctity, displayed in his unperturbed faith with the phrase 'intrepida mente' ('with fearless mind').[29] While the lexical choice of the adjective *intrepidus* to describe Cuthbert's saintly fortitude is unique, occurring in no other antecedent hagiography, the concept of unshakable faith as a litmus test for sanctity is widespread.[30] The ensuing

27 *VCA*, p. 90.

28 *VCA*, pp. 88–89. The titles of each chapter in the *VCA* appear in most manuscripts and will be considered authorial.

29 *VCA*, pp. 90–91.

30 For example, G. B. Ladner, '*Homo Viator*: Mediaeval Ideas on Alienation and Order', *Speculum* 42 (1967), 233–59, at 240–42, discusses the idea of an internalised *stabilitas* as evidence of a true monk, in connection with chapter one of Benedict's *Regula* and the wandering monks called Gyrovagues. See St Benedict, *La règle de saint Benoît*, vol. 2, ed. A. de Vogüé and J. Neufville

miracle is framed as a command, contingent upon Cuthbert's sanctity, and is displayed here in his unperturbed response to the physical conflagration.

The *VCA*'s description of the miraculous event itself is also given in uniquely naturalistic terms. Unlike the *Vita Martini*, where it is the saint's presence that pushes the flames back against the driving wind,[31] Cuthbert is here described as humbling himself before God in silent prayer. The *VCA*'s narrative highlights the theme of obedience: 'Et statim etiam deprecante eo *uentus ingens extitit ab occidente*, et omnem magnitudinem flamme a domibus innocenter abegit' ('and immediately, even while he was praying, *there arose a mighty wind from the west* and drove away the whole volume of flame harmlessly from the houses').[32] Despite the miracle that has taken place, the *VCA*'s depiction of Cuthbert's actions and Creation's response is subdued, presenting a picture not focused on allegory for exegesis or demonic confrontations to display the power of God over pagan deities, but instead a vision of the world acting in a prelapsarian manner. The wind and fire act as the elements do in the physical world, a series of physical events related causally, in no way inimical or functioning as a thinly veiled signifier. While the Genesis narrative does not explicitly depict elemental Creation as being under Adam's dominion, the expansion of Adam's authority to include it can be found in wider Anglo-Saxon literature, for example in *Genesis A*:[33]

(Paris, 1972), Ch. I.10–11, pp. 438–40: 'Quartum vero genus est monachorum quod nominatur gyrovagum [...] semper vagi et numquam stabiles' ('The fourth kind are called Gyrovagues [...] always wandering and never stable'). All quotations from this text are from this edition, which will hereafter be referred to as 'Benedict, *Regula*'; also, T. D. Hill, 'The Middle Way: Idel-Wuldor and Egesa in the Old English *Guthlac A*', *RES* 30 (1979), 182–87, at 182, in relation to *Guthlac A*. I address this more fully in chapter 5.

31 *Vita Martini*, Ch. XIV.2, p. 36.

32 *VCA*, pp. 90–91.

33 This can also be linked to the wider tradition of Anglo-Saxon depictions of Edenic spaces as perpetually pleasant, where the elements, particularly the weather, never function in a way to harm humanity. See below, pp. 279–80. All quotations from *Genesis A* taken from *Genesis A: A New Edition, Revised*, ed. A. N. Doane (Tempe, 2013).

Temað nu and wexað, tudre fyllað
eorðan ælgrene, incre cynne,
sunum and dohtrum. Inc sceal sealt wæter
wunian on gewealde and eall worulde gesceaft.
(ll. 196–99)

(Now be fruitful and multiply, fill the all-green earth with offspring, the progeny of you two, with sons and daughters. The salt water and all the worldly Creation shall remain under your rule.)

Cuthbert's power over the wind displays the saint's ability to usher in prelapsarian existence within the postlapsarian present.

Obedience and Holiness

In what is now Cuthbert's most well-known Creation miracle, where two sea animals dry his feet after a night of prayer (*VCA* II.3), the *VCA* presents the saint restoring Creation into a prelapsarian moment through a further emphasis on obedience. The chapter begins with the hagiographer's familiar physical framing, tying the miracle to a specific place and time, when Cuthbert went to the monastery of Coldingham in response to a request from the nun Æbbe. The *VCA* describes how, while there, Cuthbert continued his habit of keeping vigils by walking along the sea shore and singing.[34] At a certain point he enters the sea in a moment that echoes Irish ascetic tradition:

> Ille uero homo Dei Cuðberht, inobstinata mente adpropinquans ad mare usque ad lumbare in mediis fluctibus iam enim aliquando usque ad ascellas tumultuante et fluctuante tinctus est. Dum autem de mare ascendens, et in arenosis locis litoris flectens genua orabat, uenerunt statim post uestigia eius duo pusilla animalia maritima humiliter proni in terram, lambentes pedes eius, uolutantes tergebant pellibus suis, et calefacientes odoribus suis. Post seruitium autem et ministerio impleto accepta ab eo benedictione, ad cognatas undas maris recesserunt.[35]

34 *VCA*, p. 80.
35 *VCA*, pp. 80–81.

(But that man of God, approaching the sea with his mind made resolute, went into the waves up to his loin-cloth; and once he was soaked as far as his armpits by the tumultuous and stormy sea. Then coming up out of the sea, he prayed, bending his knees on the sandy part of the shore, and immediately there followed in his footsteps two little sea animals, humbly prostrating themselves on the earth; and, licking his feet, they rolled upon them, wiping them with their skins and warming them with their breath. After this service and ministry had been fulfilled and his blessing had been received, they departed to their haunts in the waves of the sea.)

It is immediately apparent, as Crane points out, that the sea animals' actions of rendering service to the visiting Cuthbert are a mirror of the faithful hospitality he rendered to the angel, described in the preceding chapter.[36] In that episode, the disguised angel arrives during a wintry morning and is received by Cuthbert, who obediently carries out his monastic duty as the guest-minister, not only washing the angel's hands and feet, but also rubbing them to get them warm.[37] There are further parallels, however, with the angelic guest miracle, which are displayed in the interaction between the saint and Creation. Not only are the sea animals paralleled with Cuthbert, but also the littoral landscape is paralleled with the guesthouse of the monastery. The shore is a similarly liminal place, both of the sea and of the land, just as the guesthouse is both of the monastery and of the secular world, often physically overlapping the monastic and mundane worlds by its physical separation from the cloisters.[38] This pairing of the created world with the monastic displays how the imitative nature of monastic ordering participates in the restored divine order of the universe. Cuthbert's parallel service to the angels and sea animals functions within this structure, and serves to highlight the restoration of a portion of Creation

36 See Crane, *Animal Encounters*, pp. 26–31, who argues that the presentation of these collocated chapters displays a specifically insular concept of hospitality. See *VCA*, pp. 76–78.

37 *VCA*, p. 76.

38 See, for example, L. M. Bitel, *Isles of the Saints: Monastic Settlement and Christian Community in Early Ireland* (Ithaca, NY, 1990), who describes the particular liminal qualities of the guesthouse, pp. 201–02. See also C. M. O'Sullivan, *Hospitality in Medieval Ireland, 900–1500* (Dublin, 2004), pp. 141–46.

through his saintly obedience. This focus on obedience, Creation, and holiness is furthered by the reaction of the only witness to the miracle: the sinful brother hiding among the rocks.

The *VCA* depicts this monk as a kind of spiritual pedant, following Cuthbert to spy on his activities and discover any illicit behaviour.[39] It is an action of sinful hubris, and when he beholds the miracle of the world as it should be, of the saintly man restoring the prelapsarian world, he is terrified by this confrontation with holiness: 'clericus uero familiae supradictus in scopulosis locis latens uisu *pauidus et tremebundus,* tota nocte coangustatus prope mortem accederat' ('[t]he above-mentioned cleric of the community lay hidden amid the rocks, *frightened and trembling* at the sight and, being in anguish all night long, he came nigh to death').[40] The *VCA*'s presentation of the sinful brother's response to the miracle, fear and trembling ('pauidus et tremebundus'), resembles not only one of the central biblical encounters between humanity and the holiness of God, but also similar episodes in hagiography. In the former, Heb. 12:21 describes Moses' reaction to his theophany on the mountain while receiving the Decalogue in similar terms: 'Et ita terribile erat quod videbatur. Moyses dixit: *Exterritus sum, et tremebundus*' ('And so terrible was that which was seen, Moses said: *I am frightened, and tremble*').[41] The latter can be seen, for example, in Adomnán's *Vita Columbae,* in which Columba's foster-father, Cruithnechan, returns to find the whole house illuminated with clear light and a ball of fire over the sleeping child Columba.[42] Cruithnechan's response

39 *VCA,* p. 80.

40 *VCA,* p. 80–81.

41 For Moses' theophany see Deut. 9 and Ex. 19–20.

42 Adomnán, *Adomnán's Life of Columba,* ed. A. O. Anderson and M. O. Anderson (Oxford, 1991), p. 184. All quotations and translations are from this edition, which will be referred to as the *Vita Columbae*. Cf. *Vita Antonii,* in P. H. E. Bertrand, 'Die Evagriusübersetzung der Vita Antonii. Rezeption, Überlieferung, Edition. Unter besonderer Berücksichtigung der Vitas Patrum-Tradition', Ph.D. diss. (University of Utrecht, 2006) [available at http://igitur-archive.library.uu.nl/dissertations/2006-0221-200251/index.htm], Ch. 35, p. 172, which highlights the kind of fear experienced by encounters with the holy: 'Metus enim non tantum ex pauore animi; quantum ex magnarum rerum saepe concititur aspectu' ('For their fear often arose not so much from timidity of the spirit, but as much from the recognition of the presence of

involves the same physical effect as the sinful brother, trembling: '*Qui uiso* statim *intremuit*, et prostrato in terram uultu' ('*Seeing this*, he began at once *to tremble*; and [...] bowed his face to the ground').[43] Columba's foster-father is confronted with a vision, the emphasis once again on something seen ('uisio'), and he trembles ('intremuit') with terror in response.[44] Adomnán's depiction also echoes the hagiographical trope of fantastic portents, which declare the singularity of the saint-to-be. The *VCA*, however, uniquely connects this physiological response to holiness with the restoration of Creation.

The *VCA*'s emphasis on the confrontation of the sinful human with holiness is furthered by its use of Scripture to substantiate and authorise this miraculous moment. First the author draws a parallel with Daniel and the lions to support the relationship between holiness and the restoration of Creation: 'narrans seruitionem animalium, *sicut leones in ueteri legimus Danihelo servire*' ('the animals ministered to the saint, *just as we read in the Old Testament that the lions* ministered to *Daniel*').[45] When taken together with the second scriptural reference in the *VCA*, a more complete interpretative framing comes into focus. Cuthbert's visionary ability to know of the brother's spying is paralleled with the perception of Peter in the New Testament, in which he detects the lies of Ananias and Sapphira: '*sicut uiderat Petrus* Annaniam et Saphiram spiritum sanctum temptantem' ('*just as Peter detected* Ananias and Sapphira when

---

superior beings'). All quotations will be from this edition, which will be referred to as *Vita Antonii*.

43 *Vita Columbae*, pp. 184–85.

44 For another example, see *Vita Antonii*, Ch. 35, p. 172, which describes how even good men may fear the appearance of holy beings: 'si quis pro conditione fragilitatis humanae miro fuerit eorum fulgore perterritus, omnen continuo ex corde auferant metum [...]. Metus enim non tantum ex pauore animi; quantum ex magnarum rerum saepe concitiur aspectu' ('If, being men, some fear the vision of the good, those who appear immediately take fear away [...]. For their fear arose not from timidity, but from the recognition of the presence of superior beings').

45 *VCA*, pp. 82–83. Cf. *VCA* and *Fontes*. See B. Ward, 'The Spirituality of St. Cuthbert', in *Cuthbert and Community*, pp. 65–76, at 72, for a discussion of this passage in relation to notions of Christological typology.

they were tempting the Holy Spirit').[46] The brother in the *VCA* is not slain as they were, but instead admits his fault to Cuthbert and seeks pardon in an obedient way,[47] thereby reinserting himself into his proper position within the hierarchy of the monastery, and saving himself from a death brought about by disobedience. The most famous of all Cuthbert's Creation miracles in the *VCA* is therefore focused on the saint's ability to restore a portion of Creation into its prelapsarian state, specifically by means of obedience. The deliberate pairing of Cuthbert's saintly obedience with the sea animals' subsequent obedience serves to create a moment of restored Creation whose holiness is terrifying for the disobedient monk. Through his obedient penance the restored created world and the monastic world are further united, portraying obedience within a monastery as part of the transformation of the greater world.[48]

Obedience, *Praecipio*, and Restored Creation

The majority of miracles involving provisions in the *VCA* are described as the operation of angelic or divine hands, including the bread and meat found by Cuthbert and his horse in the shelter, as well as the dolphin flesh cut and washed as if by hand on the seashore. The miracle of the eagle that brings Cuthbert and his companion a fish as they travel in the mountains is, however, distinct. This miracle is presented in keeping with the *VCA*'s focus on obedience, and the virtue of Cuthbert's obedience to restore Creation into prelapsarian relationship with him. Within this frame, the *VCA* explores here not only Cuthbert's obedience but, for the first time in this text, the obedience of one of his subordinates, an emphasis that again highlights the text's monastic focus. This shift to include Cuthbert's subordinate focuses on the term *praecipio*, which the *VCA* employs in only specific circumstances, and

46 *VCA*, pp. 82–83. See Acts 5:1–10.
47 *VCA*, p. 80.
48 Cf. Crane, *Animal Encounters*, p. 28.

only with imperative force.[49] This lexical choice is used to unite the restored created world with the human and monastic world in a similar manner to the holy moment of the sea animals.

For this interpretative paradigm to function, the *VCA* once again establishes that Cuthbert's sanctity, as exemplified specifically through his (monastic) obedience, is the factor upon which the miraculous restoration of Creation is contingent. The hagiographer achieves this by way of biblical framing, which, as Benedicta Ward cautions, is not merely 'pious trimming',[50] but is employed with the expectation that the context of each verse is implicitly understood. This is clear in the conversation between Cuthbert and his companion as they evangelise amidst the mountain peoples along the river *Tesgeta*.[51] Prompted by Cuthbert, his companion assesses their situation in the following terms: that they do not know anyone in the vicinity, and therefore he does not expect any sort of kindness or provision from the strangers around them.[52] Cuthbert's faithful response, that the Lord will provide for them, is given via three biblical references; the first invokes the providential Mtt. 6:33, spoken verbatim by the saint: 'Quaerite ergo primum regnum Dei et iustitiam eius, et haec omnia adicientur uobis' ('Seek first the Kingdom of God and his righteousness and all these things shall be added unto you').[53] This verse comes towards the end of Mtt. 6 amidst a series of examples of how God will provide for the faithful: how the birds of the air do not sow or reap, but are provided for by God;[54]

49 *DMLBS*: 3, 'To instruct, order, command'. While the dictionary editors have placed a usage from the *VCA* in definition 2, 'to instruct, advise, direct', given the context and use both biblically and within the *VCA*, the most applicable definition here is one with imperative force, hence I am using the third definition. Both verb and noun forms (*praecipio* and *praeceptum*) are used interchangeably as needed in the following discussion, as their meanings and use, other than grammatical function, are equivalent in the *VCA*. See below, pp. 34–36.

50 Ward, 'The Spirituality of Cuthbert', p. 68.

51 While the *VCA* does not describe the people Cuthbert visits as pagans, using instead the term 'rusticanos' ('rustics'), it does describe Cuthbert baptising them, which indicates at least some of them were in need of conversion. See *VCA*, pp. 84–85: 'docens rusticanos et baptizabat eos' ('teaching the country people [...] and baptising them').

52 *VCA*, pp. 84, 86.

53 *VCA*, p. 86.

54 Mtt. 6:26.

and how the lilies of the fields are clothed in raiment more splendid than Solomon's.[55] These statements of God's provisions are all encapsulated in the verse's position as a part of the Sermon on the Mount, and thus represent a key aspect of Jesus' teachings on how to live a holy life. Thus the invocation of the passage from Matthew, to seek first God's kingdom, has a direct connection to the idea of God divinely providing earthly subsistence. Cuthbert connects this New Testament quotation with what he terms a prophetic portion of Ps. 36, which Jesus' words are fulfilling: 'Et ut impleatur prophetae dictum, *Iunior fui, etenim senui, et non uidi iustum derelictum*, et reliqua' ('in order that the saying of the prophet may be fulfilled: *I have been young and am now old, yet I have I not seen the righteous forsaken*, and so forth').[56] While the portion of the psalm referenced is clear enough in its application here, it is only within the context, and implicit recognition, of the rest of the psalm that the full meaning can be ascertained. Ps. 36 is focused on the providential hand of God supporting the faithful and bringing judgment upon the sinful, and just three verses later this support is tied to the idea of holiness: 'Quia Dominus amat judicium, et non derelinquet *sanctos suos*; in aeternum conseruabuntur' ('For the Lord loves judgment, and will not forsake *his saints*: they shall be preserved forever') (Ps. 36:28). Cuthbert's faith that God will grant them provision in their missionary quest is contingent upon his own saintly status, defined contextually by his obedience.

The third and final biblical reference is from Luke 10, the chapter describing Jesus sending out seventy-two of his disciples in pairs as missionaries to preach, teach, and convert. Much like the use of the passage from Ps. 36, the *VCA* relies on the expected interpretative practice of the audience to fill in the context of the quoted passage and bring the significance of the reference for the *VCA* into proper focus. Cuthbert finishes his first speech with a portion from Luke 10:7: '*Dignus est* namque *operarius mercede sua*' ('For the *labourer is worthy of his hire*').[57] Contextually this verse speaks of the divine support granted to the faithful in their missionary journeys. When

55 Mtt. 6:28.
56 *VCA*, pp. 86–87. Italics show biblical quotation.
57 *VCA*, pp. 86–87. Italics show biblical quotation.

the three biblical references are drawn together, even before Cuthbert's prophetic utterance concerning the actions of the eagle, the *VCA* frames the upcoming miracle in the following manner: Cuthbert and his companion are missionaries travelling into a potentially pagan and antagonistic land, but God will support them with physical provisions because of Cuthbert's sanctity (echoing Adam's prelapsarian non-agrarian existence); and Cuthbert's sanctity is tied to his obedient actions, how his seeking God first makes his subsequent labour worthy. The ensuing miracle, which fulfils Cuthbert's prophetic statement, can be understood in these terms, which presents a moment equivalent to that of the sea animals: a portion of the world operating in a prelapsarian manner.

With Cuthbert's sanctity firmly supported by biblical reference for the contingent miracle of provision while engaged in evangelism, the *VCA* adds to this miracle the obedience of Cuthbert's subordinate. Through this addition, the eagle episode is transformed into another exploration of the intimate connection between the obedience of saints and their miracles, and the role that monastic obedience has in bringing these miracles about. The effects of the Fall are temporarily repaired through Cuthbert, and in this miracle his subordinate takes part in a moment whose holiness is akin to that which terrified the sinful brother:

> Post paululum autem iter agentibus illis, uiderunt aquilam super ripam fluminis sedentem. Currens etiam ad aquilam puer *secundum praeceptum* serui Dei, hesitans, inuenit piscam grandem, portantique ad eum integrum, dixit puero, Cur piscatori nostro ieiunanti partem ad uescendum non dedisti? Tunc uero puer, sicut *praeceperat* homo Dei, partem piscis aquilae dedit. Alteram autem secum portantes inter homines assauerunt et manducauerunt.[58]
>
> (After a short time, as they went on their way, they saw the eagle settling on the bank of the river. The boy ran towards the eagle *in accordance with the command* of the servant of God, and stopping, he found a large fish. The boy brought the whole of it to him, whereupon Cuthbert said: "Why did you not give our fisherman a part of it to eat since he was fasting?" Then the boy, *in accordance with the commands* of the man of God, gave half of the fish to the eagle while

58 *VCA*, pp. 86–87.

they took the other half with them, and broiling it in the company of some men, they ate it.)

The focus here is upon the restoration of proper relational hierarchy, where each element functions in righteous obedience to that which is above it. The prophetic utterance and biblical reference have already established Cuthbert's saintly obedience and God's contingent blessing, yet here the *VCA* adds his companion's obedience to God through Cuthbert as participating in God's support. After the eagle is seen alighting on the bank of the river the young man runs towards it, but does so only in accordance with Cuthbert's commands: '[c]urrens [...] puer *secundum praeceptum* serui Dei'. Even in the boy's misunderstanding of the proper response to such a miracle, forgetting the care of the fasting eagle, he immediately follows Cuthbert's second command: 'sicut *praeceperat* homo Dei'. The lexical choice of the verb *praecipio* and its noun form *praeceptum* echoes the biblical usage of the term, which is most commonly employed for God's commands to humanity, whether via his own voice, that of Jesus, or those of his elected servants, like the apostles.[59] Importantly, the first use of *praecipio* in Scripture is found in Gen. 2:16–17, the great command to Adam and Eve to eat of every tree in Eden except the Tree of the Knowledge of Good and Evil: '*Praecepitque* ei, dicens: Ex omni ligno paradisi comede; de ligno autem scientiae boni et mali ne comedas' ('And he *commanded* him, saying: Of every tree of paradise you shall eat: but of the tree of knowledge of good and evil, you shall not eat'). This prohibition is, of course, broken, and the effects of the Fall ensue. Augustine interprets this moment in his *De Genesi ad litteram* as focused on obedience to God's command (*praecipio*), rather than the actual moral status of the prohibited tree itself:

> Sed ab eo ligno quod malum non erat prohibitus est homo ut ipsa per se *praecepti* conseruatio bonum illi esset et transgressio malum [...] Hoc expertus est homo *contemnens praeceptum Dei*, et hoc experimento didicit quid interesset inter bonum et malum, *bonum* scilicet

59 For example, Acts 1:4; 1 Sam. 2:29; 1 Thes. 4:2, 11.

> *oboedientiae, malum* autem *inoboedientiae,* id est superbiae, contumaciae, peruersae imitationis Dei, et noxiae libertatis.[60]
>
> (But man was prohibited from that tree which was not evil, so that the keeping of the *command* was good for him, and the transgression evil [...]. Man, *despising the command of God,* experienced this, and from this experience learned what the difference was between good and evil, the *good of obedience,* and the *evil of disobedience,* which is pride, obstinacy, the perverse imitation of God, and the damage of freedom.)

Thus the great prohibition, the fulcrum upon which the fate of humanity turned, is based upon the obedience of Adam and Eve to God's initial command (*praeceptum*). The *VCA* is careful and consistent in its use of the term *praecipio,* each of which centres on the role of obedience in relationships within the divine hierarchy (inherently connected to the Fall narrative and Adam's initial disobedience). It is therefore unsurprising that the term is employed in the *VCA* to depict restoration miracles like that of the eagle. Even as the saintly Cuthbert is obeying God so too must his monastic followers obey him – figuring an ideal monastic community – and only in so doing might the miracle occur and the world function as it should have, had humanity not broken the original *praeceptum.* Thus every actor in this scene (from Cuthbert to boy to eagle) functions in proper obedience, contributing to the prelapsarian moment in the postlapsarian world. The *VCA* re-emphasises the centrality of obedience immediately after the eagle miracle by describing how the men, now granted provisions, continue onwards into the mountains according to God's will.[61]

The *VCA* is careful in its employment of the term *praecipio,* using it only thirteen times, and in all but three instances in the same framework: Cuthbert gives an order, his subordinates either obey or partially obey, and miracles occur. The three divergent uses of *praecipio* in the *VCA* nonetheless reinforce the primacy of obedience.[62]

60 Augustine, *De Genesi ad litteram,* pp. 251–52; the passage is used unaltered by Bede in *In Genesim,* pp. 51–52.

61 *VCA,* p. 86.

62 *VCA,* pp. 65, 120, 128. While these three uses of *praecipio* do in fact follow a similar pattern, not all of them specifically involve the restoration

The text begins, in the manuscripts which Colgrave identifies as closest and most faithful to the archetype,[63] with a chapter titled 'De prohemio oboediendi'. As Colgrave notes, both this and the following chapter, 'De prefatione scribendi', are taken wholesale from one chronological source, as well as three common hagiographical sources: the *Epistola Victorii Aquitani ad Hilarum de cursu paschali* and the *Vita Antonii* for Ch. I, and the *Vita Sancti Martini* and *Actus Silvestri* for Ch. II.[64] Despite the precedent of such borrowings in hagiography, especially with prefatory material, it is telling that in this highly derivative first section the *VCA* not only introduces the chapter with an original title, 'De prohemio oboediendi', but also chooses to begin with *praeceptum* as the first word in the *vita*: '*Praeceptis* tuis utinam sancte episcope Eadfride' ('I would that the result of my obedience to your *commands*, O holy bishop Eadfrith').[65] The hagiographer has cast the *VCA* itself within the paradigm that he uses for most of the miracles, connecting the monastic virtue of obedience with the term *praecipio*. Towards the end of Ch. I he repeats this dynamic, re-emphasising the relationship between his obedience and the commands he is following: 'fiducia sit peragere posse quod *praecipitis*' ('I can accomplish the behest which you so confidently *command*').[66] While this emphasis on faithful obedience in the creation of a *vita* is in keeping with hagiographical precedent, the framing of his task in the composition of the *VCA* as an act of obedience to a command (*praeceptum*) creates a parallel between the faithful hagiographer and those who are obedient to Cuthbert's

of Creation. The first is in reference to the prophecy by a three-year-old child concerning Cuthbert's predestined election to sanctity (p. 65). The other two are focused on another expected miraculous power of the saint, healing (pp. 120, 128).

63 *VCA*, p. 44. Colgrave's X-group – O1, O2, and A – all have the prologue and *Prefatio*. His Y-group – H, B, and the connected T and P – omit both. Interestingly, D. A. Bullough highlights, in 'A Neglected Early-Ninth-Century Manuscript of the Lindisfarne *Vita S. Cuthberti*', *ASE* 27 (1998), 105–37, at 108, that Colgrave's manuscript M, now Arras, Bibliothèque municipale, MS 1029 (orig. Arras 812), omits the *capitula* and both prefaces. Nonetheless this discussion accepts the evidence presented by Colgrave for the inclusion of both.

64 *VCA*, p. 310.

65 *VCA*, pp. 60–61.

66 *VCA*, pp. 62–63.

commands. The author's obedience can therefore be seen as taking part in the miraculous transformation of the world within the *VCA*, as well as in the continuing miracles in his present-day Lindisfarne.

The other notable divergent use of *praecipio* in the *VCA* occurs during Cuthbert's eremitic time on Farne. Here, in a moment of foreshadowing, Cuthbert not only prophesies the death of King Ecgfrith but also the length of his own future bishopric. In keeping with hagiographical precedent, as well as similar moments in the *VCA* itself,[67] Cuthbert orders Ælfflæd not to reveal the miraculous prophecy until he has died: 'Et tu quoque audi quod ego *praecipio* tibi in nomine Domini nostri Iesu Christi, ut quamdiu uixero, nulli hoc indicaueris' ('And you too harken! I [*command*] you in the name of our Lord Jesus Christ, tell this to no one while I live').[68] While this interaction between Cuthbert's command and the abbess's obedience results in no direct miracle, the emphasis on obedience and its relationship to the exercise of prophetic power is just as clear as that between obedience and miracles in the other chapters. In the lead-up to this prophecy, Abbess Ælfflæd is depicted as humbly asking for the saint to visit her, and then immediately upon his arrival physically humbles herself on bended knee as she abjures him for prophetic utterance concerning her brother, King Ecgfrith.[69] It is in direct response to this compounding of humility and obedience exemplified in Ælfflæd's actions that Cuthbert begins his prophetic utterance.[70] Thus the prophetic miracle itself is just as dependent on obedience as most of the miracles within the *VCA*; the obedience of the abbess in not revealing Cuthbert's prophecy is simply the last in a series of obedient actions on her part.

67 For example, *VCA* II.3, where Cuthbert tells the healed brother (after the sea animal miracle) never to relate the story until the saint has died.

68 *VCA*, pp. 104–05. Colgrave translates *praecipio* and *praeceptum* variously as 'bid', 'behest', 'exhortation', and 'command'. As argued above, however, the *VCA* consistently employs the term with the specific force of 'command'. I have therefore emended the translation here and throughout to 'command'.

69 *VCA*, p. 102.

70 *VCA*, p. 102.

Eremitism and Pilgrimage

As Simon Coates notes, the *VCA* is divided into four sections following the structure of the metrical *Vita Sancti Martini* by Fortunatus and Gregory of Tours' *Liber de uirtutibus Sancti Martini episcopi*, as well as the thematic separation of chapters from Sulpicius' writings.[71] These divisions are employed chronologically, presenting specific moments in Cuthbert's career. While there is no internal spiritual progression, as in the Antonian tradition, the sections are nonetheless distinct.[72] In Book III, the *VCA* draws the events of Cuthbert's life into the present and familiar landscape of Farne Island, arguably the most important from the Lindisfarne community's perspective. Cuthbert's miracles during his eremitic stay are distinct within the *VCA* as they are not only concerned with restoring sections of Creation via his sanctity, but also are specifically concerned with their continuing presence for Lindisfarne. This secondary focus creates textual sites for potential pilgrimage, reflecting the continuing relationship of the monastic community at Lindisfarne with their increasingly popular saint. The impetus behind such emphasis can be connected with the propagandistic intention of the *VCA* identified by Joseph McMullen, who argues that the proliferation of place names and specific details were meant to 'strengthen the connection between Lindisfarne and churches or territories already in its diocese' and also 'to claim an extension of the Lindisfarne *paruchia* beyond its traditional diocesan bounds'.[73] As Nancy Edwards suggests in discussing Celtic saints in general, this emphasis on portions of the landscape is indicative of the process by which such secondary sites also became part of the veneration of each saint, and 'became part of the pilgrimage ritual culminating in a visit to the most important site associated with the saint'.[74] Within

71 S. Coates, 'The Construction of Episcopal Sanctity in Early Anglo-Saxon England: the Impact of Venantius Fortunatus', *Historical Research* 71 (1998), 1–13, at 4–5.

72 See Stancliffe, 'Pastor and Solitary', p. 25.

73 A. J. McMullen, 'Rewriting the Ecclesiastical Landscape of Early Medieval Northumbria in the Lives of Cuthbert', *ASE* 43 (2014), 57–98, at 90.

74 N. Edwards, 'Celtic Saints and Early Medieval Archaeology', in *Local Saints and Local Churches*, pp. 225–65, at 226.

this cultural context the *VCA*'s consistent focus on the present efficacy and evidence of these miracle sites on Farne unites them with the monastery at Lindisfarne. The sites, in McMullen's structure, would serve to expand the influence of Lindisfarne in the ecclesiastical landscape. The type of pilgrimage envisioned here, however, is specifically non-elite, short-term pilgrimage, usually for the purpose of attaining physical healing. This union of the world restored via the saint's obedience with the present-tense physical landscape of Farne creates a unique but connected exposition with the previous portions of the text, highlighting how the monastic world could partake in the restoration of Creation.

The focus of any pilgrimage to Lindisfarne after the elevation of Cuthbert would be, of course, his uncorrupt corpse, the true *locum sanctum*, placed on the south side of the altar.[75] The author of the *VCA* recognises this, particularly through his depiction of various post-mortem miracles associated with Cuthbert later in the *VCA*.[76] In an original way, however, the author seeks to unite the sites of Cuthbert's eremitic miracles with Cuthbert's physical *locum sanctum*, as places where the saint's power could be both glimpsed and, potentially, experienced. Alan Thacker notes this process in connection with the creation of secondary relics of Cuthbert, in particular the dust from the pit where Cuthbert's body had been washed.[77] Furthermore, the kind of pilgrimage that is implied by this unification of external landscape sites and Cuthbert's *locum sanctum*, and which is described in the post-mortem healing miracles in the *VCA*, is of the kind most closely linked with local saints: the pilgrimage to a holy site, usually the saint's burial place, for

75 Thacker, 'Making of a Local Saint', p. 46.

76 *VCA*, pp. 133–39.

77 Thacker, 'Making of a Local Saint', p. 46: 'In addition to the shrines of imported secondary relics [...] subsidiary miracle-working sites in and around the monastery included Cuthbert's hermitage on Farne and the pit near the monastic church into which water used in the funeral washing had been poured away'. In addition to these, Thacker highlights the specific secondary relics attested in the Latin *vitae* of Cuthbert: 'Also credited with miracle-working properties were Cuthbert's funeral vestments, removed at the elevation and displayed by the shrine, and hairs cut from his incorrupt head, placed in small portable reliquaries and given to friends of the brethren', p. 46.

miraculous healing. As Diana Webb points out, Bede discusses this type of lay pilgrimage when distinguishing his three types of pilgrimage, only one of which he would have consistently called *peregrinatio*.[78] Whereas the first two kinds of pilgrimage, self-exile for God (the true *peregrinatio*) and lengthy journeys to distant holy sites, most notably Rome, were undertaken by people of the higher classes in Anglo-Saxon society, this third kind of pilgrimage was 'undertaken by humble people'.[79] The various holy battle sites connected with St Oswald provide a helpful parallel. Catherine Cubitt, for example, points out that Oswald's cult focused on the battlefields Heavenfield and Maserfield, and it was to these locations that people journeyed 'for healing, and removed fragments of the cross or earth from the King's place of death'.[80] Furthermore, she posits that the pilgrims who came to such sites were non-elites in search of physical healing, and that such a depiction is representative of the *loci* for low-status lay piety: 'its cult places were centered not upon corporeal relics and shrines but rather upon holy places in the landscape'.[81] The miracles performed at these sites can be characterised as mundane, in that the focus is predominantly on a physical ailment or difficulty within the physical world. In Bede's depiction of the posthumous miracles at specific locations associated with St Oswald, the parallels with the *VCA*'s depiction become evident:

> Cuius quanta fides in Deum, quae deuotio mentis fuerit, etiam post mortem uirtutum miraculis claruit. Namque in loco ubi pro patria dimicans a paganis interfectus est, usque hodie sanitates infirmorum et hominum et pecorum celebrari non desinunt. Vnde contigit ut puluerem ipsum, ubi corpus eius in terram conruit, multi auferentes et in aquam mittentes suis per haec infirmis multum commodi adferrent.[82]

78 D. Webb, *Pilgrimage in Medieval England* (London, 2000), p. 4.

79 Webb, *Pilgrimage*, p. 4.

80 C. Cubitt, 'Universal and Local Saints in Anglo-Saxon England', in *Local Saints and Local Churches*, pp. 423–53, at 425.

81 Cubitt, 'Universal and Local Saints', p. 426.

82 Bede, *HE*, III.9, pp. 242–43. All quotations and translations are from this edition. The text will be abbreviated as *HE* for reference, and *Historia* in the main text.

(His great faith in God and his devotion of heart were also made clear after his death by certain miracles. Indeed in that place where he was slain by the heathen fighting for his fatherland, sick men and beasts are healed to this day. It has happened that people have often taken soil from the place where his body fell to the ground, put it in water, and by its use have brought great relief to their sick.)

While Oswald represents traditional red martyrdom, death for the sake of Christ, as opposed to Cuthbert's white martyrdom, the daily sacrifices of the ascetic,[83] the emphasis on physical healing at a cult site not directly connected to the saint's physical relics is telling, as is the familiar hagiographical trope of water mixed with dirt as an endlessly renewable secondary relic.[84] The primary connection to the *VCA* is the way in which the physical location becomes a locus for a specific kind of lay pilgrimage, focused on miraculous physical healings. As Cubitt notes, both of Bede's subsequent examples of miracles occurring at the site are centred upon physical healing of the laity.[85] First, a travelling man's horse falls deathly ill, until it comes to the place where Oswald had died and is cured.[86] Likewise the second miracle relates the healing of an innkeeper's daughter who visits the site due to reports from the man with the healed horse, and is cured from palsy.[87] In each case the saint's virtue is experienced not in direct connection with the relics of his physical body, but instead through points in the landscape, which function as similar *loci* where such potency could be experienced and such healing hoped for.

In *VCA* III.1, the author unites the *foci* of obedience and potential pilgrimage sites in his depiction of Cuthbert's arrival and life at Lindisfarne. This is evident in the summative description of Cuthbert's initial arrival to Lindisfarne at the request of Bishop Eata: '*praesens et absens* demoniacos sanauit, et alios uarios languores

83 C. Stancliffe, 'Red, White and Blue Martyrdom', in *Ireland in Early Mediaeval Europe: Studies in Memory of Kathleen Hughes*, ed. D. Whitelock, R. McKitterick, and D. Dumville (Cambridge, 1982), pp. 21–46, at 21–23.

84 Thacker, 'Making of a Local Saint', p. 71.

85 Cubitt, 'Universal and Local Saints', p. 426.

86 Bede, *HE*, III.9, p. 242.

87 Bede, *HE*, III.9, pp. 242, 244.

curauit' ('where, both *present and absent, he healed* those possessed of devils and cured various other infirmities').[88] While the subsequent wonder-working narrative makes no mention of the saint casting out demons or healing infirmities during his time as a hermit (prefiguring instead the miracles described in Book IV after he took up the bishopric), the emphasis on Cuthbert's ability to work miracles in the physical space of Lindisfarne, whether or not he was actually there, is telling; in so doing the *VCA* is positing the saint's continued potency in present-day Lindisfarne. This emphasis on Cuthbert's miraculous power being centred on a location unites with the *VCA*'s overall focus on the physical landscape, in that the central feature is the location, a place ripe with potential for pilgrimage. The initial summative frame serves, therefore, to emphasise the continued presence and potency of Cuthbert in the anonymous author's own Lindisfarne. The *VCA* also unites this focus on Cuthbert's continuing efficacy with the greater focus of the *VCA*, delineated lines of obedience:

> Viuens quoque ibi secundum sanctum scripturam, contemplatiuam uitam in actuali agens, et nobis regularem uitam primum componentibus constituit, quam usque hodie cum regula Benedicti obseruamus.[89]
>
> (He dwelt there also according to Holy Scripture, following the contemplative amid the active life, and he arranged our rule of life which we composed then for the first time and which we observe even to this day along with the rule of St Benedict.)

Cuthbert is depicted, in keeping with hagiographical tradition, as an ideal, capable of joining together the contemplative and active lives, while also providing a means of rule for the monastery, a list that strictly delineates obedience within a hierarchy. This *regula mixta,* potentially involving a synthesis of traditions from the respective Rules of Columbanus and Benedict,[90] would have

88 *VCA*, pp. 94–95.
89 *VCA*, pp. 94–97.
90 *VCA*, p. 324. Colgrave puts forward the example of Gaulish monasteries which used just such a synthesis of the Benedictine and Columban rules. He posits that the 'more practical as well as the more moderate Benedictine rule was added to that of Columban as early as the

emphasised obedience within a hierarchy,[91] from God to bishop to monk, and so on. That Cuthbert is the centre from which this order is delineated immediately establishes not only his sanctity, but also his status as the closest to God within it.

The *VCA* explores the dual focus on obedience and pilgrimage sites in the first visible evidence for Cuthbert's miracles on Farne: the miraculous construction of his hermitage. The depiction of this construction in two related miracles in Book III points to their inclusion as evidence of Cuthbert's dedication to the eremitic ideal, and therefore provides the necessary contingency for the ensuing moments of Creation's restoration, as well as the secondary focus of their continued presence on Farne. Cuthbert does not simply find a cave and inhabit it like Saint Benedict,[92] or take over existing ruins like Saint Antony,[93] but sets out physically to alter the landscape and create an eremitic space. He does this by constructing a hut that combines the round dug-out floor of early Anglo-Saxon

seventh century and finally supplanted it. It is therefore very possible that the traditional rule of St Cuthbert, probably that also in use at Melrose and Ripon (*V.P.* c. 9), was observed, together with the Benedictine rule, tending to make the latter stricter and more ascetic than was the original intention of its founder.'

91 This is most clearly seen in the *Regula monachorum* of Columbanus, which begins with a chapter on obedience, explicitly uniting unquestioning obedience within a monastic setting with general obedience to God; St Columbanus, 'Regula Monachorum', in *Sancti Columbani opera, Scriptores Latini Hiberniae II*, ed. and trans. G. S. M. Walker (Dublin, 1957): 'Ad primum verbum senioris omnes ad oboediendum audientes surgere oportet, quia oboedentia deo exhibetur' ('At the first word of the senior, all on hearing should rise and obey, since their obedience is shown to God'), p. 122. It also concludes by reemphasising obedience regardless of the task: 'Nihil itaque recusandum est oboedientibus veris Christi discipulis, quamvis durum et arduum sit, sed cum fervore, cum laetitia arripiendum est, quia si talis non fuerit oboedientia, non erit acceptabilis domino' ('Thus nothing must be refused in their obedience by Christ's true disciples, however hard and difficult it be, but it must be seized with zeal, with gladness, since if obedience is not of this nature, it will not be pleasing to the Lord'), p. 124.

92 See Gregory the Great, *Grégoire le Grand: Dialogues*, ed. A. de Vogüé, 3 vols (Paris, 1978–1980) vol. 2, p. 132. All references will be from this edition using volume and page number(s). The text will be referred to as Gregory's *Dialogues*.

93 *Vita Antonii*, Ch. 12, pp. 164–65.

dwellings[94] with the encircling cashel, or wall, typical of those built around their huts by hermits in Ireland.[95] Cuthbert, however, is no desert saint from an imagined antiquity, but is the saint of Lindisfarne in the present, whose works are still visible, and whose power is therefore all the more present. In III.1 and III.2 the *VCA* gives two nearly identical miracles involving Cuthbert's moving of seemingly immovable stones, joining together the two focal points of this section: obedience and the continued presence of his miracles. *VCA* III.1 describes Cuthbert delving through unforgiving stone to the depth of a human, before proceeding to build his enclosing wall.[96] The act of digging here functionally recalls that of Adam after his expulsion from Eden, and serves to highlight the reparative power of Cuthbert's sanctity, as tied to his obedience. Where Adam digs the earth as a result of disobedience, Cuthbert's actions here are part of his obedience to the eremitic ideal;[97] likewise the expulsion of the demons from Farne by Cuthbert contrasts with the Fall of the first couple, who succumbed to the devil's influence. This obedience is connected with Cuthbert's path towards eremitism, as he is following the proper trajectory for such a life through extensive training and time within the monastic life, before setting out to become a hermit.[98] The miraculous assistance given by God in this foundational moment is fitting, as Cuthbert is not only fulfilling the eremitic ideal, but also, in his labour, redeeming the postlapsarian

94 *VCA*, p. 326.

95 See *VCA*, where Colgrave highlights the similar construction of Irish eremitic sites, p. 326: 'and such too was the regular plan of Irish monasteries, as can still be seen at Skellig Michael and elsewhere'.

96 *VCA*, p. 96.

97 See Gen. 3:17, 'Ad Adam vero dixit, "Quia audisti vocem uxoris tuae et comedisti de ligno ex quo praeceperam tibi ne comederes, maledicta terra in opere tuo; in laboribus comedes eam cunctis diebus vitae tuae"'; ('And to Adam he said, "Because you have heeded the voice of your wife and have eaten from the tree which I commanded you not to eat, cursed is the earth in your work; with labour and toil will you eat of it, all the days of your life."').

98 The delineated progression towards the eremitic life via the cenobitic life was widely accepted in Anglo-Saxon England, in great part due to the influence of texts like Benedict's *Rule*; see Benedict, *Regula*, Ch. I.3–5, pp. 436–48; see also p. 124, for a discussion of this progression in Bede's *VCM*.

labour of Adam. The *VCA* therefore describes the actual construction of his cashel in miraculous terms:

> Alterum uero cubitum mirabilem desuper cum lapidibus *incredibilis magnitudinis nisi scientibus tantum Dei uirtutem in eo esse*, et terra commixtis constructum aedificauit.[99]
>
> (He also built a marvellous wall another cubit above it by placing together and compacting the earth, *stones of such great size as none would believe except those who knew that so much of the power of God was in him.*)

The emphasis on the visibly wondrous size of the stones, and how no one would believe unless they knew of Cuthbert's miraculous life, unites his saintly obedience with the working of a miracle, as well as with the continuing evidence of that miracle in the wall. The *VCA*'s wording implies the visitation of others to see the wondrous hermitage, and points to its focus as a place of pilgrimage already within present-day Farne.

The second, nearly identical miracle, found in *VCA* III.2, presents another incident of Cuthbert's divine strength, this time demonstrated by his moving of a stone which four of his brethren had been unable to move, even with the help of Cuthbert's own cart. Cuthbert is said to command the four brethren to bring him the stone, so that he could use it in the construction of his cell: 'Fuit namque lapis in interiore parte insule, quem uehere in uehiculo suo iiii fratribus uisitantibus eum in adiutorium edificii sui *praecepit*' ('Now there was a stone in the interior of the island which he *commanded* four brethren who were visiting him to carry in his cart for the use of his building').[100] As discussed in the previous section concerning the eagle-and-fish miracle, the focus falls on the saint's order (*praecipio*) to the brothers. The saint is clearly in complete obedience to God, as evidenced by his occupation of Farne, and is here giving a direct command to those under him. Their response echoes that of Cuthbert's subordinate in the eagle episode: 'Illi sine mora *statim obedientes*, uenerunt ad lapidem' ('With *prompt obedience*

99 *VCA*, pp. 96–97.
100 *VCA*, pp. 96–97.

they came at once to the stone').[101] Here the brethren are depicted as swiftly obeying the saint's commands, and yet they are unable to complete them by human effort. While the ensuing miracle is left ambiguous, whether a continuation of his divine strength or another instance of angelic ministration, each individual functions appropriately within the hierarchy of obedience.

It must be noted that there is no condemnation of the brothers in this chapter, as they were obedient to the extent their mortal frames allowed. Thus when they return and see the stone already moved and placed in Cuthbert's building, they simply praise God.[102] The *VCA* closes this episode with a biblical reference, chosen solely to give verity to the claims that God can, and therefore does, work miraculously through his saints: 'Mirabilis Deus in *sanctis suis*, et reliqua' ('God is marvelous in *His saints*, etc.').[103] The unwritten rest of the psalm further develops this connection: 'Mirabilis Deus in sanctis suis; Deus in Israel ipse *dabit uirtutem et fortitudinem plebi suae*. Benedictus Deus!' ('God is wonderful in his saints: the God of Israel is he *who will give power and strength to his people*. Blessed be God') (Ps. 67:36). Thus the first two miracles on Farne foreground the structured, hierarchical obedience of everything in relation to that which is above them, God to Cuthbert, Cuthbert to the brethren, as well as the continued presence of physical evidence for those miracles, creating a site for potential pilgrimage.

Water from Stone

The *VCA* transforms Cuthbert's first restoration of Creation on Farne, the bursting forth of a well from the stone floor of his cell in III.3, to focus once again on his saintly obedience and the continued evidence of his miraculous deed. This twin focus not only further highlights the relationship between monastic obedience and prelapsarian miracles, but also further unites the potential lay pilgrimage sites on Farne with the monastery at Lindisfarne.

101 *VCA*, pp. 96–97.
102 *VCA*, p. 96.
103 *VCA*, pp. 96–97.

While the author draws from the rich tradition of biblical exegesis and hagiographical precedent for this miracle, it is clear that he has adapted the episode in order to set Cuthbert apart from other saints. The foundation of all antecedent and parallel hagiographical accounts of well miracles is the biblical narrative in Ex. 17 and Num. 20, where Moses strikes a rock with his staff and produces a well to alleviate the thirst of the Israelites in their desert wandering.[104] Tomás O'Sullivan highlights how this relatively straightforward narrative was given a 'wide range of hermeneutic possibilities' by Paul's assertion that the rock from which Moses obtained a well is Christ,[105] as well as through John's connection in his gospel between the physical water produced for the Israelites and the 'living water' of the Holy Spirit.[106] O'Sullivan further highlights how the Church Fathers, including Augustine and Gregory the Great, as well as later exegetes like Bede, produced a variety of interpretations of these allusions: 'the waters of baptism [...] the four rivers of the gospels flowing out to irrigate the world, or [...] the spiritual sustenance [...] in the mystery of the eucharist'.[107] While the four rivers mentioned here are interpreted as the gospel flowing from Christ, they biblically refer to the river in Eden which splits into four heads: Gen. 2:10, 'Et fluvius egrediebatur de loco voluptatis ad inrigandum paradisum, qui inde dividitur in quattuor capita' ('And a river went out of the place of pleasure to water paradise, which from thence is divided into four heads'). The river itself comes from a single spring in Eden: Gen. 2:6, 'sed fons ascendebat e terra inrigans universam superficiem terrae' ('but a spring rose out of the earth, watering all the surface of the earth'). Bede's *In Genesim* interprets this motif in Gen. 2:10 as providing an image of

104 Num. 20:6, which includes Moses' and Aaron's prayer: 'Dominus Deus, audi clamorem huius populi, et aperi eis thesaurum tuum *fontem aquae vivae*, ut satiati, cesset murmuratio eorum' ('O Lord God, hear the cry of this people, and open to them your treasure, *a fountain of living water*, that being satisfied, they may cease to murmur').

105 T. O'Sullivan, 'The Miraculous Production of Water from Rock and the Impact of Exegesis on Early Irish Hagiography', *Eolas: The Journal of the American Society of Irish Medieval Studies* 3 (2009), 19–50, at 24–25. Also, see 1 Cor. 10:1–4.

106 O'Sullivan, 'Exegesis', 26. See John 4:10–14; 7:38.

107 O'Sullivan, 'Exegesis', 49.

the prelapsarian world (in relation to the description of the land of Sodom and Egypt as verdantly watered like Eden), much like the miraculous interaction of saints and birds:[108]

> Et prouida utique dispositione Dominus ac conditor rerum in nostro orbe uoluit habere similitudinem nonnullam patriae illius ad quam possidendam in primo parente creati sumus, *ut ad promerendum eius reditum* de uicino nos admoneret *exemplo*.[109]
>
> ('And certainly by this careful arrangement the Lord and Creator of things wished to have some likeness of that fatherland in our world, for the possession of which we were created in our first parents, in order to suggest, by a nearby *example, to deserve its return*.')

A miraculous well can recall, then, Eden before Adam's Fall.

From this wealth of interpretative options hagiographers from Sulpicius Severus to Adomnán composed scenes wherein their saints participated in this tradition, and presented their saints engaging the narrative and received exegesis in interrelated and often highly allegorical ways. The author of the *VCA* had ample material upon which to base his depiction of Cuthbert's well miracle, but his use of this common hagiographical topos is distinctive even when compared with the hagiographies he demonstrably knew, or whose ideas were common currency within his contemporary monastic milieu. The *VCA*, for example, does connect the well miracle to biblical precedent through references to Moses, who received water from a rock when he struck it with his staff (Ex. 17:6), and to Samson, who was given water in the jawbone of an ass (Jud. 15:16). Neither scriptural precedent, however, lends more than a touch of legitimacy in the *VCA* to the act of water being given from an unexpected place. The *VCA* makes no attempt to explore the exegetical interpretation of these two Old Testament narratives, and includes no extended engagement with the wealth of potential symbolic meaning. These verses are instead invoked to provide precedent for the physical miracle itself, that act of obtaining water from dry rock when in desperate need. This same process is in play

108 See above, pp. 12–14 for discussion of Bede's theology; see also below, pp. 49, 115, 159, 163, 278–84.
109 Bede, *In Genesim*, p. 48.

with the description of the sweetness of the water from Cuthbert's well, 'in ea aqua a Deo donata omnis liquoris sibi esse suauitas' ('he enjoyed in that God-given water the sweetness of every kind of drink'),[110] which echoes Christ's miraculous transformation of water into wine as an implicit legitimising reference.[111] The *VCA*'s emphasis on Cuthbert's ability to restore portions of Creation is, however, supported by the overarching Edenic motif in the well miracle (especially given the context of Farne and the ensuing restoration miracles that follow), and contributes to the focus on obedience and such restorations.

The central role of obedience in the depiction of well miracles has hagiographical precedent, as can be seen in the *Vita Sancti Colmani*. The exegetical interpretation of the Mosaic well miracle as connected with Christ and the Holy Spirit is implicit here, with the miraculous production of water from dry ground acting to heal bodily destitution, even as the act of baptism heals internal sin. The miracle occurs when a man implores Colman to heal his son. Colman's response comes in the form of an order: '*Iube* terram prope *fodi*, ubi uolueris fontem habere' ('*You should dig* near this ground, and there you will have the fountain you wish').[112] The father's immediate obedience to this order leads to the miraculous well's appearance, and it is from this obedience that the primary miracle of healing occurs when the boy is washed therein:

> Et facta fossa, fons lucidus erupit. Aitque sanctus Colmanus illi homini: 'Laua filium tuum in ipsa aqua'. Et lauato puero, ualidus corpore et sanus sensibus ipse de fonte surrexit. Hoc uiso, omnes repleti sunt gaudio, tam propter aquam, quam propter sanitatem pueri.[113]
>
> (And when he made a ditch, a gleaming fountain burst forth. And saint Colman said to the man: "wash your son in this water". And he washed the son, and the son rose from the fountain strong of body

110 *VCA*, pp. 98–99.

111 John 2:1–11.

112 *Vita Sancti Colmani*, in *Vita Sanctorum Hiberniae partim hactenus ineditae*, ed. C. Plummer (Oxford, 1968), vol. 1, p. 272. The edited text as a whole will hereafter be referred to as *VSH*, and the individual *Lives* by their abbreviated titles: *Vita Colmani, Vita Maedoc*, etc.

113 *Vita Colmani*, p. 272.

and sound in mind. This was seen, and all filled with joy, on account of the water, which gave health to the child.).

The father's faith reveals his righteous obedience, first in humbly asking for the saint's help, and second by obeying Colman and digging the ditch. It is through these acts of obedience that the miracle occurs. It must be stressed, however, that the well miracle is not the focus of this episode in the *Vita Colmani,* but functions in a subsidiary role to the primary miracle: the bodily healing of the boy. By contrast the miracle in the *VCA* centres on the miraculous well itself.

A similar depiction of the role of obedience in the manifestation of a well miracle occurs in the *Vita Sancti Maedoc.* Similar to the *VCA,* St Maedoc, obedient to God, orders a subordinate to delve into the earth physically (achieved here through the uprooting of the tree), and through that subordinate's obedience the miracle occurs.[114] The miracle is closer in its purpose, however, to the *Vita Colmani,* as the miraculous well functions in the *Vita Maedoc* primarily as the catalyst by which a second miracle may occur: to display Maedoc's power in protecting the monastery's lands from worldly temptation. The water from his newly-created well flows along the border between the monastery's land and that of the farmer Becc, whose daughters become temptations by washing their clothes in it.[115] Within both of these hagiographical parallels the focus on obedience as a functional portion of the enactment of the well miracle is paired with its subsidiary status: the wells are important, but are not at the heart of their respective episodes.

The *VCA* clearly engages with this tradition, including elements from the biblical narrative as well as echoing exegetical and hagiographical material, and yet the function, focus, and contextual application of the well miracle in the *VCA* are all distinct. The depiction is grounded upon the *VCA*'s continual focus on the transformation of the physical landscape over the spiritual, and Cuthbert is therefore presented not battling demons, as Antony, Martin, Columba, or later Guthlac did, but instead overcoming the highly physical

[114] *Vita Maedoc,* in *VSH,* vol. 2, ed. Plummer, p. 151.
[115] *Vita Maedoc,* p. 151.

difficulties of his chosen eremitic site. The *VCA* frames the miracle by focusing on Cuthbert's manner of saintly obedience. In his action of receiving the pilgrim brethren and immediately preaching to them he unites the active and contemplative lives: 'Quibus ille secundum morem eius primum uerbum Dei predicauit' ('According to his custom, he first preached the word of God to them').[116] It is from this established obedience that Cuthbert orders the brothers to start digging: 'et fodite in medio pauimento domus meae hanc saxosam terram, quia potens est Deus de rupe petrina petenti aquam suscitare' ('and do you dig this rocky ground in the middle of the floor of my dwelling, because God is able from the stony rock to bring forth water for him who asks').[117] The miracle occurs, however, only after the brothers follow the saint's command, signalled once again by the noun *praeceptum*:

> Fratres uero *secundum praeceptum* eius foderunt terram. Et orante eo, *statim fontem aquae uiue*, sursum in obuiam eius, de saxosa terra erumpere manantem inuenerunt.[118]

> (So the brethren, *according to his command*, dug the earth and as he prayed, at once they discovered *a fountain of living water* which broke out of the rocky ground and poured forth before him.)

The unification of Cuthbert's obedient prayer with the brothers' unhesitant and unquestioning obedience to the saint results in the miraculous transformation of a portion of Creation into a state which functions at the service of Cuthbert. The depiction, echoing the miracles discussed previously, of each individual acting in obedience within the divine hierarchy, is central to the *VCA*'s alteration of the common hagiographical trope. This leads directly to the most substantial change the *VCA* makes to its well miracle – the focus on the continuing presence of the well and its potential as a site for pilgrimage in connection with the hagiographer's present moment.

In the final section of *VCA* III.3, for the first time in the narrative, the author explicitly unites the miracles with his present moment: 'Cuius nos magnam suauitatem dulcedinis *usque hodie* degustantes

116 *VCA*, pp. 98–99.
117 *VCA*, pp. 98–99.
118 *VCA*, pp. 98–99.

cum gratiarum actione probauimus' ('The great sweetness of its flavour we have proved and still thankfully prove by tasting it, *even until the present day*').[119] The use of the second-person plural for the verbs is telling because it is the first time the author has employed such a self-reflexive statement in describing Cuthbert's miracles. Not only is this miracle still efficacious in the present day, but also its sweetness has been tested by the hagiographer himself. At first glance the phrasing for the continued presence of the well is prosaic enough, *usque hodie*, a common idiomatic Latin phrase used by authors such as Jerome to indicate something that is simply still occurring in the present.[120] The application of the idiom in hagiography, however, points to something more meaningful, as it is commonly used to describe locations connected to sanctity and pilgrimage, as can be seen in Bede's description of St Oswald:

> Namque in loco ubi pro patria dimicans a paganis interfectus est, *usque hodie* sanitates infirmorum et hominum et pecorum celebrari non desinunt.[121]
>
> (Indeed in that place where he was slain by the heathen fighting for his homeland, sick men and beasts are healed *to this day*.)

Here the phrase is used to describe a location that now functions as a site for pilgrimage, where lay pilgrims journey for the purpose of obtaining healing through the continued efficacy perceived to be present in the physical location. Jerome himself was not unfamiliar with this use, and in his *Vita S. Hilarionis* employs the phrase to discuss a dispute between the Palestinians and the Cypriots:

119 *VCA*, pp. 98–99.

120 Jerome, *Epistulae*, ed. I. Hilberg and M. Kamptner, 4 vols, CSEL 54–56.2 (Vienna, 1996), *Epistola XXXIX*, p. 302: '[F]lent *usque hodie* iudaei et nudatis pedibus in cinere uolutati sacco incubant' ('The Jews weep *to this day*, and with naked feet they lie in sackcloth and roll in ashes'). See also *Epistola LXV*, p. 636: 'βάϱις uerbum sit ἐπιχώϱιον palaestinae et *usque hodie* domus ex omni parte conclusae et in modum aedificatae turrium ac moenium publicorum, βάϱεις appellentur' ('Baris is the word epichorion of Palestine: and to this day the place is enclosed in all parts, and in the manner of a built tower with ramparts of the state, is called Baries').

121 Bede, *HE*, III.9, pp. 242–43.

> Cernas *usque hodie* miram inter Palaestinos et Cyprios contentionem, his corpus Hilarionis, illis spiritum se habere certantibus. *Et tamen in utrisque locis magna quotidie signa fiunt.*[122]
>
> (Even *to this day* there is a surprising rivalry between those of Palestine and those of Cyprus, the first contending they have the body of Hilarion, and the other his spirit. *And yet in both places great miracles are done daily.*).

The phrase is also employed by medieval hagiographers to describe the continued presence of wells associated with their saints. For example, Adomnán's depiction of Columba's well miracle, performed for the purpose of baptising a child, concludes with passing mention of its continued presence: '*ubi hodieque* fonticulus sancti nomine Columbae pollens cernitur' ('there *even today* a spring is seen, that is potent in the name of Saint Columba').[123] Likewise in the *Vita Sancti Albei*, after the saint has caused water to pour forth from a stone to supply a newly built church, the *vita* connects the well with a recognisable landscape feature in the present: 'Cui Albeus ait: "Quamuis modica est, eterna tamen erit." Unde *usque hodie* nomen fluminis istius *Buanan Cille Ruaid*' ('Ailbe said to him: "Although it is little, it will nonetheless be eternal". Thus *up to this day* the name of that river is *Buanan Cille Ruaid*').[124] In both of the early Irish examples, however, the phrase is employed without specific pattern, and in both *vitae* the wells are not connected to the specific notion of lay pilgrimage, as their wondrous appearances function as subsidiary miracles. In the *Vita Columbae*, the primary miracle is the prophecy concerning the fate of the baptised child

[122] Jerome, *Vita S. Hilarionis* in *Jérôme: Trois vies de moines: Paul, Malchus, Hilarion*, ed. P. Leclerc, E. M. Morales, and A. de Vogüé, Sources chrétiennes 508 (Paris, 2007), p. 298. All references to these three *Lives* from Jerome will be from this edition, cited by title and page number.

[123] *Vita Columbae*, pp. 108–09.

[124] 'Vita Sancti Albei', in *Vitae Sanctorum Hiberniae: ex codice olim Salamanticensi nunc Bruxellensi*, ed. W. W. Heist (Brussels, 1965), p. 124. See also O'Sullivan, 'Exegesis', 39, n. 64. Compare with Plummer, *VSH*, p. 54: 'Tunc sanctus Colmanus ad Albeum dixit: "Modica est aqua ista." Cui Albeus ait: "Quamuis modica est, semper non deficiet, et riuulus eternus erit usque ad finem mundi huius." Vnde nomen fluuii dicitur Buanann Chille Ruaid, id est indeficiens riuulus Chelle Ruaid.'

and,[125] as O'Sullivan highlights, the entire narrative echoes that of Exodus: 'It presents a picture of the Christian life as a journey through the desert, during which we are saved and nourished by the waters of grace which flow, through the conduit of the saint, from the rock that is Christ.'[126] The miracle is unconcerned with the physical well, and presents no attempt to focus on the site's potential for pilgrimage. Similarly, Ailbe's well functions as a signifier pointing to exegetical interpretation, rather than as a future location for pilgrimage. Ailbe's miracle focuses on the perpetuity of the well ('eterna tamen erit') in such a way that the stream, in O'Sullivan's words, 'fulfill[s]' the promise of John 4:13:[127] 'qui autem biberit ex aqua quam ego dabo ei, non siteit in aeternum' ('but he that shall drink of the water that I will give him, shall not thirst forever'). The Irish *vitae* function differently than the *VCA*, and have distinct purposes for their well miracles.

The use of the phrase *usque hodie* (and slight variations thereof) is not unique to the *VCA*, but builds on hagiographical tradition. What is distinctive, however, is the *VCA*'s consistent pattern of employing the term, using it only when describing the miracles on Farne with a single important exception:[128] in reference to Cuthbert's translation and the interment of his bones and clothing as relics in Lindisfarne. Here the continued efficacy of the miraculous sites is connected explicitly to the primary *locum sanctum* and, I would suggest, to a concerted attempt to create points of pilgrimage in connection with Lindisfarne:

125 See *Vita Columbae*, pp. 108–09: 'Sanctus, ad proximam declinans rupem flexis genibus paulisper orauit; et post orationem surgens eiusdem rupis frontem benedixit. De qua consequenter aqua abundanter ebulliens fluxit, in qua continuo infantulum baptizauit. De quo etiam baptizato haec uaticinans intulit uerba [...] Quae omnia eidem uiro iuxta sancti contigerunt uaticinium' ('The saint turned aside to a rock close by, bowed his knees, and prayed for a little while. And rising after his prayer, he blessed the face of the rock, from which thereupon water flowed in an abundant cascade; and he immediately baptized the infant. And also, concerning the baptized child, he pronounced these prophetic words [...] All these things happened to the man according to the saint's prophecy.').

126 O'Sullivan, 'Exegesis', 37.

127 O'Sullivan, 'Exegesis', 40.

128 See *VCA*, pp. 95–107.

> Nam sudarium reuoluentes quo capud eius cingebatur, pristine candiditatis pulchritudinem custodiens, et ficones noui quibus calciatus est, in basilica nostra contra reliquiis *pro testimonio usque hodie* habentur.[129]
>
> (They unwound the headcloth in which his head was wrapped and found that it kept all the beauty of its first whiteness; and the new shoes, with which he was shod, are preserved in our church over against the relics, *for a testimony, up to this present day*).

The *VCA* closes this episode with an implicit invitation for pilgrimage in the description of the sweetness of the water Cuthbert drank from his well, 'in ea aqua a Deo donata omnis liquoris sibi esse suauitas' ('he enjoyed in that God-given water the sweetness of every kind of drink'),[130] which echoes Ps. 33:9, '[g]ustate, et uidete quoniam suauis est Dominus; beatus uir qui sperat in eo' ('O taste, and see that the Lord is sweet: blessed is the man that hopes in him'). It is a call to come and taste the sweet water, directly connected to seeing the saint in Lindisfarne. Thus the *VCA* has adapted the well miracle to align both with its focus on the role of obedience in the restoration of Creation, and with its novel desire to present the miracles on Farne as sites for potential pilgrimage where such restorations once took place.

Restored Creation and Pilgrimage

In III.4 and III.5, the *VCA* brings to fruition its exploration of the role of obedience in Cuthbert's ongoing restoration of Creation, as well as its related focus on the literary creation of sites for potential pilgrimage. The two miracles found in these chapters form a distinct pair, with the first presenting the restoration of elemental Creation, and the second the animate world. These two function together as a microcosm of all that has gone before, and are in many ways the

[129] *VCA*, pp. 132–33.

[130] *VCA*, pp. 98–99. Also, in the final chapter of the *VCA* summarising miracles omitted, the author relates how Cuthbert later did turn water to wine, pp. 138–39.

culmination of the hagiographer's exploration of the saint's miraculous interaction with Creation.

VCA III.4 places the connection between obedience and the restoration of Creation at the heart of the ensuing miracle through the framing device of its title: 'De ligno quod *mare seruiens* ei detulit' ('How the *sea served* him by bringing him wood').[131] The miracle is focused upon the sea's obedient service to the saint and, as with the other miracles depicted on Farne, this service is contingent on Cuthbert's saintly obedience. In keeping with the *VCA*'s focus on the physical landscape of Farne, the text here carefully describes the rocks hollowed out by the waves over which Cuthbert wishes to place a twelve-foot beam to build his *domunculus* ('toilet'). Akin to the stone miracles and the well miracle, the *VCA* presents Cuthbert here employing his brethren in the construction of his habitable eremitic landscape and thereby engaging directly with Creation. This interaction is distinct, however, from those discussed above. Here the *VCA* changes the lexis by which Cuthbert gains assistance, from its consistent and particular use of the term *praecipio*,[132] to the weaker terms *peto* and *deposco*.

In the initial depiction of Cuthbert's request for a twelve-foot beam, the *VCA* employs the verb *peto*: 'lignum xii pedum in longitudine ad fundamentum alicuius domunculi *petiuit*' ('some wood twelve feet in length, which *he desired* as a foundation for a certain small building').[133] After the brothers forget, the question is phrased as an earnest request (*deposcere*): 'Quod uero a fratribus *deposcens* non perpetrasset' ('He *had asked* the brethren [for this beam] but would not have obtained it').[134] Both requests are a departure from the consistent use of *praecipio* to describe Cuthbert's commands to his brothers in the previous miracles. The first term, *peto*, has a range of meanings whose unifying feature is supplication,[135] and is rarely invoked as an authoritative command. This term is used consistently in Scripture for the act of imploring someone else for

131 *VCA*, pp. 98–99.
132 See above, pp. 34–36.
133 *VCA*, pp. 100–01.
134 *VCA*, pp. 100–01.
135 *DMLBS*, 4: 'to ask (someone) to do something'; 5: 'to try to obtain by asking, request'.

something.[136] The term therefore carries no imperative stress or sense of an authority giving an order that is expected to be obeyed.

The second term, *deposco*, carries more of an imperative stress than *peto*, but is likewise used most often in the context of putting forth a need, rather than as a directive.[137] The word is never used in Scripture, and apart from the *VCA* does not appear to be part of the common hagiographical lexis; for example, Bede does not use the word in either of his *vitae* of Cuthbert. The *VCA*, however, is consistent in the application of this term as well. In each of its seven uses the meaning is the same: a request stemming from a need. At the end of Book I, for example, a woman tormented by a devil to madness is healed when her husband asks Cuthbert to send a priest to help her, '[s]ed tantum presbiterum aliquem secum mittere, et requiem sepulture *deposcebat*' ('but he only *asked* Cuthbert to send a priest with him and that she might find peace in the grave').[138] This term is used as an equivalent act of imploring in *VCA* IV.17, where a paralytic boy is healed by asking to wear the shoes from Cuthbert's uncorrupt body: 'Ideo namque *deposco* ab abbate calciamenta que circumdederunt pedes sancti martyris Dei incorruptibilis' ('So *I ask* the abbot for the shoes which were on the feet of the holy and incorruptible martyr of God').[139] In the *VCA*, therefore, the verb *deposco* functions as a form of supplication, and carries with it little sense of an authoritative command; the failure to respond to such a request carries no consequence as severe as the Fall from breaking

[136] For example, John 4:10, in which Jesus explains to the woman about the living water he is able to give, he tells the woman that if she knew, she would ask (*peto*) him for it: 'Si scires donum Dei, et quis est qui dicit tibi: Da mihi bibere, tu forsitan *petisses* ab eo, et dedisset tibi aquam vivam' ('If you knew the gift of God, and who it is that says to you, "Give me to drink"; you perhaps would *have asked* of him, and he would have given you living water'); also, Ps. 20:5, in which David is described as asking God for life: 'Vitam *petiit* a te, et tribuisti ei longitudinem dierum in saeculum, et in saeculum saeculi' ('He *asked* life from you: and you have given him length of days for ever and ever'). For other examples see Est. 5:3; Acts 3:2; and Ex. 10:11.

[137] *DMLBS*, 1: 'to request, seek, pray (for); 2, to entreat, beseech'. Augustine uses the term with an equivalent meaning, see *Epistola XC*, in *Sancti Aurelli Augustini epistulae LVI–C*, ed. K. D. Daur, CCSL 31A (Turnhout, 2005), p. 153.

[138] *VCA*, pp. 92–93.

[139] *VCA*, pp. 136–37.

a *praeceptum*. The change is slight but telling: the apparent disobedience of the brothers' forgetfulness is no longer a breaking of a *praeceptum*, but is instead a much more forgivable trespass, as it was only a request, not a command. The *VCA* highlights this distinction by declaring that the monks should not be judged too harshly.[140]

The brothers' forgetfulness becomes instead the catalyst through which the miraculous takes place. Cuthbert's saintly obedience, exemplified in prayer, is rewarded by the restoration of the wildest of all elemental forces, the sea:

> Nam cum eadem nocte *mare fluctibus undans in honorem serui Dei*, stipitem xii pedum detulens specialiter, iam ad [ostium] scopuli ubi ponendus erat in aedificium natantem deportauit.[141]
>
> (For that same night *the sea, uplifting its waves in honour of the servant of God*, landed a floating timber exactly twelve feet in length, just at the opening by the rock where it was to be placed for the building.)

The author's deliberate use of the noun *servus* connects the sea's service to Cuthbert as given in the title of the chapter, 'mare *seruiens*', with the saint in his complete obedience as the ultimate servant of God ('*serui* Dei'). This unity of service, of God served by Cuthbert who is subsequently served by restored Creation, is given in direct contrast to the forgetfulness of the brothers, who marvel at the sea's service in obedience to the saint: '*admirantes* quod *mare* in *honorem* Christi magis *obediens anachorite* quam homines parauerunt' ('*marvelling* that *the sea* in *honour* of Christ had accomplished more than men, in *obedience* to *the hermit*').[142] The sea uses its waves to honour Cuthbert by placing the needed timber in exactly the right spot, contrasting the idle hands of the brothers who lifted nothing in response to the saint's requests. This depiction of perfect order, of the world operating as it should in prelapsarian harmony, is emphasised by the author's declaration of the perfect actions performed by the sea. Not only does the sea bring a floating timber of exactly the right

140 *VCA*, p. 100.

141 *VCA*, pp. 100–01. Colgrave leaves *hostium* when MS P gives *ostium*, which is the word he uses for translation, and which makes the most contextual sense. I have chosen to emend to *ostium*, therefore, for reasons of sense.

142 *VCA*, pp. 100–01.

size ('specialiter'), but also places it exactly where it needed to be for the construction of the little chamber. While the scene does not display the level of blinding holiness of the sea animal miracle, the same theological framing is on display: the saint, by virtue of his obedience and his position as the paragon servant of God, restores Creation so that it serves him as it would have served Adam in the prelapsarian world.

In keeping with the Farne section as a whole, the *VCA* moves directly from the depiction of the miracle itself to the continued efficacy of the site for potential pilgrimage: 'et *adhuc usque hodie* nauigantibus domus super lignum transuersum aedificata apparet' ('and *even until this day* the house, built upon the cross-timbers, is still to be seen by mariners').[143] The employment of the phrase *usque hodie* ties the continued presence of the chamber to the stones and well that can be seen on Farne itself. This visibility functions as a beacon for pilgrimage to Farne, as it existed as a point of reference for those travelling by sea from nearby Bamburgh. The emphasis on its visibility is not simply descriptive of its present state, but is brought to the fore to further the depiction of Farne as a space where miracles have, and potentially still could, occur.

The second chapter of this pair, III.5, describes a miracle which echoes the earlier sea miracle, with Cuthbert engaging animate instead of elemental Creation in the form of thieving ravens. This episode has its roots in hagiographical precedent, and scholars such as Thacker have connected it to the story of the penitent wolf in Sulpicius Severus' *Dialogi*, as well as the narrative of Columbanus banishing the raven that steals a glove in the *Vita Columbani*.[144] In the former, the depiction resembles the *VCA* insofar as it contains an animal seeking repentance from a hermit; otherwise the depictions have little in common, as there is a pre-established relationship between the wolf and the hermit, and there is also no rebuke or

143 *VCA*, pp. 100–01.

144 See Thacker, 'Social and Continental Background', pp. 90, 101. Stancliffe, 'Pastor and Solitary', p. 26, sees a connection to the raven who takes away St Benedict's poisoned loaf in Gregory's *Dialogues*, vol. 2, pp. 160–70. See also Cavill, 'Some Dynamics of Story-telling', at 4–5. The various hagiographical examples have their source in the biblical narratives of Noah and Elijah; see below, p. 116.

interaction between them, with the wolf independently recognising its trespass.[145] In the latter, the depiction in the *Vita Columbani* aligns with the *VCA* only in so far as Columbanus also banishes a thieving raven by verbal command after the bird has returned.[146] The *VCA*'s depiction is distinct in many ways, firstly in its linking of this episode to the sea's obedience: 'Sicut ergo diximus mare seruientem homini Dei, ita et aues coeli obedierunt ei' ('We have told how the sea served the man of God; so also the birds of the air obeyed him').[147] Bede later uses the same pair of miracles to explicate his exegesis concerning the restoration of Creation in his two *vitae* of Cuthbert.[148] The *VCA* is not as straightforward in its interpretation of these miracles, but contextually the connection between saintly obedience and this theology brings this joining of elemental and animate Creation in service to the saint in line with Augustinian exegesis of the Fall.

The *VCA* presents Cuthbert as once again diligent in his eremitic life, here via agrarian labour, when the two ravens tear apart the thatch from Cuthbert's guesthouse to build their nest.[149] While scholars have interpreted Cuthbert's ensuing reaction (shooing the birds away) as a depiction of the defence of the human and constructed,[150] the focus here rests not on the privileging of the humanly wrought building and a delineated sense of property rights, but instead on Cuthbert's ability to restore Creation into a proper relationship, functioning as it should within the now restored divine hierarchy.

The author of the *VCA* is once again careful here in his lexical employment when describing obedience and disobedience,

145 See Sulpicius Severus, *Dialogi*, Book I.14, *PL* 20, cols 192C–193C.

146 See Jonas, *Ionae Vitae Sanctorum Columbani, Vedastis, Iohannis*, ed. B. Krusch (Hanover, 1905), I.15, pp. 178–79. Hereafter the *Vita Columbani*.

147 *VCA*, pp. 100–01. Thacker, 'Social and Continental Background', p. 101, notes a verbal correspondence with the *Vita Columbani*, but once again the parallels are minor and not particularly relevant to the discussion at hand: compare *VCA*, '*aues* coeli *obedierunt* ei', with *Vita Columbanus*, '*avium oboedientia* clarescant'. See *Vita Columbani*, I.15, pp. 178–79.

148 See *VCA*, p. 225, and *VCM*, Chs. XVII–XIX; See also, Bede, *In Genesim*, pp. 68, 94, for similar discussions.

149 *VCA*, p. 100.

150 For example, Crane, *Animal Encounters*, pp. 34–35; and Alexander, *Saints and Animals*, pp. 46–47.

distinguishing between two different kinds of authoritative interactions in the banishing of the ravens and their subsequent miraculous contrition. Mirroring his care in the previous chapter, where Cuthbert never employs the stronger term *praecipio* in his directions to the forgetful brothers, the author of the *VCA* presents Cuthbert's initial prohibition in a similar manner: '*Prohibuit* autem eos leni motu manus, ne hanc iniuriam fratribus nidificantes facerent' ('*He bade* them, with a slight motion of his hand, not to do this injury to the brethren, while building their nests').[151] Here instead of the verb *deposco* the action is described with *prohibeo*, and while *prohibeo* is used with a similar force to *praecipio* elsewhere, including by Augustine in reference to the prohibition in Eden,[152] the term has specific meaning within the *VCA*. The verb *prohibeo* is used only three times in the text including this one, and in the other two instances is applied as a word of description rather than prescription or directive. *VCA* II.4 employs *prohibeo* to describe the actions of the raging sea that stopped Cuthbert and his brothers from continuing their sea-journey in the lands of the Picts.[153] The term is used to describe the hindering power of the storm, its preventative ability. Equivalently in *VCA* III.6, Cuthbert uses *prohibeo* as a description of the hindering power of the illusory fires, which try to prevent the villagers from hearing the word of God.[154] In light of these previous instances, the *VCA*'s employment of *prohibeo* here to describe Cuthbert's actions in response to the ravens is telling. There is little force, then, in either the word *prohibeo* or Cuthbert's action, where he tries to banish the ravens with a perfunctory wave of his hand: '*leni* motu manus' ('a *slight* motion of his hands').[155] The birds, unsurprisingly, are unmoved. Cuthbert must invoke his

151 *VCA*, pp. 100–01.

152 See, for example, Augustine, *De Gen. con. Man.*, p. 143.

153 *VCA*, pp. 82–83: 'tempestas maris potestatem iterum nauigandi *prohibuit*' ('[T]he tempestuous sea *prevented* them from continuing their voyage').

154 *VCA*, pp. 88–89: 'nec foras currentes a uerbi Dei auditione tardamini, per inlusionem *proibiti*' ('do not run out and be prevented from hearing the word of God, nor be *hindered* by an illusion').

155 *VCA*, pp. 100–01.

authority over animate Creation with a true, firm command, by which the subordinate obedience of the ravens is displayed:

> postremo motato spiritu, austere *praecipiens* in nomine Iesu Christi de insula discedere exterminauit. Illis igitur nec requies, nec mora patriam secundum *preceptum* eius deserentibus.[156]
>
> (at last his spirit was moved and sternly *commanding* them in the name of Jesus Christ to depart from the island, he banished them. Without any pause or delay, they deserted their homes according to his *command*.)

There is, therefore, a contrast of the two terms in the *VCA*: *prohibeo* and *praecipio*. The former implies a descriptive and preventative use, and the latter, a use that echoes the great command in the Garden that was broken. Within such a paradigm Cuthbert's second action, described as a *praeceptum*, represents a depiction of the world as it should be. The miracle that follows his rebuke is therefore causally dependent upon Cuthbert's continuing obedience to the eremitic ideal, evidenced by his attempt to provide his own sustenance,[157] together with his unification of the active and contemplative lives, which allow his command (*praeceptum*) to be received and followed by animate Creation, here represented by the ravens.

The *VCA*'s focus on the relationship between the divine obedience of the saint and the ensuing restoration of animate Creation is furthered in its depiction of the subsequent miraculous return and penitence of the ravens:

> post triduum alter e duobus reuertens ante pedes hominis Dei fodienti iam ei, terram supra sulcum expansis alis, et inclinato capite, sedens et merens humili uoce ueniam indulgentie deposcens, crocitare cepit.[158]
>
> (after three days, one of the two returned to the feet of the man of God as he was digging the ground, and settling above the furrow

156 *VCA*, pp. 100–01.

157 See *VCA*, pp. 100–01: 'primum enim duobus uel tribus annis de opere manuum suorum antequam clausus obstructis ianuis intus maneret, laborans cotidianum uictum acceperat' ('For at first, for two or three years before he shut himself in behind closed doors, he laboured daily and gained his food by the work of his own hands').

158 *VCA*, pp. 102–03.

with outspread wings and dropping head, began to croak loudly, with humble cries asking his pardon and indulgence.)

Scholars have noted the formulaic nature of the birds' contrition,[159] with the prostration and head bowing echoing both Scripture, more generally, and passages in works like Benedict's *Regula*, which though not a direct source here, is an illuminating correspondence nonetheless. In Ch. 71 of the *Regula*, Benedict focuses on obedience and how a brother might seek reconciliation when he has failed to observe it:

> Si quis autem frater pro quavis minima causa ab abbate, vel a quocunque priore suo corripiatur quolibet modo [...] mox sine mora tamdiu *prostratus in terra ante pedes eius iaceat satisfaciens*, usque dum benedictiones sanetur illa commotio.[160]
>
> (If, however, a brother is corrected in any way by the Abbot or by any of his superiors [...] let him immediately, without delay, *prostrate himself on the ground at his feet*, and lie there making satisfaction until that agitation is healed with a blessing.)

While the description of monastic obedience in Benedict's *Regula* is not directly equivalent to that of the *VCA*, the echo in the *VCA* of such expectation in the actions of each individual involved is nonetheless striking: a person who has breached the rules of obedience must come before their spiritual better and bow to the earth before their feet and ask for pardon. To conclude, however, that the ravens are nothing but signifiers,[161] a subtle mimetic critique of bad monastic practice, is to miss the central focus of the passage in the context of the *VCA* as a whole. While there is didactic potential in this depiction (Bede expands this in his later iterations), the fundamental focus of this episode is on the actual interaction of the saint with physical Creation, with real ravens on the island of Farne. The premise by which the miracle works is that while ravens do indeed make nests and would have used a thatched roof

159 For example, see Crane, *Animal Encounters*, p. 35, in relation to her argument about vernacular forms of hospitality.

160 Benedict, *Regula*, Ch. LXXXI.2, p. 668.

161 See S. Duncan, '*Signa de Caelo* in the Lives of St Cuthbert: the Impact of Biblical Images and Exegesis on Early Medieval Hagiography', *The Heythrop Journal* 41 (2000), 399–412, at 404–05.

as building material, they would not, under the normal causal circumstances of the postlapsarian world, leave when commanded, and then return to ask for pardon. The echo of monastic praxis in the penitent raven's actions is therefore not primarily about that praxis, but points to something more fundamental: that with Cuthbert's sanctity the prelapsarian hierarchy has been restored, and within such a relationship, humanity has regained true stewardship over Creation. His sanctity is once more defined in relation to obedience here, as the *VCA* calls him the servant of Christ ('Seruus [...] Christi').[162] The miracle is complete, and the saint pardons the ravens' 'sin'.[163]

The miracle is not left, however, in the past tense of the primary narrative, but is brought into present-day Lindisfarne. Not only did the saint pardon the 'sins' of the ravens, he allowed them to stay and make a home on Farne, where they remain to the present day: '*usque adhuc* illic manent' ('and they remain there *until today*').[164] This is the third time the *VCA* has used this construction with *usque adhuc*, which unites this miracle with the general focus on the present evidence for the wonders Cuthbert performed in places familiar and traversable to the audience of the *VCA*. When the author reports that he has verified this by trustworthy witness, as hagiographical precedent demands, it is in a sense collapsing all times into one: the moment of the miraculous, its evidence in the witnesses whose boots were greased by the lard, and finally, the ravens who still live on Farne in the present. The implication, via the repeated use of *usque adhuc* in combination with the verified witnesses, is of another site for potential pilgrimage, where any raven that is seen living on Farne becomes a reminder of their miraculous prelapsarian interaction with Cuthbert.

162 *VCA*, pp. 102–03.
163 *VCA*, p. 102.
164 *VCA*, pp. 102–03.

Conclusion

The *VCA* stands near the beginning of Anglo-Latin hagiography, and as such sets a precedent for the later *vitae* produced in England. It centres on Cuthbert's ability to restore Creation by means of his sanctity (evidenced in his perfect obedience in a monastic setting), which echoes and redeems Adam and Eve's disobedience at the Fall in Eden. This focus is most clearly seen in the author's persistent emphasis on the role of obedience within the divine hierarchy as necessary for the temporary restoration of portions of Creation. The strict obedience to monastic life (whether Columban, Cuthbertine, or even Gregorian) exemplified in Cuthbert's actions, and those of his subordinates, reflects the wider process by which the world is transformed. The *VCA* draws this discussion of obedience and transformation down to earth by delineating a physical landscape, not only through highly physical descriptions, as well as the inclusion of numerous place names, but also by means of creating textual sites for potential lay pilgrimage. These *foci* in the landscape, including Cuthbert's hut on the sea-cliff on Farne, the fountain on Farne, and Cuthbert's tomb on Lindisfarne, not only display the saint's ability to restore Creation, but also function to unite the continuing efficacy of Cuthbert's miracle-working on Farne with the monastic community at Lindisfarne.

2

Ruminative Poetry and the Divine Office: Bede's Metrical *Vita Sancti Cuthberti*

Bede's metrical *Vita Sancti Cuthberti* (*VCM*)[1] occupies a unique place in hagiography. Chronologically it is the second life of Saint Cuthbert and, if the extant manuscripts are any indication, it was widely influential, especially in the mid-tenth century.[2] It is also the only extant verse saint's life composed by Bede. Scholars have discussed its relation to Latin metrics,[3] its place in the *opus geminatum* tradition in England,[4] and its relation to the development of Bede's ideas concerning Cuthbert as they relate to his later *VCP*.[5] The *VCM* has, however, received remarkably little critical attention as a text in its own right. Bede's *VCM*, I will argue here, anticipates his reshaping of Cuthbert in the *VCP* into an idealised Gregorian monk-pastor, by emphasising Cuthbert's monastic obedience in observing the Divine Office. Further still, the *VCM*, in form and content, was not only a ruminative and poetic exercise for Bede himself, but also was designed as a text for a monastic audience as an aid for rumination and meditation.

1 All quotations from *Bedas metrische Vita Sancti Cuthberti*, ed. W. Jaager (Leipzig, 1935).

2 See M. Gretsch, *Ælfric and the Cult of the Saints in Late Anglo-Saxon England* (Cambridge, 2005), pp. 103–10, concerning the *VCM* in particular, and pp. 75–110, for the popularity of the Cuthbertine cult well into the eleventh century.

3 For example, Lapidge, *Anglo-Latin Literature, 600–899*, esp. pp. 313–48.

4 B. Friesen, 'The *Opus Geminatum* and Anglo-Saxon Literature', *Neophilologus* 95 (2011), 123–44; P. Godman, 'The Anglo-Latin "Opus Geminatum": from Aldhelm to Alcuin', *MÆ* 50 (1981), 215–29; G. Wieland, '*Geminus stilus*: Studies in Anglo-Latin Hagiography', in *Insular Latin Studies: Papers on Latin Texts and Manuscripts of the British Isles, 550–1066*, ed. M. W. Herren (Toronto, 1981), pp. 113–33.

5 See Newlands, 'Bede and Images of Cuthbert'; G. H. Brown, *A Companion to Bede* (Woodbridge, 2009), pp. 82–85.

The *VCM* was likely composed during the reign of Osred, between 705 and 716. On the basis of Bede's positive descriptions of the king in lines 552–55, such as the reference to him as 'venerabile',[6] and as a new Josiah in lines 554–55, Michael Lapidge posits a date of composition during the early period of Osred's reign, between 705 and 707.[7] These positive references to Osred are in direct contrast to the portrayal of the later part of his reign, during which he becomes, in Lapidge's phrase, 'a wicked Ahab, defiling nuns and murdering noblemen'.[8] Lapidge has also identified two distinct recensions of the poem. The first recension, which Lapidge categorises as the B-recension after the single manuscript witness, Besançon, Bibliothèque municipale 186, represents an early draft of the poem by Bede, potentially close to the initial version produced between 705 and 707.[9] The second recension, comprising nearly twenty manuscripts, which Lapidge terms the vulgate version, represents a revised text produced by Bede in connection with his subsequent composition of the *VCP*, sometime after 721.[10] Lapidge's observations allow a unique comparative approach whereby Bede's poetic development can be viewed, as the vulgate revision was almost entirely concerned with perfecting issues of verse and metre.[11] They also highlight the importance of seeing the production of the *VCM* as comprising the second half of a conceived *opus geminatum* with the anonymous prose *VCA*; this is despite the later reception history

6 Lapidge, *Anglo-Latin Literature, 600–899*, p. 34; Cf. *VCM* l. 552.

7 Lapidge, *Anglo-Latin Literature, 600–899*, p. 34. More recently Alan Thacker has suggested a range of 705–06, also in reference to King Osred. See A. Thacker, 'Bede and History', in *The Cambridge Companion to Bede*, ed. S. DeGregorio (Cambridge, 2010), pp. 170–89, at 181.

8 Lapidge, *Anglo-Latin Literature, 600–899*, p. 34. Also see Newlands, 'Bede and Images of Cuthbert', 76, n. 19.

9 Lapidge, *Anglo-Latin Literature, 600–899*, pp. 345–46, posits that this MS 'was copied from a (lost) exemplar in Anglo-Saxon miniscule originating from the area of the Anglo-Saxon mission and dating from the second half of the eighth century. It thus takes us back to the earliest continental phase of the transmission of Bede's writings to the continent, to the period of Lul's request to the abbot of Wearmouth/Jarrow for *libellos de uiro Dei Cudberceto metro et prosa conpositos*'. The *VCM* appears on folios 1r–24r.

10 Lapidge, *Anglo-Latin Literature, 600–899*, p. 347. For dating of Bede's *VCP*, see Colgrave, *VCP*, p. 16.

11 Lapidge, *Anglo-Latin Literature, 600–899*, p. 341.

where Bede's *VCP* and *VCM* appear together in eight manuscript witnesses, whereas the *VCM* never appears with the *VCA*.[12]

The *VCM* as *Opus Geminatum*

As many scholars have noted, Bede's *VCM* was created in the tradition of the *opus geminatum*;[13] though taking a number of forms, it is helpfully defined by Friesen as 'a pair of texts, one in verse and the other in prose, which ostensibly treat the same subject', but which 'does not require that the same writer compose both halves, nor that they be written or read in a particular order'.[14] The *opus geminatum* tradition does, however, often rely on intertextual relationships, where the meaning of one half is enriched by knowledge of the other half. This twin form has its root in classical paraphrase, most notably in the works of Cicero, Quintilian, and Theon,[15] but which came to be applied to biblical content by Juvencus in his verse paraphrase of the life of Christ, *Evangelia*,[16] in 329 or 330.[17] As Friesen explains, in Juvencus the 'paraphrase brings together two contrary realms: the glory of the divine law and the ornament of terrestrial language'.[18] Juvencus's *Evangelia* was an attempt to unite the established literary traditions of high classical culture with material dealing with divine truth, for the purpose of engaging a cultured audience more familiar with classical rhetoric and metre.[19]

12 See Friesen, '*Opus Geminatum*', at 139: 'Bede's insistence upon the unity of these two works carried somewhat more currency than Aldhelm's: of the thirty-eight manuscripts which contain the prose life, eight also contain the verse (Colgrave and Mynors 1985, p. 54)'. Colgrave's list, *VCP*, pp. 20–39, is equivalent to that found in *Anglo-Saxon Manuscripts: a Handlist*, ed. Gneuss and Lapidge, p. 897.

13 See for example, Wieland, '*Geminus stilus*', pp. 113–33; Friesen, '*Opus Geminatum*', 137–39; E. Knibbs, 'Exegetical Hagiography: Bede's Prose *Vita Sancti Cuthberti*', *Revue bénédictine* 114 (2004), 233–52, at 236–37.

14 Friesen, '*Opus Geminatum*', 124.

15 See Friesen, '*Opus Geminatum*', 124–29, for historical overview.

16 See Juvencus, *Evangeliorum Libri Quattuor*, ed. J. Huemer, CSEL 24 (Vienna, 1891).

17 Friesen, '*Opus Geminatum*', 130.

18 Friesen, '*Opus Geminatum*', 130.

19 See Friesen, '*Opus Geminatum*', 131; and E. Thornbury, *Becoming a Poet in Anglo-Saxon England* (Cambridge, 2014), pp. 50–51.

Lapidge argues that Juvencus was the first Christian-Latin poet, and that he created in the *Evangelia* a diction 'which would be imitated and adopted by all subsequent Christian-Latin poets'.[20] The *opus geminatum* found its complete and most influential form in Caelius Sedulius' fifth-century twin works, the verse *Carmen Paschale* and the prose *Opus Paschale*.[21] Sedulius here follows Juvencus in his creative endeavour, seeking, as both Friesen and Godman note,[22] to provide a learned audience with biblical material of technical and metrical sophistication.[23] The verse portion of the *opus geminatum* as conceived here becomes a pragmatic evangelistic tool. This is due in great part to the pleasure derived through its poetic form which is connected to the Roman rhetorical appeal to the virtue of *voluptas* or pleasure, notably discussed by Quntillian in *Institutio Oratoria* I.28.[24] The final late antique flowering of the tradition is the *De actibus apostolorum* by Arator,[25] the first verse paraphrase of the book of Acts, which was widely read and from which Bede quotes extensively,[26] especially in his *VCM*.[27] Much of the difficulty in reading the *VCM* stems from Bede's imitation of Arator, whose diction, as Lapidge notes, is extremely compressed, a feature he also notes of Bede's *VCM*.[28]

20 M. Lapidge, 'Versifying the Bible in the Middle Ages', in *The Text in the Community: Essays on Medieval Works, Manuscripts, Authors, and Readers*, ed. J. Mann and M. Nolan (Notre Dame, IN, 2006), 11–40, at 13–14.

21 Godman, 'Anglo-Latin "Opus Geminatum"', 217; Thornbury, *Becoming a Poet*, p. 51.

22 See Friesen, '*Opus Geminatum*', 217–18; Godman, 'Anglo-Latin "Opus Geminatum"', 132.

23 Lapidge, 'Versifying the Bible', p. 17, notes that Sedulius 'opened up the possibility of using [the form] for figural and typological interpretation'.

24 Noted by Godman, 'Anglo-Latin "Opus Geminatum"', 218.

25 *Aratoris subdiaconi De actibus apostolorum*, ed. A. P. McKinlay, CSEL 72 (Vienna, 1951).

26 See L. T. Martin, 'The Influence of Arator in Anglo-Saxon England', in *Proceedings of the PMR Conference* 7 (Villanova, 1982), pp. 75–81, esp. 77–80; Friesen, '*Opus Geminatum*', 134; G. Bonner, 'Bede and Medieval Civilization', *ASE* 2 (1973), 71–90, at 86–87; for influence on Bede specifically see M. Lapidge, 'Bede's "Metrical Vita S. Cuthberti"', in *Cuthbert and Community*, pp. 77–93, at 90; and A. G. Holder, 'Bede and the New Testament', in *The Cambridge Companion to Bede*, ed. S. DeGregorio (Cambridge, 2010), pp. 142–55, at 145.

27 See Thacker, 'Social and Continental Background', pp. 114–15.

28 Lapidge, 'Versifying the Bible', p. 20; Lapidge, *Anglo-Latin Literature, 600–899*, pp. 332–33.

The first Anglo-Latin author to engage with the *opus geminatum* tradition was Aldhelm with his twin works, the prose *De Laude Virginitatis* and the verse *Carmen De Virginitate*, both written sometime after 680.[29] What is relevant to the discussion here is Aldhelm's metaphorical explanation for the *opus geminatum* form itself, as noted by Friesen and Godman,[30] where the tradition is depicted as comprising distinct and interrelated portions of a building. The prose constitutes the firm foundation and walls of the structure, while the verse becomes the roof: 'cum tegulis trochaicis et dactilicis metrorum imbricibus firmissimum culmen caelesti confisus suffragio imponam' ('as if the rhetorical foundation stones were laid and the walls of prose were built, so I shall – trusting to heavenly support – build a sturdy roof with trochaic slates and dactylic tiles of metre').[31] This 'vertical dichotomy'[32] is concerned with its metaphorical height rather than its function as a roof image of covering and protection, corresponding to the implicit view of poetry as *voluptas*, high ornament available for pleasurable interaction. Friesen argues that Aldhelm's ascription of 'elegance to verse' and 'eloquence to prose' displays how he perceived the categories, where 'elegance is understood primarily as an aesthetic quality, whereas eloquence focuses on speech that is "well done" […] through its replete articulateness'.[33] The interdependent relationship depicted here functions, in keeping with the building metaphor, with the prose as functional tool and the verse as pleasurable experience. Despite this clear delineation Aldhelm's *opus geminatum* functions poorly within these parameters, as the *De Laude Virginitatis* is in fact much more complex than the *Carmen De Virginitate*, and can hardly be said to possess a more utilitarian function.[34] Nonetheless the *Carmen De Virginitate* necessarily entails knowledge of the prose

29 On the difficulty of dating the twin works see *Aldhelm: The Poetic Works*, trans. M. Lapidge and J. L. Rosier (Cambridge, 2009), pp. 12–14.

30 Friesen, '*Opus Geminatum*', 135, and Godman, 'Anglo-Latin "Opus Geminatum"', 220.

31 *Aldhelmi Opera*, ed. R. Ehwald, MGH AA 15 (Berlin, 1919; repr. 1961), p. 321; see also Friesen, '*Opus Geminatum*', 135. Trans. from *Aldhelm: The Prose Works*, trans. M. Lapidge and M. Herren (Cambridge, 1979), p. 131.

32 Friesen, '*Opus Geminatum*', 135.

33 Friesen, '*Opus Geminatum*', 135.

34 Friesen, '*Opus Geminatum*', 135.

De Laude Virginitatis for its full meaning to be ascertained,[35] granting the twin texts a modicum of unity as an *opus geminatum*. The manuscript transmission of these texts, however, runs counter to Aldhelm's intentions, as they are not copied together in any extant manuscript,[36] suggesting they were, in fact, primarily read as independent texts.

What is most relevant to the present discussion, however, is Aldhelm's development of the function of the *opus geminatum* form itself, wherein the verse is perceived as less utilitarian and more for pleasure. Bede's first foray into this tradition is his prose rendering of Paulinus of Nola's verse *Vita Sancti Felici*, composed c. 700–05.[37] Bede considers the prose half of an *opus geminatum* as meant for utilitarian consumption, in contrast with Aldhelm's essential sense of the unity of a single work in two parts.[38] Bede emphasises this in both the clarity of diction in the prose text itself,[39] as well as in his explicit framing in the *praefatio*:

> qui quia metricis potius quam simplicibus sunt habiles lectoribus, placuit nobis ob plurimorum utilitatem eandem sancti confessoris historiam planioribus dilucidare sermonibus.[40]
>
> (They are more understandable to readers learned in metrics than to simpler ones; it has pleased me, for the utility of the majority, to make clear the life of the holy confessor in plainer discourse.)

Bede shifts, however, the manner in which this utilitarian function was achieved from what Friesen calls the 'expansiveness and minute detail', which characterises the style of Aldhelm and Sedulius, to one of 'brief clarity'.[41]

35 Godman, 'Anglo-Latin "Opus Geminatum"', 221.

36 Godman, 'Anglo-Latin "Opus Geminatum"', 221. As Godman notes, a full list of the relevant manuscripts is found in *Aldhelmi Opera*, pp. 329–31.

37 Godman, 'Anglo-Latin "Opus Geminatum"', 222.

38 Godman, 'Anglo-Latin "Opus Geminatum"', 222; Wieland, '*Geminus stilus*', pp. 116–17.

39 See Friesen, '*Opus Geminatum*', 137.

40 *Vita S. Felixi, Venerabilis Bedae Opera IV*, ed. J. A. Giles (London, 1893), p. 174; Godman, 'Anglo-Latin "Opus Geminatum"', 222, and Wieland, '*Geminus stilus*', p. 116, both quote this section of the *Praefatio* in their discussions.

41 Friesen, '*Opus Geminatum*', 137.

Bede's next hagiographical endeavour took him into the second portion of the *opus geminatum* tradition. The *VCM* is, by scholarly consensus, a difficult and often abstruse text,[42] even when considered contextually with Sedulius' *Carmen Paschale* and other verse texts in the tradition. Indeed, Janie Steen describes it as 'dense with rhetorical devices – a tribute to a saint famed for his eloquence',[43] and identifies seventeen distinct rhetorical devices in just the first thirty-six lines, with a further eight distinct devices throughout the rest of the *VCM*.[44] This density of rhetorical schemas and tropes, in combination with what Lapidge describes as a style of 'compressed understatement',[45] creates a text that is almost unintelligible outside of the conceptualised interdependence of the *opus geminatum* tradition.[46] The necessity of including the *VCA*, and later Bede's own *VCP*, in any reading of the *VCM* is key to any examination. It is here in the *VCM* that Bede furthers the poetic side of the *opus geminatum* tradition, following in the footsteps of Arator's highly allegorical *Historia Apostolica*, and presenting what is essentially a 'meditation on the spiritual significance of the events described prosaically by the Lindisfarne author'.[47] In the *VCM* Bede is certainly engaging with the nature of a verse text in the *opus geminatum* tradition, but he also fuses *voluptas* with *ruminatio*, combining the pleasures of complex rhetorical verse with the meditative impulse underlying monastic life.[48] This synthesis produces a poem whose complex layers of meaning can only be ascertained by extended rumination, especially on the many biblical resonances and echoes.[49] The obtuseness of the *VCM* is meant to inspire rumination,

42 See Lapidge, *Anglo-Latin Literature, 600–899*, pp. 332–33; Friesen, '*Opus Geminatum*', 138; and Godman, 'Anglo-Latin "Opus Geminatum"', 222.

43 J. Steen, *Verse and Virtuosity: the Adaptation of Latin Rhetoric in Old English Poetry* (Toronto, 2008), p. 23.

44 Steen, *Verse and Virtuosity*, pp. 25–27. She uses Bede's own *De schematibus et tropis* for her definitions.

45 Lapidge, *Anglo-Latin Literature, 600–899*, pp. 32, 333.

46 See Lapidge, *Anglo-Latin Literature, 600–899*, pp. 332–33.

47 Lapidge, *Anglo-Latin Literature, 600–899*, p. 333.

48 See Thornbury, *Becoming a Poet*, pp. 51–52, for a brief discussion of the inherently meditative nature of the verse Anglo-Latin hagiographical portions of the *opus geminatum* tradition.

49 Lapidge, *Anglo-Latin Literature, 600–899*, p. 336.

which necessitates knowledge of at least the general Cuthbertine tradition in the form of the *VCA*.

The practice of *ruminatio* and its connection to monastic life has received a considerable amount of criticism and scholarly attention. The most influential conception of the practice for early Anglo-Saxon England comes in Augustine's interpretation of Ps. 46 involving the image of a cow chewing the cud.[50] This metaphor defines *ruminatio*, in André Crépin's words, as the 'spiritual process of eating, savouring, and meditation upon the Word of God'.[51] *Ruminatio* was most commonly associated with the Psalter, which was the foundation of all monastic learning.[52] As Lapidge has noted, however, monastic education also favoured meditation and rumination on the verse paraphrases of Scripture that make up the *opus geminatum* tradition, including Juvencus, Sedulius, Avitus, and Arator.[53] The preference for such poetic renditions of biblical material for rumination is based on both the mnemonic function of poetry, and the inherent concentration and reflection needed for poetry's full meaning to be ascertained.[54] This ruminative potential stems primarily from the types of rhetorical devices and techniques of such poetry, like those identified by Steen in the *VCM* above, for example: alliteration (whose inherent aural repetition can function mnemonically); the Golden Line (which requires time to unpick); and polyptoton (which requires a knowledge of the multiple senses of terms and their relationship to one another). The mnemonic

50 See Augustine, *Enarrationes in Psalmos*, ed. E. Dekkers and J. Fraipont, 3 vols. CCSL 38–40 (Turnhout, 1956; repr. 1990), p. 529. All further references from this edition, using page number.

51 A. Crépin, 'Bede and the Vernacular', in *Famulus Christi: Essays in Commemoration of the Thirteenth Centenary of the Birth of the Venerable Bede*, ed. G. Bonner (London, 1976) pp. 170–92, at 172.

52 For example, Crépin, 'Bede and the Vernacular', p. 171.

53 Lapidge, 'Versifying the Bible', pp. 11–12. See also A. Orchard, 'The Word Made Flesh: Christianity and Oral Culture in Anglo-Saxon Verse', *Oral Tradition* 24 (2009), 293–318, who likewise emphasises the influence of the Christian-Latin poets.

54 For an in-depth discussion of the manner in which the Old English verse translation of the Psalter may have functioned mnemonically, see F. Leneghan, 'Making the Psalter Sing: the Old English *Metrical Psalms*, Rhythm and *Ruminatio*', in *The Psalms and Medieval English Literature: From the Conversion to the Reformation*, ed. T. Atkin and F. Leneghan (Cambridge, 2017).

aural quality of poetry aids in the internalisation of knowledge, and poetic devices aid in the reflection and meditation of that knowledge; these actions are at the heart of *ruminatio*. If the reading of poetry is an inherently ruminative act, the creation of poetry is an equivalently meditative one. The most famous instance of an explicit act of poetic creation as *ruminatio* in Anglo-Saxon literature is Bede's story of the cowherd Cædmon who, through divine aid, creates a poem in praise of God in the act of creation: *Cædmon's Hymn*. The narrative involves the humble Cædmon internalising biblical knowledge, spending the night ruminating on it, and then weaving the praise poem afterwards.[55] As Crépin and others have noted, Bede uses the term *ruminatio* in his depiction of this story, explicitly aligning Cædmon's activity with that of rumination.[56] The poetic act can therefore be seen as part of a faithful monk's life, functioning both as an act of *ruminatio* in the making, as well as an act of meditation in the later reading, both for the author and for his audience.

The act of creating a verse text on biblical themes as part of a monk's meditative life speaks directly to the composition of the *VCM*. The poem also reveals a further layer of meditation, as its position in Bede's canon points to how it functioned for him as a practice in the act of poetry. This can be discerned, as Lapidge notes, by the existence of the B-recension, which allows a rare glimpse into Bede's development as a poet. When he returned to the *VCM*, nearly twenty years later while composing his *VCP*, the vulgate revision is primarily concerned with perfecting aspects of verse, with only a few substantive changes.[57] I would suggest that the early date and relative imperfection of Bede's metre and rhetorical technique in the *VCM* points to the work being, in part, a

55 Bede, *HE*, III.24 (22), pp. 415–21.

56 Crépin, 'Bede and the Vernacular', p. 181; A. Orchard, 'Poetic Inspiration and Prosaic Translation: The Making of *Cædmon's Hymn*', in *Studies in English Language and Literature: 'Doubt Wisely', Papers in Honour of E. G. Stanley*, ed. M. J. Toswell and E. M. Tyler (London, 1996), pp. 402–22, at 404, highlights how Bede utilises the biblical figures from which the mastication metaphor is drawn, Lev. 11:3 and Deut. 14:6, 'nearly a dozen times in his works'.

57 See Thornbury, *Becoming a Poet*, p. 193; and Lapidge, *Anglo-Latin Literature, 600–899*, p. 341.

practice in poetic technique by the future writer of *De arte metrica*. While the *VCM* was later copied and sent to the continent, it may have been originally intended as a personal devotional exercise for Bede, both in meditation on the themes but also on the form. This is suggested by Bede's very uncharacteristic biographical aside in the vulgate's prefatory letter to John, where he describes a miracle he experienced while writing the *VCM* (presumably referring to a text closer to the B-recension), where Cuthbert heals his tongue: 'Ex quibus unum est, quod in me ipso, sicut jam tibi dixit, per linguae curationem, dum miracula eius canerem, expertus sum' ('One of these I experienced myself, as he told you, in the healing of my tongue while I was composing this verse about his miracles').[58] The complexity and abstruse nature of the early B-recension, in combination with the later precision of the vulgate version of the *VCM*, as well as the prefatory letter to John, suggest that the original composition was a personal exercise for Bede, which was only afterwards circulated widely.

Meditative Landscapes and Monastic Obedience

Bede's *VCM* functions as a meditative tool in its conceptual position as the poetic second part of a hagiographical *opus geminatum*. This meditative construction is likewise present in Bede's depiction of Cuthbert's ability to restore Creation due to his sanctity. While Bede follows the sense of his main source, the *VCA*, connecting the nature of Cuthbert's sanctity to his perfect obedience within a monastic setting, he locates Cuthbert's sanctity in a more specifically monastic context. This monastic obedience is explored in the *VCM* not, however, through the narrative and the careful use of the verb *praecipio* as it was in the *VCA*, but instead is highlighted in two distinct ways. First, Bede adds a speech in which Cuthbert describes his sanctity as found within a monastic hierarchy. Second, and in direct connection to the meditative purpose of the *VCM*, Bede repeatedly depicts Cuthbert observing the Divine Office, crucially

58 *VCM*, p. 57.

both before his entry into the monastery and in his life thereafter. For example, in *VCM* XX, a section with no source in the *VCA*, Bede presents Cuthbert giving an extended speech to those who had sought him out. During this exposition on how the 'weapons of faith [...] overcome empty deceits without battle' (ll. 469–70), Bede firmly sets Cuthbert's achievements within a monastic structure and focuses on the necessity of obedience within it:

> Artior est monachi quae semper subdita iussis
> Vita sub imperio servit devota parenti
> Excubiasque famemque preces manuumque laborem
> Ad votum gaudent proni frenare regentis.
>
> (ll. 479–82)

> (The life of a monk is strict, always subservient to commands; it obeys the elder devoutly in inclining itself under his dominion. Monks rejoice in their vigils, fasting, prayers, and manual labour, subjecting themselves to the rule of their superior.)

Here Bede highlights the centrality of obedience to the monastic life not only via the content of the statement, but also through the emphasis created by the structure of lines 479–80, where the sense carries directly from 'semper subdita iussis' to 'Vita', and the subsequent alliterative repetition of the two words for subjugation, 'subdita' and 'sub', where the noun *vita* is syntactically placed after the commands (*iussis*) which it is to obey. The creation of this novel speech by Bede speaks to the central role of monastic obedience within his depiction of Cuthbert's sanctity.

Similarly, and in keeping with the focus on rumination throughout the text, Bede foregrounds Cuthbert's distinctly monastic obedience in a series of passages in which the saint implicitly follows the monastic hours of the Divine Office. The most developed of these instances occurs in *VCM* VI, where Bede relates the miraculous provisions granted to Cuthbert when he and his horse turn aside to shelter from a storm. Bede is faithful to the *VCA* here in depicting Cuthbert's sanctity while still an untrained youth, but departs from his main source in presenting this sanctity as fundamentally monastic.[59] In connection with this focus, as will be demonstrated below, Bede expands the *VCA*'s description of the storm that prompts Cuthbert to seek shelter in a shepherd's hut, the catalyst for the miracle, in keeping with his general process of descriptive expansion in the *VCM*. Bede's description of the storm's fury in the *VCM* is considerably longer than in the *VCA*, taking up nearly half of the chapter's length, seven out of sixteen lines. More importantly, Bede alters Cuthbert's reaction to the wintry storm, from the saint only 'waiting for the storm to cease' and 'praying to God' while doing so,[60] into a statement about God's position as ruler of elemental Creation, as well as an implicit depiction of Cuthbert keeping the Divine Office. The underlying motivation for the passage, emphasising Cuthbert's devotion to monastic observance via the Divine Office, is further suggested by one of the revisions made to the B-recension in the final vulgate text:

B-recension:	Vulgate recension:
Expectansque udos dominum conponere flatus Parieti conectit equum, solitisque retentus Carminibus.[61]	Parieti et adnectit quo venerat ipse caballum, Expectansque udos dominum conponere flatus. *Divinis horam dum sacrat laudibus* almus. (ll. 169–71)
(And waiting for the Lord to calm the wild wind, he tied the horse to the wall, and held his accustomed songs.)	(And the holy man bound the horse to the wall where he arrived, waiting for the Lord to calm the wet wind, while *hallowing the hour with heavenly praise*.)

The first statement in the B-recension, describing Cuthbert's acknowledgment of God's position as ruler over elemental Creation, is identical in the vulgate. The other two statements are similar

59 For an alternative view, see Gretsch, *Ælfric and the Cult of Saints*, p. 90, who argues that the *VCM* presents him as already a monk during this miracle.

60 *VCA*, p. 70.

61 *VCM*, p. 70, note to l. 171.

in sense but have been revised and expanded in the vulgate. While the description of Cuthbert's nightly activity in the B-recension, his 'accustomed songs' ('solitus [...] carminibus'), clearly possesses the general sense of the vulgate, the lexical changes represented in the later revision point instead to Bede's desire to emphasise the ruminative potential of the passage by linking the moment to the Divine Office. This revision allows Bede to embed ruminative material in the vulgate revision, and further connects Cuthbert's knowledge of God's power over elemental Creation with his obedient observance of the monastic hours. This is displayed in the collocation of the terms 'divinis', 'horam', and 'laudibus' in the vulgate, which echo the lexis of the Divine Office.[62] Bede is using the first two of these terms to bring to mind the Office in general, with the choice of the verb 'laudibus' meant to bring a specific temporal designation to the scene. Bede's poetic description of the clouds covering the stars ('percurrit sidera nimbus') in connection to the practice of Lauds, which during Bede's time was recited sometime before and into dawn,[63] creates a scene that functions to aid monastic rumination by implicit reference to the Divine Office. Bede's revision of the scene from the *VCA* depicts Cuthbert travelling through the night and now singing Lauds at dawn as the storm passes.

As with most of the *VCM*, however, the literal level of the text is being used as a ruminative pathway towards deeper meaning, here connecting with the actual form of the Divine Office that would have structured not only Cuthbert's, but also Bede's, daily life. According to Jesse Billett, the office of Lauds in use during the early Anglo-Saxon period under discussion included Psalms 148–50 daily as the penultimate portion, directly before the *Benedictus*.[64] Given the regularity in practice ushering in each new dawn, these three psalms' connection to Cuthbert's present circumstance has a firm foundation. Ps. 148 is primarily concerned with all the created world praising God, and includes in verse 8 a list of specific sections of elemental Creation and their position under God's

62 See, for example, Benedict, *Regula*, Ch. XLIII.1, p. 586, which begins with the phrase 'Ad horam divini Officii' ('At the hour of the Divine Office').

63 J. D. Billett, *The Divine Office in Anglo-Saxon England, 597–c.1000* (London, 2014), p. 38.

64 Billett, *The Divine Office*, pp. 38–39.

authority: 'Ignis, grando, nix, glacies, spiritus procellarum, quae faciunt verbum ejus' ('Fire, hail, snow, ice, stormy winds which fulfill his word') (Ps. 148:8). Other than the first term, 'ignis', the others resonate with the wintry storm howling outside Cuthbert's shelter, and the declaration of God's power over the elements at the end of the psalm, 'que faciunt verbum ejus', provides biblical support for Bede's statement that Cuthbert waits for God to calm the storm.[65] Further still, this verse was interpreted by Augustine in his *Enarrationes in Psalmos*, a text which Bede knew well, as relating to God's direct providential control over Creation, in which the elements are not moved 'by chance': 'Dei in omni motu suo deseruiunt' ('in every motion they obey God').[66] Within this exegesis on verse 8, Augustine contrasts those who understand the true nature of Creation, all things praising God by functioning in their ordained manner, with foolish ('multi stulti') men, who instead believe that the elemental world is below God's purview:

> uisum est illis quia superiora omnia Deus gubernat, inferioria uero despicit, abjicit, deserit, ut nec curet ista, nec gubernet, nec regat.[67]
>
> ([They] think that because God rules all things above, those below he despises, casts aside, abandons, so that he neither cares for them, nor guides them, nor rules them.)

Cuthbert, therefore, is depicted as implicitly understanding this exegesis, faithfully awaiting God to calm the wild winds He rules.

The second action Cuthbert takes, the praise of God through the implicit singing of Lauds, resonates with the second two psalms,

65 There is a parallel with Bede's later depiction of St Chad in *HE*, which presents the saint responding to storms by going into the chapel to pray, using portions of the Psalter. P. Darby, *Bede and the End of Time* (Farnham, 2012), p. 102, highlights how for Chad storms function eschatologically, and that in Bede's exegesis such meteorological signs function 'as overt reminders from God, sent to make us think about the final judgement'. See also F. Leneghan, 'Preparing the Mind for Prayer: *The Wanderer*, *Hesychasm* and *Theosis*', *Neophilologus* 100 (2015), 121–42, who notes the eschatological function of the weather in the Chad episode, and argues that it is meant to arouse compunction, a ruminative exercise.

66 Augustine, *Enarrationes in Psalmos*, p. 2173. For Bede's familiarity with the text see Lapidge, *Anglo-Saxon Library*, p. 201.

67 Augustine, *Enarrationes in Psalmos*, p. 2172.

149 and 150, which are both concerned with humanity's praise of God. Ps. 149 takes the Church's call to sing praises to God as its topic, and twice speaks in general of saints included in this call: 'Alleluja. Cantate Domino canticum novum; laus ejus in ecclesia sanctorum' ('Sing to the Lord a new canticle: let his praise be in the church of the saints') in verse 1, and '[e]xsultabunt sancti in gloria, laetabuntur in cubilibus suis' ('The saints shall rejoice in glory: they shall be joyful in their beds') in verse 5. Augustine's exegesis on this psalm once again lies behind the figural significance of the moment in the *VCM*. He interprets verse 5 in great depth, and contrasts foolish secular glory, 'gloria stultorum, popularis illa quae dicitur' ('the glory of fools, popular glory as it is called'),[68] with the humble spiritual glory of saints. Augustine focuses in particular on the second portion of verse 5, the location in which this glorying is to take place, 'in cubilibus suis' ('in their closets'), which he interprets as speaking to the humble seclusion of the saintly *cubile*, ultimately representing their hearts.[69] This not only resonates with the eremitic tradition in general, but also Cuthbert's situation where, huddled in a freezing hut with only his horse for company, the propriety of his praise is unequivocal, and the implicit meditative function is highlighted through the focus on internal action. Ps. 150 functions similarly, consisting of a five-verse exhortation to praise God, constructed by means of ten repetitious phrases each beginning with the imperative *laudate*, as seen in verse 1: 'Alleluja. *Laudate* Dominum in sanctis ejus; *laudate* eum in firmament virtutis ejus' ('*Praise* the Lord in his holy places: *praise* him in the firmament of his power').[70] Here Augustine expounds the nature of the saints referred to, as they are glorified by God, which resonates with Cuthbert's monastic form of sanctity, the focus on the internal glorying occurring in his heart of Ps. 149, and the external reality of his solitary prayer in the shepherd's hut. Taken together these moments leading up to the miraculous provision of sustenance for Cuthbert and his horse stem directly from his monastic sanctity, displayed in his

68 Augustine, *Enarrationes in Psalmos*, p. 2184.

69 Augustine, *Enarrationes in Psalmos*, p. 2185. Cassiodorus gives an equivalent interpretation in *Expositio Psalmorum*, Ps. 149, ll. 68–69.

70 See also Ps. 150: 2–5 for an equivalent construction.

obedience to the ordering of the hours in the Divine Office. Bede thus emphasises the role of monastic obedience, and presents a scene whereby the implied monastic audience, through meditation and rumination, might be ushered towards a similar obedience, as well as be reminded of the potency of God's authority over the created world. Likewise, the causal link between Cuthbert's monastic obedience and the calming of the storm foreshadows the later miracles in which his sanctity allows for the restoration of Creation.

Exegesis and Restored Creation

Explicit exegesis concerning the restoration of Creation first enters the Anglo-Saxon hagiographical tradition in the *VCM*. Whereas the *VCA* works implicitly within the Augustinian tradition, interpreting the Fall as ontological and fundamentally reparable through Cuthbert's extraordinary sanctity, centred on his perfect obedience, Bede, the consummate exegete, supplies his own interpretation for the restoration of Creation, which he later expands in both *In Genesim* and the *VCP*.[71] The exegesis occurs in *VCM* XIX during Cuthbert's tenure as a hermit on Farne, in an explicatory framing for the miracle in which the sea brings Cuthbert the beam that the brothers had forgotten:

> Quid referam aequoreas iusto famularier undas
> Obsequiumque illis elementa impendere, qui se
> Imperiis subdunt devota mente supernis?
> (ll. 451–53)

> (What shall I relate of the sea's service to the righteous man, and of how the elements serve those who subject themselves devoutly to heavenly commands?)

While the lexis does not include the use of the term *praecipio* from Augustine and the *VCA*, the sense is very much the same: Cuthbert's

[71] While the dating of the two versions of *In Genesim* is still contentious, even Kendall's early date of 717–18 would place the work at least ten years after the composition of the B-recension of the *VCM*; see Bede, *On Genesis*, ed. Kendall, p. 45.

sanctity and connected ability to restore Creation are based upon his complete obedience to God's commands. As noted by Lapidge, however, this chapter includes one of the few substantive revisions in the vulgate recension.[72] In the B-recension, the chapter begins instead with an allusion to a miracle otherwise unattested in the Cuthbertine tradition, in which a group of pregnant seals will not give birth until they receive a blessing from Cuthbert:

> Quid referam uitulas foetus sub fasce grauatas
> Non ausas illic uetri deponere pondus
> Ni prius ipse sacra dextra permitteret illis?
> Quarum illi patrium seruit cum fluctibus aequor.[73]

> (What shall I say of the seals, weighed down with the burden of pregnancy, who did not dare to let down their offspring until the old man [Cuthbert] had first blessed them with his holy right hand? The seals' watery homeland also served the man with its waves.)

The removal of the allusion in the vulgate version appears to aid the cohesion of the chapter as a whole, and while theologically viable within the Cuthbertine tradition, the miracle of the pregnant seals adds little to a chapter focusing on the service rendered to Cuthbert by the sea. This is evident in the authorial titles for the passage: in the *VCA*, 'De ligno quod mare seruiens ei detulit' ('How the sea served him by bringing him wood'); and in the *VCM*, 'Qualiter eidem mare servierit' ('How the sea served the same saint').[74] Likewise, Bede's primary addition to the miracle is, in keeping with the work as a whole, his exegesis on the purpose for the restoration miracles, specifically in their emulative function. Bede transforms the *VCA*'s depiction of the brothers' awe at the sea's honouring of Cuthbert[75] into a reproach of the brothers' disobedience:

72 Lapidge, *Anglo-Latin Literature, 600–899*, pp. 342–43. Lapidge also points to a potential analogue in Adomnán's *Vita Columbae*, p. 41.

73 *VCM*, p. 92, note to 455; compare with ll. 451–53.

74 *VCA*, pp. 98–99; *VCM*, p. 92. The titles to the individual chapters are represented in most of the manuscripts and will be considered authorial.

75 See above, p. 59; compare *VCA*, pp. 100–01: 'Fratres [...] et admirantes quod mare in honorem Christi magis obediens anachorite quam homines parauerunt' ('[T]he brethren [...] marvelling that the sea in honour of Christ had accomplished more than men, in obedience to the hermit').

Pandere, deposuit, hominumque oblivia pontus
Increpat et nautas sua concutit unda pudore.
(ll. 460–61).

(The sea spread out, placed [it], rebukes the men's forgetfulness, and the wave shakes the sailors through their shame.)

This alteration serves to focus the chapter on the obedience of the sea, and the theological support and pedagogical purpose of such obedience: the miracle occurs only in response to Cuthbert's sanctity, and functions as an exemplum from which the brothers are meant to reform their behaviour.

Ch. XIX is a microcosm of Bede's approach to the restoration of Creation in the *VCM* as a whole, whereby his exegetical framing (ultimately stemming from Augustine) provides the controlling mechanism by which the miracles are theologically justified. This approach functions as follows: Cuthbert's sanctity, connected to his perfect obedience, restores Creation into prelapsarian relationship with him, but these miracles function primarily as elaborate exempla for other souls. This familiar pattern points towards the *VCM*'s primary function as a meditative tool, where the imitative function of these miracles is less concerned with praxis, and more with rumination and the spiritual progression of the reader.

Triptych 1 (Chapters III, IV, and V): Pre-Monastic Sanctity

The first Creation miracle in the *VCM* occurs in one of Bede's original chapters, which he inserts as Ch. III, directly after his rendering of the episode where the angel appears and heals Cuthbert's diseased knee.[76] This chapter forms part of a group of three connected chapters (III–V) which function as a unit to establish Cuthbert's pre-monastic sanctity, and tie that sanctity to his ability to restore Creation. Bede sandwiches the *VCA*'s account of Cuthbert's vision of Aidan's soul ascending to heaven, here in Ch. IV, between Ch. III, which relates how Cuthbert changed the winds and saved a group of monks swept out to sea, and Ch. V, which describes Aidan

[76] See *VCA*, pp. 66–69.

working a similar miracle, whereby a stormy sea is calmed through the ministrations of the saint. The grouping unites Cuthbert's power over the winds with Aidan's, authorising the saint by connection to his illustrious predecessor through the prophetic vision of Aidan's death. Further still, Bede connects Cuthbert's vision with the overarching emphasis on the saint's keeping of the Divine Office, once again joining a specifically monastic form of obedience to the restoration of Creation. In each of the mirrored miracles where the saints calm storms, the elements implicitly enact the exegesis in Ch. XIX: Creation acts thus only in response to the sanctity of the two men.

While narrative progression is not a central focus of the organisational structure of the *VCM*, the positioning of this tripartite group highlights Cuthbert's providential sanctity long before he enters the monastic life. This is made clear by Bede's framing for the miracle in Ch. III, which is united with Ch. II, and presents the upcoming miracle in Ch. III as a rare moment of progression: 'Hinc sacra maiori firmatus robore corda / Celsithronum didicit precibus pulsare tonantem' ('Thereafter, strengthened in his heart with greater focus, he learned to batter the lofty-throned Thunderer with his prayers') (ll. 95–96). Cuthbert's interaction with the angelic healer in Ch. II is here a strengthening force for his convictions, and the ensuing miracle is meant to illustrate his progression in sanctity. The episode in Ch. III relates how Cuthbert's fervent prayers were able to change the course of strong winds that were driving a company of monks on five rafts further and further out to sea. Bede follows his overarching methodology of abstraction, obliquely referring to the actual location where this took place via, as Steen notes, ekphrasis: 'Est locus insignis fluvii super ostia Tini, / Eximio iam tunc monachorum examine pollens' ('There is a distinguished place above the mouth of the river Tyne, which at that time was the thriving location of an extraordinary group of monks') (ll. 97–8).[77] Bede is here using the traditional opening for ekphrasis, 'est locus',[78] to indicate the exemplary nature of the monastic establishment without drawing out the description. He follows this by

[77] See Steen, *Verse and Virtuosity*, p. 27.
[78] Steen, *Verse and Virtuosity*, pp. 40–41.

presenting the episode in highly descriptive terms, relating how the monks were suddenly caught by the swift current and strong wind on the Tyne, and swept out to sea:

> Qui veherent dum ligna feri per terga fluenti,
> Fluminis et venti subito feriuntur ab ictu.
> Quinque fuere rates, rapido quae gurgite cunctae
> Oceani canum pronae labuntur in aequor.
> Iamque oculis abstracta procul velut aliger undis
> Mergulus innaret, paret per caerula puppis.
>
> (ll. 99–104).

> (Once while transporting wood on the surface of this wild river, they were suddenly struck by the wind-blown river. There were five rafts, all of which were swept headlong by the fierce swirling water out into the foam-crested ocean. Now, pulled far from men's eyes, the boats appear through the deep-blue, like a diving sea-bird floating on the waves.)

The profusion of descriptive terminology employed here is representative of Bede's approach to rendering aspects of Creation throughout the *VCM*, and can best be characterised as providing imaginative detail to aid in the meditative process on the miracles, whereby their vivid physicality provides a point of connection for the reader.[79] Bede's focus on detail for *ruminatio* is likewise aided by the aural effects of Bede's poetic technique: for example, the alliteration of initial 'f' in the first two lines (*f*eri, *f*luenti, *F*luminis, and *f*eriuntur), which centres on the descriptive content. Here the description functions to highlight the plight of the central narrative: the monks are swept not simply into the sea, but into a tumultuous expanse of water frothing with the white foam of breaking waves

[79] For a discussion of this process and its ties to *ruminatio* see R. Wehlau, 'Rumination and Re-Creation: Poetic Instruction in *The Order of the World*', *Florilegium* 13 (1994), 65–77, at 72. Also, M. Carruthers, *The Book of Memory: A Study of Memory in Medieval Culture*, 2nd edn (Cambridge, 2008), pp. 202–17, for the connection between *ruminatio* and reading. Lapidge, *Anglo-Latin Literature, 600–899*, p. 355, gestures towards this in relation to the inherent difficulty of the *VCM*, connecting it back to the distinction between the prose and verse portions of the *opus geminatum* in Alcuin's *Vita S. Willibrordi*, as well as Arator's *De Virginitate*: 'As Arator's poem was intended as a meditation on the meaning of the acts of Peter and Paul, so Bede's poem is intended as a meditation on the life and significance of Cuthbert'.

('Oceani canum pronae [...] in aequor'), all of which establishes the ferocity of the sea into which the monks have been tossed. Likewise, Bede adds further detail for the imagination to ruminate on by way of a naturalistic simile, that they have been carried so far they seem tiny, like sea-birds, appearing and disappearing with every rise and fall of the ocean ('velut aliger undis / Mergulus innaret, *p*aret *p*er caerula *p*u*pp*is'). This rich depiction (again utilising alliteration) allows for Cuthbert's subsequent miracle to have greater impact, as the imaginative peril has been grounded in the physicality of a scene filled with realistic detail that would have been familiar to both Bede and to most monastic readers. They would be used to the dangers of water-travel, whether river or sea, and the sudden shifting of winds would be a common occurrence. It is with this realistic detail that Cuthbert's ensuing speech to the mocking crowd, and his subsequent miraculous changing of the wind, attain true meditative force, both within the narrative itself and in the ruminative act of reading. Cuthbert's exhortation to the crowd to 'sympathise' with the monks and 'pray to God' also, however, includes a declaration of God's position as creator and ruler over elemental Creation:

> Condoleamus et his, quos verbera maesta fatigant,
> Vel dominum potius, qui flabra creavit et undas
> Oremus, dignetur iter donare salutis.
>
> (ll. 112–14).

> (Let us suffer with those afflicted by lashes of sorrow, or let us pray to the Lord, who created the winds and the waves, that He may deign to grant [us] a safe journey).

It is this very understanding that allows him to 'batter the lofty-throned Thunderer with his prayers', which is meant as a demonstration of that understanding. God's power over elemental Creation is manifested only through Cuthbert's obedient prayer and sanctity, and functions within the narrative of the *VCM* as an evangelistic tool: the scoffing throng is ashamed ('erubuit') at the power displayed and in response glorifies God who answers the prayers of his followers ('firmet qui vota suorum') (ll. 118–19).

Bede connects this point of progression, where Cuthbert's prayers are miraculously efficacious, with Ch. IV, through its general chronological setting: '*Haec inter* teneros laetis *dum* collibus agnos / Pasceret' ('*During this time, while* he was feeding gentle lambs in the joyful hills') (ll. 120–21). While Carole Newlands takes the phrase to be an 'abrupt temporal clause' which 'does not link the preceding episode of the monks' rescue' with the ensuing vision, translating '[h]aec inter [...] dum' as 'in the meantime',[80] the Latin phrase could equally be translated with the more connective 'during this period', and given its position as the central link from the *VCA* for Bede's two novel chapters, it is likely that its meaning is closer to the latter. The vision occurs, therefore, during the same period, that is, during Cuthbert's youth before he had entered the monastery, and is equally connected to his miraculous changing of the winds in Ch. III. Bede expands and alters this vision in two distinct ways: first, he connects it with Cuthbert's implicit observance of the Divine Office (and therefore obedience to monastic discipline); second, he fleshes out Cuthbert's description of the vision in a direct speech to the other shepherds, imaginatively depicting what the *VCA* only alludes to: 'ut uiderat indicauit' ('he described [the vision] [...] just as he had seen it').[81] The first alteration of the vision can be seen in Bede's rendering of the *VCA*'s description of Cuthbert's nightly custom of prayer, 'pernoctans in uigiliis secundum morem eius, mente fideli, pura fide, uberrimis orationibus' ('spending the night in vigils according to his custom, offering abundant prayers with pure faith and with a faithful heart'),[82] into an implicit reference to the Divine Office: 'vigil nocturnis [...] in hymnis' ('keeping vigils with nocturnal hymns') (l. 121). The collocation of the terms 'vigil', 'nocturnis', and 'hymnis', once again draws in the monastic practice of the Divine Office, here the Night Office in particular. He has taken the *VCA*'s 'vigiliis' and expanded it into the monastic practice of vigils, beginning his implicit depiction of Cuthbert's obedience via his adherence to the monastic hours. The ensuing vision, following the *VCA*, is therefore causally connected to Cuthbert's nightly custom; it is his monastic form of obedience, driving back the 'shadows of the night' with psalmody, that allows the vision to occur. Bede expands the description of the

80 Newlands, 'Bede and Images of Cuthbert', p. 89.
81 *VCA*, pp. 68–69.
82 *VCA*, pp. 68–69.

vision itself in the narrative, enlarging the *VCA*'s image of angels ascending and descending while carrying the soul in a globe of fire ('in globo igneo ad coelum') into a scene of intense radiance:

> cernit [...]
> Ignea sidereis fulgescere castra maniplis
> Atque polis sanctam rutilae per gaudia pompae
> Ferre animam.
>
> (ll. 121–24)

(he perceives [...] fiery hosts gleaming in celestial legions and a holy soul being carried to heaven amidst this radiant procession's joy.)

This expansion is furthered by Bede's addition of Cuthbert's direct address to his fellow shepherds. The imaginative scope of the passage is striking, encompassing the 'twin gates' opening to allow the soul entry into the 'ethereal hall', as well as the soul soaring beyond 'the stars' to see the 'high-throned king', and emphasises the meditative function of the miracle, both within the narrative and for the audience. In keeping with Bede's transformation of the saint, the descriptive speech is closed by a didactic section that points to praxis associated with the monastic life, the Divine Office in particular, and thus highlights Cuthbert's pastoral role. The speech achieves this by focusing on Cuthbert's instructions for the shepherds within the narrative, and, by extension, for the meditative reader as well:

> Discite, pastores, vigili tutamine mandris
> Insidias noctis furvosque cavere leones,
> Vobis ut angelicae pateant sacra carmina laudis
> Bethleaque deum cernatis in arce potentem.
>
> (ll. 135–38)

(Learn, shepherds, from the guarding of your sheep, to beware nocturnal ambush and shadowed lions, so that the sacred hymns of angelic praise may be open to you, and you may see the God of Bethlehem, strong in His fortress.)

The imperative 'discite', echoing biblical use and foreshadowing that of Bede himself in later works,[83] grants Cuthbert a pastoral authority not found in the *VCA*. The content of his sermon simultaneously reinforces Bede's depiction of Cuthbert's pre-monastic obedience to monastic ideals, as well as providing further meditative material for the reader. The miraculous vision is causally connected to Cuthbert's practice of nocturnal psalmody, and his declaration to the shepherds delineates how this monastic practice not only defends them against the wiles of the devil, envisioned here as shadowed lions,[84] but allows the faithful to be given similar visions of God in heaven: 'Vobis *ut* angelicae pateant sacra carmina laudibus' ('*so that* the holy hymns of angelic praise may be accessible to you'). The meditative function of this didactic extension is clear: by adhering to the ruminative monastic practice of the Divine Office, the faithful might also be given luminous visions. The passage as a whole highlights the twin pillars of Cuthbert's sanctity: his role as saint, and his obedience to a specifically monastic habit of life, even before entering the monastery. The emphasis of this chapter is altered from the *VCA*'s focus on prophecy, to focus on establishing Cuthbert's sanctity.

Bede concludes his first triptych with his second original chapter (V), in which he relates the miracle of Aidan calming the stormy sea. The position of this chapter is clearly meant to provide a mirror

83 See esp. Mtt. 9:13, 11:29, 24:34; Mark 13:28; Is. 1:17; and Bede, *HE*, II.2, p. 138, where he presents Augustine referencing Mtt. 11:29.

84 Given the late hour, the image of 'dusky lions' here evokes 1 Peter 5:8, as the verse was part of the Night Office at the hour of Compline: 'sobrii estote vigilate quia adversarius vester diabolus tamquam leo rugiens circuit quaerens quem devoret' ('Be sober and watch: because your adversary the devil, as a roaring lion, goes about seeking whom he may devour'). Its employment here is meant to foreshadow Cuthbert's monastic obedience even before he has entered the monastery. While there is no extant evidence for the use of this verse as a *Capitula* during the early Anglo-Saxon period, its inclusion in the Roman Breviary tradition speaks to the likelihood of its inclusion in the Anglo-Saxon Divine Office. See *Breviarium Romanum ex decreto Sacrosancti Concilii Tridentini Restitutum S. PII V. Pontificis Maximi jussu editum Clementis VIII et Urbani VIII. Auctoritate recognitum, cum officiis sanctorum novissime per Summos Pontifices usque ad hanc diem concessis; in Quatuor Anni Tempora divisum* (Rome, 1568; repr. facs. Libreria Editrice Vaticana, Città del Vaticano, 1999), p. 135, which begins Compline with 1 Peter 5:8 as the *lectio brevis*. This is also supported by a similar evocation in Felix's *VSG*. See below, p. 205, in relation to the *VSG*.

to Ch. III and further support the depiction of Cuthbert, even in his youth, as a model for English sanctity. The narrative involves a monk asking for a blessing of protection from Aidan as he and his companions are about to undertake a sea journey; Aidan agrees, prophesies a storm, and gives him a blessed chrism to pour out to calm the waves and wind; the sailors are beset by a storm, and after pouring out the oil are given safe passage through a calmed sea. Bede once again provides his own exegesis of the miracle as a conclusion to the chapter:

> Virtus sic una gemello
> Effulget radio: dictis qui tristibus ante
> Terruerat, prompto relevat solamine maestos,
> Quique prophetali praefatur mente futura,
> Ipse elementa suis refrenat turbida jussis.
> (ll. 159–63)

> (Thus, one miracle shines forth with a twin beam: he who had before frightened with his sorrowful prophecy quickly lifted up the sad sailors with swift relief; and the very one who had predicted future events with prophetic spirit restrained the wild elements with his own commands.)

In this summative exegesis, Bede highlights the two important facets of Cuthbert's saintly activity by way of mirroring those of Aidan: the speaking of prophecy and the restoration of Creation. Regarding prophecy, Bede employs the descriptive term *vates* ('prophet') for Aidan which, as Newlands notes, is Bede's favourite term for Cuthbert throughout the *VCM*.[85] While Bede does not specifically term Cuthbert's vision of Aidan in Ch. IV a prophecy, instead declaring that the vision occurs simultaneously with Aidan's actual death and ascension,[86] the prophetic function of the vision is nonetheless apparent; this is doubly so here, as the *VCA* terms the vision

85 Newlands, 'Bede and Images of Cuthbert', 103, highlights that Bede uses the term thirty-two times in the *VCM* for Cuthbert. Newlands also notes that the emphasis on his prophetic powers, via *vates*, is unique to the *VCM*, and is substantially reduced in the *VCP*, where Bede most often refers to Cuthbert as *pater* (103–04).

86 *VCM*, ll. 140–41.

prophecy ('prophetans').[87] Regarding Creation, Aidan is described as enacting the exegesis of Ch. XIX, whereby the elements obey the saint, here following his commands ('jussis').[88] It is in the extended descriptions of Creation that the potency of this restoration can be found. Consonant with the depictions of the wild sea in Ch. III, Bede renders the stormy sea in this chapter with poetic force in three separate accounts, including a general description, Aidan's prophecy, and the subsequent narrative description:

> Presbyter undisoni quidam vaga caerula ponti
> Scandere iussus.
>
> (ll. 145–46)

> (A certain priest was ordered to go forth on the wandering blue of the wave-sounding sea.)

> Petis aequor ut altum
> Obvius adverso insurget septentrio flatu;
> Venti sed fremitus tempestatesque sonoras.
>
> (ll. 148–50)

> (As you make for the deep ocean, the north wind will rise against you with its opposing gust; but [remember to calm] the blasts of the wind and the roaring storm.)

> Cum subito gravis instat hiems, furit undique pontus.
>
> (l. 155)

> (when suddenly harsh winter comes, and the sea rages on all sides.)

The extended emphasis on the various sensory stimuli of the storm is striking, including the tactile rising of the north wind and its 'blasts', the aural 'wave-sounding sea' and the wind's 'howling gales', as well as the visual 'wandering blue' of the raging sea. Together this profusion of descriptive elements serves to aid the

87 *VCA*, pp. 68–69. Bede also moves prophecy towards the later portions of Cuthbert's life to represent his development.

88 This can be compared with the *VCP*, which is much more explicit and uses terms of command, 'imperiis' for example, which represents a development on Bede's part in the *VCP*.

meditative function of the *VCM*, equivalent to that found in Ch. III, by providing a complex depiction whose meaning only becomes apparent through extended rumination. When Aidan is capable of not only prophesying the storm's appearance but also subduing its fury, the exegesis is given due force. Here a small and representative portion of Creation is subordinated to the saint, just as later depicted in Ch. XIX.

Bede's first triptych functions as a unit, presenting Cuthbert as a saint of equal standing with Aidan, despite his youth and, crucially, before he has entered the monastic life. While Aidan's sanctity is presented as a given ('sacerdotis', 'vatis', 'antistes [...] celsus'),[89] Cuthbert's sanctity is founded not only upon his providential place as a saint, but also his inherent knowledge of, and obedience to, monastic praxis. Cuthbert's ability to restore Creation, mirrored and authorised in Aidan's miracle, is theologically framed by Bedan exegesis. Ch. III provides meditative material on the true source of, and reason for, the restoration of Creation, whereby the knowledge of God's omnipotence as creator of the elemental world allows Cuthbert to intervene. Ch. IV exhorts the reader to keep monastic hours: obedience to the Night Office is depicted as one of the reasons why Cuthbert is granted a vision of God, which points to the potential benefits for faithful monastics who adhere to such practice. Ch. V not only provides support for Cuthbert's sanctity, but also allows for reflection on the power of sanctity to bring the world back into its prelapsarian state. In each of these chapters, the level of descriptive detail serves to grant force to the ruminative material, grounding the imaginative process of meditation in its physicality. This structure is mirrored in each of the subsequent restorations of Creation in the *VCM*.

Triptych 2 (Ch. VII, VIII, and IX): Monastic Sanctity

Bede's focus on the connection between Cuthbert's sanctity and his monastic obedience, especially to the Divine Office, and the subsequent restoration of Creation, is further evident in one of Cuthbert's most prominent miracles, the ministration of the sea animals. This

89 *VCM*, ll. 142, 146, and 148 respectively.

miracle is part of a second triptych of connected chapters: VII, VIII, and IX. Bede retains the collocation from the *VCA* of the miracle of the angel and bread while Cuthbert is serving as 'guest-master' (*VCM* VII), with the following miracle of the sea animals in *VCM* VIII, and likewise highlights the mirroring of Cuthbert's obedient ministration to the angelic visitor with the sea animals' service. Further still, these three chapters are connected chronologically (all occurring in the same general time), indicated not only by the emphasis on the hardships of winter in each, but also through the repetition of *interea* in the beginning of each chapter. The function of this triptych is equivalent to the first triptych, and highlights the causal connection between Cuthbert's obedience, depicted in both his actions of service as well as continuation of obedience to the Divine Office, and the drawing of Creation into prelapsarian relationship.

Bede begins Ch. VII with another chronological division and subtle gesture towards Cuthbert's progression: 'Hinc maiora petens monachis Hrypensibus almus / Corpore, mente, habitu, factisque adiungitur' ('From this time [Cuthbert], seeking greater things, is joined in mind, body, habit, and deed to the monks of Ripon') (ll. 180–81). The chronological division of *hinc* points to a new portion of Cuthbert's life centred upon his entry into monastic life at Ripon. This progression is necessary for Bede's (and Benedict's) vision of the holy life,[90] whereby the active and communal life are the precursors for the contemplative, and thus monastic training is essential for eventual eremitic seclusion. Bede's list of the various sections of Cuthbert's life that the saint seeks to yoke to the monastic life is related to a number of similar lists, including Pope Damasus' *Carmen XV, De Sancto Felice*.[91] The miracle itself closely follows the

90 See Benedict, *Regula*, Ch. I.3–5, pp. 436–38. Also, see above, p. 45 for this structure in the *VCA*.

91 Noted by Jaager, *VCM*, p. 71; see Pope Damasus, *Carmen XV. De Sancto Felice*, *PL* 13, col. 389A: 'Corpore, mente, animo, pariter de nomine Felix' ('The body, mind, and spirit are equally (in the man) named Felix'). This is related to other lists, for example, Paulinus of Nola, *Epistulae*, ed. G. de Hartel, CSEL 29 (Vienna, 1894; repr., ed. with suppl. M. Kamptner, 1999), p. 279, l. 21: 'Corpore mente fide castissimus incola Christi' ('The body, mind, and faith, are in pure habitation with Christ').

VCA in most respects, highlighting the recognition of Cuthbert's sanctity in his immediate appointment to the post of 'guest-master' ('hospitibus famulus'),[92] and depicts Cuthbert receiving the angel in disguise, compelling him to stay and eat, only to find the angel vanishing and supplying heavenly bread. Once again Bede elaborates the *VCA*'s depiction of physical Creation, here the cold, winds, and snow of December, to heighten the imaginative physicality of the scene as an aid for rumination:

> Hospitis hic habitu mediae sub tempore brumae
> Advenit, vatisque humili de more receptus
> Artus abluitur sacros, quique algida blando
> Palmarum adclinis mulcet vestigia fotu
> . . . .
> Ne frigusque famesque nivosaque flabra Decembris
> Ingeminent itiner longo sub calle rigentis,
> Qui matutinis advenerit algidus auris.
>
> (ll. 185–88; 190–93)

> (This stranger arrived in the appearance of a guest in the midst of the shortest day [winter], and having been received by the prophet's humble customary ministrations, his holy limbs are washed and Cuthbert's palms rub his frozen feet with gentle warmth [...] so that the cold and hunger and snowy gusts of December would not redouble the guest's difficulties, already numb with cold from the length of his journey, who had arrived in the cold morning air.)

The typological connection with Christ as well as with monastic tradition is evident, and Bede grants further force to the subservience of Cuthbert by his elevated reference to the saint as *vates* in the very act of washing the visitor's feet. The *VCM* further highlights the extraordinary cold of the scene, not simply repeating the *VCA*'s comment of the cold feet of the visitor warmed by Cuthbert's attention but adding to the overall description by enumerating the hardships of winter travel via polysyndeton,[93] including *frigus, famem,* and *nivosum,* as well as the detail of the specific month, December. This emphasis on the frigidity of the moment brings

92 *VCM,* l. 182.
93 See Steen, *Verse and Virtuosity,* p. 27.

imaginative potency to the clearly contrastive gift of the heavenly loaves. Bede expands the *VCA*'s bare description of the holy bread ('panis suauissimi' ['the sweetest bread']),[94] into gleaming loaves ('fulgentes polline panes'), whose fragrance fills the building with the scent of roses ('roseo nidore').[95] He also greatly expands Cuthbert's speech by utilising the inexpressibility topos[96] within which the saint elaborates on the heavenly nature of the guest and his gift through a series of hyperbolic comparisons where the earthly pales before the heavenly:

> Pascere, non pasci veniens et fercula portans,
> Qualia non surgunt nostrae de germine messis,
> Lilia nec candent nec sic rosa fulgida flagrat
> Nostraque mannifluo spernuntur mella sapore.
>
> (ll. 206–09)

> (He came to feed not to be fed, carrying with him the food of a kind that does not grow from the sprouts of our harvest. Our lilies do not shine nor our roses burn so brightly, and our honey is worthy of scorn compared with this manna-like scent.)

The reversal of hospitality is a clear indication of Cuthbert's saintly status and is presented through a line from Sedulius employing polyptoton,[97] '[*p*]*ascere* [...] *pasci*', and which leads directly into descriptive hyperbolic comparisons. The frozen world is suffused with the sensorial wonders of heaven, illuminating the darkness of early morning in the bitter winter with the light of the loaves, gleaming brighter than lilies or roses, and the fragrance of roses combines with the aroma of honey that recalls the blessed manna granted to the Israelites (Ex. 16:31). There is a deliberate echoing between Cuthbert's humble obedience in service to the visitor, and the subsequent blessing he receives in the form of a heavenly vision, which follows the overall structure for miracles in the *VCM*, and presents further grounding for the restoration of Creation via the sanctity of the person involved in Ch. VIII.

94 *VCA*, pp. 78–79.
95 *VCM*, ll. 202, 200.
96 See Steen, *Verse and Virtuosity*, p. 28.
97 See Steen, *Verse and Virtuosity*, p. 28.

The famous sea animal miracle of Ch. VIII is linked chronologically to the previous miracle through the use of *interea* in the opening line (l. 220). While *interea* covers a wide lexical range expressing varying levels of contemporaneity,[98] including its weaker use expressing 'a vague resumptive idea',[99] it is here being used in its primary chronological function, to denote events that happen more or less within the same period, in this case, during Cuthbert's time of training within the monastery. It is through this connective term that Bede glosses over Cuthbert's transition from Melrose to Repton in the *VCA*, as well as the location of the miracle near Coldingham; the miracle occurs instead within the scope of Cuthbert's years in communal life. Bede's primary alteration of the episode continues the meditative trend of the *VCM*, teasing out the potential interpretations of the miracles themselves, including Christological typology for Cuthbert. Bede uses these miracles to reveal the saint's spiritual state, which can be connected with the exegesis in Ch. XIX, and the overarching theology of Creation's restoration via the sanctity of the individual. Bede's alterations are also framed by the previous chapter and the depiction of Cuthbert's obedient reception of the disguised angel. There are a number of substantial alterations to the presentation of the miracle, most of which serve to refine the focus of the episode on the Creation miracle itself, its connection to Cuthbert's obedient sanctity, and to Cuthbert's healing of the spying monk.

Bede opens Ch. VIII with another depiction of Cuthbert keeping the Night Office, again connecting the saint's actions to the psalmody for that section of the Office: 'Interea iuvenis solitos nocturnus ad hymnos / Digreditur' ('Meanwhile the nocturnal young man departs at night for his accustomed hymn-singing') (ll. 220–21). As Lapidge notes, Bede is utilising hypallage to transfer the adjective 'nocturnus' from its position describing Cuthbert's nightly singing

98 For a discussion of Virgil's use of the term and its general application, see O. W. Reinmuth, 'Vergil's Use of Interea, a Study of the Treatment of Contemporaneous Events in Roman Epic', *The American Journal of Philology* 54 (1933), 323–39.

99 Reinmuth, 'Vergil's Use of Interea', 328.

to the saint himself.[100] Nonetheless the lexical connection in the line between the terms 'nocturnus', 'solitos', and 'hymnos' functions similarly to the horse-and-loaf episode, providing a ruminative link to the presentation of Cuthbert's obedience within a monastic setting and the observance of the Divine Office. While not as overt as the collocation of *divinum* and *horam* in the previous passage, the use of the term 'solitus' ('accustomed') here together with the term 'nocturnus' ('nocturnal'), with its association with the Divine Office, points to a similar intention. Bede reinforces the connection four verses later through paronomasia, where he depicts Cuthbert's entry into the sea and actual singing: 'Ad mare deveniunt; collo tenus inditus undis / Marmoreo Cuthbertus agit sub carmine noctem' ('They arrive at the sea; Cuthbert spends the night singing, immersed in the waves up to his marble-like neck') (ll. 223–24). The implicit association between 'nocturnus ad hymnos' and 'carmine noctem', the first phrase being the general commentary on Cuthbert's obedient behaviour and the second being the actual description of that behaviour, focuses on the paronomasia of *nocturnus* and *noctem*. The presentation implicitly depicts Cuthbert taking part in the Night Office, or Matins, exemplifying his monastic obedience which, as the chapter continues, Bede directly connects to the restoration miracle. This connection is furthered by comparison with the specific psalms sung during this portion of the Office, as well as traditional exegesis concerning those psalms. According to Billett, the Night Office in the time of Bede would have begun daily with Ps. 94 as the Invitatory.[101] The psalm begins with another call to praise God through song: 'Venite, exsultemus Domino; jubilemus Deo salutary nostro' ('Come let us praise the Lord with joy: let us joyfully sing to God our saviour') (Ps. 94:1). The specific aspect of God which is the impetus for praise, as it is in many of the psalms, is his position as creator and sustainer of elemental Creation, which comprises verses 4 and 5 of the chapter:

[100] See Lapidge, *Anglo-Latin Literature, 600–899*, p. 353; also Steen, *Verse and Virtuosity*, p. 27.

[101] Billett, *The Divine Office*, p. 39.

> [4] Quia in manu ejus sunt omnes fines terrae, et altitudines montium ipsius sunt; [5] Quoniam ipsius est mare, et ipse fecit illud, et siccam manus ejus formaverunt.[102]
>
> ([4] For in his hand are all the ends of the earth: and the heights of the mountains are his. [5] For the sea is his, and he made it: and his hands formed the dry land.)

Ps. 94:6 immediately declares that in response to this the faithful should prostrate themselves before God: 'Venite, adoremus, et procidamus, et ploremus ante Dominum qui fecit nos' ('Come let us adore and fall down: and weep before the Lord that made us'). While not as united with the events described in the *VCM* as those of the horse-and-loaf miracle, the resonance in the progression of events is striking. Cuthbert's purpose of praising God through the singing of psalms aligns with the fundamental call of Ps. 94. His location within elemental Creation, submersion in the sea followed by prayerful kneeling on the shore afterwards, resonates with verses 4–6, which highlight God's relationship to the sea and the call for the faithful to genuflect in response to this.

Once again, the exegetical tradition of Ps. 94 further correlates with Cuthbert's nocturnal miracle. Augustine interprets the sea in this psalm as this world, which God made, and thus the waves (the temptations of the world), are bounded by his divine demarcation.[103] Augustine aligns this with the bounded nature of temptation, and instructs the faithful to be dry land, and thirst for God: 'siti gratiam Dei, ut ueniat super te imber dulcis, inueniat in te fructum' ('thirst for the grace of God: so that as a sweet shower it may fall on you, and may find fruit in you').[104] While neither Bede nor the *VCA* highlight explicitly the connection to the traditional practice of mortifying the flesh by submersion in freezing water,[105] Cuthbert's actions recall this tradition. Thus, Augustine's interpretation

[102] Ps. 94: 4–5.

[103] Augustine, *Enarrationes in Psalmos*, p. 1338: 'Est enim mare mundus iste, sed et mare Deus fecit, nec saeuire fluctus possunt, nisi usque ad littus, ubi ipse terminum posuit' ('For the sea is this world, but God made also the sea: nor can the waves rage save only so far as the shore, where He established their bounds').

[104] Augustine, *Enarrationes in Psalmos*, p. 1338.

[105] See above, p. 27.

of Ps. 94:5 as relating to the bounded nature of temptation is echoed by Bede's depiction of Cuthbert's nightly submersion, whereby he declares his thirst for God in his mortification and prayer, as well as his obedient partaking in the Divine Office.

The miraculous interaction with the sea animals occurs only after Bede's depiction of Cuthbert's obedience to monastic practice:

> Ad mare deveniunt; collo tenus inditus undis
> Marmoreo Cuthbertus agit sub carmine noctem.
> Egreditur ponto genibusque in litore fixis
> Expandit geminas supplex ad sidera palmas.
> Tum maris ecce duo veniunt animalia fundo
> Vatis et ante pedes fulva sternuntur harena;
> Hinc gelidas villo flatuque foventia plantas
> Aequoreum tergunt sancto de corpore frigus;
> Supplice tum nutu sese benedicier orant.
> Qui parens votis verbo dextraque ministris
> Impendit grates patriasque remittit ad undas
> Ac matutino tectis se tempore reddit.
>
> (ll. 223–34)

> (They arrive at the sea; Cuthbert spends the night singing, immersed in the waves up to his marble-like neck. He steps out of the sea and, with knees fixed on the shore, he spreads his two hands to the heavens in prayer. Then, behold, two animals come from the bottom of the sea and prostrate themselves on the golden shore at the prophet's feet. Then, warming his icy feet with their breath and fur they wipe the liquid cold from his holy body. Then with bowed heads they pray in supplication for a blessing. The obedient prophet gives thanks to his attendants by his word and his right hand, and sends them back to their ancestral waters; and he returns to the dwelling in the early morning.)

Bede has once again devoted greater imaginative force to the description than his main source, the *VCA*, by embellishing each portion of the miracle. While the *VCA* presents Cuthbert submersing himself to his armpits and then retreating to the shore to pray, the *VCM* transforms this into a night-long exercise ('agit […] noctem'), as well as raising the water to his neck in a Virgilian echo ('collo tenus').[106] Likewise, Cuthbert does not simply kneel on the shore to pray but, strikingly,[107] given the connection with the Night Office discussed above, fixes his knees to the shore and raises both his hands towards heaven in supplication, once again echoing Virgil.[108] The shore is depicted not simply as earthly sand but material made golden by the illumination of the miracle, as well as the implied approach of dawn. This heightening of the pictorial aspect of the scene continues Bede's poetic practice in the *VCM*, but also serves further to connect the miraculous service of the sea animals to that of Cuthbert and the disguised angel in Ch. VIII. The poem thus emphasises the difficulty of such a winter exercise ('gelidus', 'frigus'), and the sea animals' service ushers the descriptive narrative from dark winter to a warm, luminous new day, echoing the structure of Ch. VII where, after the miracle of the loaves, winter is banished.

Bede's most significant transformation of the sea animal miracle is, however, in his exegesis of this, and by extension, all of Cuthbert's miracles. Where the *VCA* frames the miracle in relation to the restoration miracle of Daniel and the lions and Peter's visionary prowess, Bede instead gives a more comprehensive interpretation of Cuthbert's miracles, which includes depicting the saint as a type of Christ. After Cuthbert orders the monk to 'remain silent' about the miracles until after his death, Bede directly links Cuthbert's actions to those of Christ: 'Summique exempla magistri / Exsequitur' ('He [Cuthbert] followed the example of the great Teacher') (ll. 244–45), a reference to Mtt. 9:27–30, after Jesus had healed two blind men. This statement is immediately followed by an implicit Christological reference, echoing Mtt. 9:5–7, in the narrative description of Cuthbert's

106 See Virgil, *Georgics, P. Vergili Maronis Opera*, ed. R. A. B. Mynors (Oxford, 1972), IV, ll. 523–25: 'tum quoque marmorea caput a cervice reuulsum / gurgite cum medio portans Oeagrius Hebrus / uolueret', ('Even then, when Oeagrian Hebros rolled the head onwards, torn from its marble neck, carrying it midstream'). Noted by Jaager, *VCM*, p. 75. All references to the works of Virgil will be from this edition, and will be referred to only by the name of the work; e.g., *Georgics*.

107 See *VCA*, p. 80.

108 See Virgil, *Aeneid*, II, ll. 152–53: 'ille dolis instructus et arte Pelasga / sustulit exutas uinclis ad sidera palmas'. Noted by Jaager, *VCM*, p. 75. Cf. Lapidge, *Anglo-Saxon Library*, pp. 226–27, on Bede's use of Virgil.

healing of the spying monk: 'Tum prece languorem pellit culpamque relaxat' ('Then with prayer he expelled the illness and forgave the sin') (l. 247). In the biblical passage referenced here, Jesus, after healing a paralytic and forgiving his sins, responds to the scribes' charges of blasphemy by declaring his authority and power over both physical ailment and sin, the fundamental ailment of the soul. Bede clearly had these verses in mind, and is gesturing towards an equivalent miraculous authority for Cuthbert. In the closing section of the chapter, Bede provides an exegesis refocusing the miraculous works of Cuthbert, both physical and visionary, upon the saint's spiritual state: 'Inque dies meritis crescenti summa tonantis / Gratia testis adest, pandunt miracula mentem' ('The highest grace of the Thunderer is present as a witness to Cuthbert's merit, growing day by day; his miracles reveal his [spiritual] state') (ll. 248–49). Cuthbert's restored relationship with the sea animals is thus further connected to his sanctity, which is founded on Cuthbert's humble obedience in washing the stranger's feet. This humble obedience is itself exemplified in his obedience to the monastic life, here via the Divine Office. The central focus of this episode for rumination is how Cuthbert's miraculous works provide evidence of his sanctity, and how that luminous behaviour, bringing warmth to the earthly winter, has its ultimate source in Christ. This image echoes one of Bede's other additions in the *VCM*, where in an introductory list of famous holy men and the lands to which they brought the light of the gospel, Bede adds Cuthbert as the luminary for Britain.[109] The illumination which Cuthbert and the other men bring, phrased as the 'lucem verbi' ('the light of the Word') (l. 13), is described as

[109] See *VCM*, ll. 1–34 for list, esp ll. 25–29: 'Nec jam orbis contenta sinu trans aequora lampas / Spargitur effulgens, huiusque Britannia consors / Temporibus genuit fulgur venerabile nostris, / Aurea qua Cuthbertus agens per sidera vitam / Scandere celsa suis docuit jam passibus Anglos' ('And this lamp, no longer content in the lap of the old world, is shed gleaming across the water, and Britain, now participating in it, gives birth in our own days to the holy splendour whereby Cuthbert, an inhabitant of the golden stars, teaches the English to ascend on high following his footsteps'). Bede's list is an adaptation from lists in two of his sources: Paulinus of Nola, *Carmina*, ed. G. de Hartel, CSEL 30 (Vienna, 1894), carmen XVIIII, pp. 118–43, esp. ll. 54–163; and Fortunatus, *De Virginitate*, in ed. F. Leo, *Monumenta Germaniae Historica, Auctores Antiquissimi* 4. (Berlin, 1881), Carminum lib. VIII, ch. III, pp. 181–91, esp. pp. 184–85, ll. 141–60.

having its ultimate source in Christ.[110] Bede's emphasis on Cuthbert as bringing illumination to this specific patch of the English landscape resonates with his wider role of illuminating the whole of Britain. The restoration of Creation displayed in the sea animal miracle is transformed to create an episode that focuses on the nature of sanctity as functioning within monastic forms of obedience. To be a saint, in this model, is to follow monastic practice. In so doing Cuthbert is granted the further grace of prophecy from God, which Bede depicts in connection with his overarching metaphor of illumination: 'Iamque prophetalis stellanti e culmine virtus / Candida praerutilo irradiat praecordia flatu' ('And now a prophetic strength from the summit of heaven, illuminates his gleaming heart with its burning radiance') (ll. 250–51). The humble *vates*, washing the feet of a frozen traveller and singing psalms while submerged in the icy sea, is lifted up through these obedient actions into the role for which his designation destines him: prophet.

Bede continues the chronological linking of the chapters in this triptych by employing the temporal marker *interea* to begin Ch. IX: 'Pictorum *interea* puppi defertur ad oras' ('*Meanwhile* he is carried by ship to the shores of Pictland') (l. 252). This chapter concludes the triptych in its unification of the three primary miraculous *foci*: the provision of sustenance, both spiritual and physical; the connection between sanctity and the calming of the elemental world; and the role of Cuthbert's distinctly monastic obedience in causally bringing these miracles about. In this chapter, Bede elaborates the description of the winter storm, echoing both the previous chapters (VII and VIII), but to a degree that surpasses them both:

> Sed reducem rigido tardant freta clausa Decembre.
> Nubila cum tonitrus cum grando polique micantes
> Cumque fames frigusque marisque pericla furentis
> Externo trepidos quaterent sub litore nautas.
>
> (ll. 253–56)

(But [the seas] hindered his return, frozen solid in December. When storm-clouds, thunder, hail, trembling skies, and when hunger, cold, and the dangers of a raging sea terrified the frightened sailors on that foreign shore.)

[110] *VCM*, ll. 16–19.

This description echoes that of Ch. VII, employing polysyndeton with an exaggerated force unique in the *VCM*, with the conjunctive use of *cum* leading into the equivalent use of *que*, the depiction of the storm building with each subsequent descriptive image; clouds come, then thunder, then hail, then the sky itself flashing or trembling, then the familiar *fames* and *frigus* that Bede used in Ch. VII, and finally the peril of the raging sea ('maris [...] furentis'). The completed depiction is of a wondrous storm whose power sends the sailors in fear to the shore. The descriptive weight of the storm grants the ensuing miracle its potency, as well as highlighting the spiritual stability of Cuthbert. In this chaotic assemblage Bede, following the *VCA*, focuses on the day of Epiphany as explanation for the anticipation of an ensuing miracle: 'Venerat alma dies, natus qua corpore signis / Enituit Christus caelorum gloria terris' ('The blessed day had arrived on which Christ, born in the flesh, shone from his marks as the glory of heaven on earth') (ll. 257–58). The illumination of Epiphany is immediately connected to Cuthbert's reaction via its position in the next sentence, which depicts the saint reacting calmly to the raging storm: 'Ille, ut erat placidus, blando sic pectore fatur' ('He (Cuthbert), being a gentle person, speaks thus from his pleasant heart') (l. 259). This emphasis on the fixity of a saint's internal state is a hagiographical commonplace, and functions further to display Cuthbert's sanctity.[111] As Lapidge notes, Cuthbert's response to the fearful sailors includes further description of the wintery scene, constructed in poetry so 'compressed as to be virtually untranslatable'.[112] Its density mirrors the building of the storm, but instead builds towards graceful resolution in Cuthbert's lengthy speech exhorting the brothers to seek God for help:

> Cernitis, aequoreo canescat ut aggere tellus,
> Aer aquas manet, glacies mare, nox tegat aethram;
> Corda fame tabent hominumque iuvamina desunt.
> Ergo deum restat precibus pulsare supinis.
> (ll. 262–64)

> (Do you see the earth whitens through mounting hoarfrost, the air flows with water, ice covers the sea, and darkness covers the

[111] See above, p. 25 n.30.

[112] Lapidge, *Anglo-Latin Literature, 600–899*, p. 335.

> sky? Our spirits melt with hunger, and help for humanity is far away. Therefore, the only thing left [to do] is to assail God with prostrate prayers.)

Lapidge argues here for a symbolic reading of the descriptive landscape,[113] in which the 'very bleakness of the wintry scene' carries 'the symbolic promise of God feeding Cuthbert as He had fed His chosen people'.[114] The imaginative physicality of the depiction here, however, echoes Bede's wider heightening of the descriptions of Creation throughout the *VCM*. Biblical parallels are used primarily as precedents for Cuthbert's faithful declaration of God's miraculous provision and protection. This scene is more concerned with Cuthbert's faith in God's provision in the physically delineated Scottish winter than with biblical precedent. Thus, the structure of Cuthbert's speech moves from enumerating the biblical parallels to the fortuitous timing of Epiphany, into a declaration of faith in God's provision based on both: 'Sic quoque nostra deo remeant duce prospera credo, / Illum si rogitans firmet fiducia pectus' ('Thus I believe prosperity will also return to us through God's guidance, if in beseeching Him faith fortifies our hearts') (ll. 276–77). Nonetheless, Cuthbert's reaction to the subsequent miraculous provisions brought about by his earnest faith in prayer points to the meditative focus on the place of monastic obedience in the enacting of the miraculous.

Bede depicts Cuthbert's obedience in this episode once again through the saint's implicit partaking of the Night Office, by adding details to the description of the saint gathering his companions at the seashore: 'Haec ubi dicta, freti socios sub litore sistens, / *Suerat ubi vigiles supplex iam ducere noctes*' ('When he had said these things and had placed his companions together on the shore *where he was accustomed to spend nights at vigil in prostrate prayer*') (ll. 278–79). While the vigil itself is not undertaken here, Bede's focus on monastic obedience is once again apparent through the collocation

[113] Lapidge, *Anglo-Latin Literature, 600–899*, pp. 335–36: 'But if we think that Bede's primary intention here was to paint a naturalistic description of the Scottish landscape in winter, we miss his point [...] In other words, the landscape in which Cuthbert and his companions finds themselves is a symbolic one.'

[114] Lapidge, *Anglo-Latin Literature, 600–899*, p. 336.

of the terms 'suerat', 'vigiles', and 'noctes'. This emphasis on the repetitive nature of this practice reflects the notion of the Divine Office as the consistent demarcating practice of each day and night, and the collocation of 'vigiles' and 'noctes' once again points to the specific portions of the Office.[115] This further reference to observing the Night Office is meant as an echo of Cuthbert's actions in Ch. VIII, and thus the same connection with Ps. 94 is implied. While not causally connected here, the depiction of Cuthbert's space on the shore as the location where he engaged in this activity in general, and the immediate discovery of God's miraculous provisions in the form of three portions of dolphin flesh, furthers the connection between Cuthbert's monastic obedience and his miracle-working. Cuthbert then prophesies the calming of the sea,[116] and they all partake of God's blessing while awaiting the dissipation of the storm. Bede adds another echo of Ch. VII here to further the numinous quality of both miracles, with his additional detail of the honeyed flavour of the fish echoing the honeyed loaves.[117] Taken together, this chapter functions as a fitting close to the triptych, emphasising once more the particularly monastic form of sanctity displayed by Cuthbert, and its relationship to the various miraculous events, from the provision of bread and fish, to prophetic vision, to the restoration of Creation in the service of the sea animals.

115 While the specific demarcations between the terms for the various portions of the Divine Office were subject to change, here Bede is once again echoing Benedictine wording, as can be seen in Ch. 8 section 3 of the Benedictine *Regula*, p. 508: 'Quod uero restat post uigilias a fratribus qui psalterii uel lectionum aliquid indigent meditationi inseruiatur' ('And the time that remains after the Night Office (Vigils) should be spent in study by those brothers who need a better knowledge of the Psalter or the lessons'). See also J. W. Mckinnon, 'The Origins of the Western Office', in *The Divine Office in the Latin Middle Ages: Methodology and Source Studies, Regional Developments, Hagiography*, ed. M. E. Fassler and R. A. Baltzer (Oxford, 2000), pp. 63–72, for an overview.

116 *VCM*, l. 284.

117 *VCM*, l. 288. As Lapidge notes, *Anglo-Latin Literature, 600–899*, p. 336, concerning the dolphin miracle discussed above, honey may also refer to the promised land, the 'land dripping with milk and honey', which can also typologically point towards heaven. The phrase appears frequently in Scripture, for example Ex. 3:8, 33:3; Deut. 31:20; Lev. 20:24; and Ezek. 20:15.

Eremitism and Restored Creation

Bede alters the *VCA*'s presentation of Cuthbert's transition into the eremitic life, during which the greater portion of the restoration miracles occur, to depict the saint in the *VCM* as an idealised Gregorian monk-pastor, who unifies the active and contemplative lives. The primary position of Gregory the Great in Bede's conception of orthodoxy and orthopraxis is well documented,[118] as is Gregory's influence on Anglo-Saxon spirituality. One of Gregory's most important contributions was his *Cura Pastoralis*, a text which, as Francis Leneghan highlights, was the 'classic portrait' of the 'ideal pastor, in equal parts teacher and contemplative'.[119] Leneghan also points out that this model of the 'mixed life' of the monk-pastor was 'imitated by early insular bishop-monks like Aidan and Cuthbert and admired by Bede and Alcuin'.[120] Clare Stancliffe, among other scholars, has noted how Bede, in his prose *VCP*, 'stylizes Cuthbert according to the monk-pastor ideal that he imbibed from Gregory the Great, explicitly pointing up Gregorian parallels to Cuthbert's miracles'.[121] This transformation of Cuthbert into an idealised Gregorian monk-pastor has its beginnings, however, here in the *VCM*. This synthesis between the active and contemplative lives is made explicit in the way Bede reorders the events depicted in the *VCA*, as well as by altering the prose frame for the chapters. Bede splits the events in *VCA* III.1 into two, creating a self-contained chapter focusing entirely on the saint's time at Lindisfarne, and a second concerned solely with the establishment of his cell on Farne, thus

118 See Thacker, 'Bede's Ideal of Reform', 140–42; Stancliffe, 'Pastor and Solitary', p. 38; and H. Appleton, 'Anglo-Saxon Landscapes: The Construction of the Environment in Old English and Related Texts', Unpub. Ph.D. thesis (University of Sydney, 2012), pp. 75–77.

119 F. Leneghan, 'Teaching the Teachers: The Vercelli Book and the Mixed Life', *English Studies* 94 (2013), 627–658 at 633.

120 Leneghan, 'Teaching the Teachers', 633.

121 Stancliffe, 'Pastor and Solitary', 28. See also H. Gittos, *Liturgy, Architecture, and Sacred Places in Anglo-Saxon England* (Oxford, 2013), pp. 32–33; Brown, *A Companion to Bede*, 84; and C. Di Sciacca, '*Concupita, quaesita, ac petita solitudinis secreta*: The Desert Ideal in Bede's Vita S. Cuthberti and Ælfric's Life of St Cuthbert', in *Hagiography in Anglo-Saxon England: Adopting and Adapting Saints' Lives into Old English Prose (c. 950–1150)*, ed. L. Lazzari, P. Lendinara, and C. Di Sciacca (Barcelona, 2014), pp. 121–81.

highlighting the saint's faithful union of the active life on Lindisfarne with the progression towards full contemplative seclusion. He also gives form to the internal balance of the saint in his attempt to live out this synthesis by providing the following prose heading for Ch. XIV: 'Qualiter anachoresim meditatus apud Lindisfarnenses monachos vixerit' ('How he lived with the monks of Lindisfarne, but desired the eremitic life').[122] Likewise in the subsequent chapter Bede centres the ensuing poetic account on the establishment of Cuthbert's eremitic cell, and the associated expulsion of demons from the island with another heading: 'Qualem sibi in insula Farne pulsis daemonibus habitationem fecerit' ('How he made for himself a habitation on Farne Island once he had expelled the demons').[123]

In Ch. XIV Bede sets out to develop the internal balance between the two sides of Cuthbert's life through an expansion of Cuthbert's impetus for seeking the solitary life, taking the *VCA*'s simple description of the saint fleeing worldly glory[124] and transforming it into a sustained depiction of Cuthbert's desire for heaven and the inherent necessity of ignoring the fame associated with his miraculous deeds:

> Talia mirantum fragili ne laude supernae
> Caelestisque exsors famae foret, abdita mavult
> Secreti lustrare, deo qua teste valeret
> Laudis ab humanae liber munirier aura.
>
> (ll. 373–76)

> (So that he might not be exempt from heavenly and celestial fame through the fleeting praise of such miracles, Cuthbert illuminates the depths of the secret location, where with God as his witness he can be free to strengthen himself against the approval of human praise.)

This description of the benefits of the eremitic life is immediately followed by an account of the saint's submission to the commands of his bishop to teach in Lindisfarne instead,[125] thus emphasising his monastic obedience even as it highlights Cuthbert's efficacy

122 *VCM*, p. 85. This corresponds to the first part of the *VCA* III.1.
123 *VCM*, p. 87.
124 *VCA*, p. 94.
125 *VCM*, l. 377.

within the active life. Here Bede condenses the depiction from the *VCA*, where Cuthbert is the author of the monastery's first written *regula*, into a general Christological image of Cuthbert revealing to the brothers, by example, the narrow path of virtue.[126] The use of the noun *callis* here echoes the familiar metaphor from Mtt. 7:13–14 in Jesus' sermon on the mount of the narrow gate and straight path that leads to life.[127] Moreover, the pastoral element of this image anticipates Bede's later verse epitaph for Wilfrid in the *Historia* V.7, which concludes with a declaration that the various deeds enumerated serve as an example by which others might follow a righteous path: 'Dona, Iesu, ut grex pastoris calle sequatur' ('Grant, Lord, his flock may tread their shepherd's path').[128] This pastoral emphasis is combined with Bede's image of Cuthbert as the great luminary for all of Britain, as he describes the saint's overall activities while living on Farne as 'gleaming miracles'.[129] The ensuing list of Cuthbert's luminous activities is concluded by Bede's novel re-emphasis on the role of external splendour as a window into the saint's inner life:

> Internam quid enim coner describere vitam,
> Quam dulcis sermone gravisve sit actibus et quam
> Accensam lacrimis acuarit ad aethera mentem,
> Cum decus externum puri sit pectoris index.
>
> (ll. 386–89)

> (Why should I attempt to describe his inner life – how sweet in his speech and how weighty in his actions, and how he urged his enflamed mind to the heavens with tears – when his external beauty is an index of his pure soul?)

In this way, Bede brings together the roles of pastor and luminary, both founded upon Cuthbert's obedience to the monastic rule of his bishop, holding his desire for eremitic seclusion until he is released to do so.

126 *VCM*, l. 378.

127 Mtt. 7:14. The phrase also recalls Is. 26:7, with its description of the path of the righteous, especially in its use of *callis*: 'Semita justi recta est, rectus callis justi ad ambulandum' ('The way of the just is right, the path of the just is right to walk in').

128 Bede, *HE*, V.19, pp. 530–31.

129 *VCM*, l. 381.

With the first portion of *VCA* III.1 adapted in *VCM* Ch. XIV, Bede is free in Ch. XV to expand the depiction of Cuthbert's conquering of Farne Island, and subsequent establishment of his eremitic retreat. Bede teases out here the implications of the *VCA*'s quick reference to the Antonian trope of expelling demons from wilderness locations – itself a reversal of Adam and Eve's weakness before the serpent in the Garden (Gen. 3) – into a full narrative. While the *VCA* has devils who cause illusions ('demonum fantasias'), Bede presents the landscape as dense ('horrens') with their presence and their threats: 'Illa prius nam horrens larvaribus insula flabris / Arcebat humana minis consortia caecis' ('For that island was previously bristling with wind-demons and closed off from human society with hidden threats') (ll. 392–93). Bede alters the *VCA*'s oblique reference to Cuthbert putting the demons to flight into a potent account of Cuthbert banishing them with the cross and the demons vanishing like smoke: 'Quam domini mox servus adit, cruce territus atrox / Turba fugit per inane vagi quasi portio fumi' ('As soon as the servant of the Lord arrived, the savage assembly was terrified by the cross and vanished into nothing like wisps of smoke') (ll. 394–95). With the banishment of the demons, Bede presents the construction of Cuthbert's retreat as an elision between heaven and earth, creating a space in which the miraculous might occur. In keeping with the stylistic density of the *VCM* this elision is accomplished through a condensing of imagery, where one sense bleeds into the other:

> Ille serena tenens, pereunte tyrannide, regna
> Terrestri aetheriam sacer aggere condidit urbem,
> Atque humiles celsis statuit sub moenibus aedes,
> E quibus astriferum tantum aspectare cacumen
> Posset et a celso secretus rege tueri.
>
> (ll. 396–400)

> (The saint, taking possession of this serene realm once [the demons'] tyranny had vanished, founded an ethereal city with terrestrial walls, and set up humble dwellings within his celestial ramparts, from which he could see only the starry sky, and could be observed in secret only by the heavenly King.)

The elision begins in Cuthbert's establishment of a heavenly city ('aetheriam [...] urbem') with earthly walls ('[t]errestri [...] aggere'), creating an image that echoes Bede's description of Cuthbert's external form revealing inner glory in the conclusion to Ch. XIV. In the next line (398) Bede furthers this elision of place through a deliberate reversal of the imagery, where the surrounding walls are now described as Latin *celsus* ('lofty'), implying something beyond the earthly, within which Cuthbert builds a structure elided by the denotations of the noun *aedes*, which can refer to either a temple or sanctuary, as well as a dwelling for humans in the plural. This structure of elision is repeated in the next two lines (399–400), where Cuthbert's intense observation of the heavens is mirrored in God's watchful observance of Cuthbert in his seclusion. The mirrored verbs of looking, *aspectare* and *tueor*, are used to depict the faithful searching of the exemplary human, Cuthbert, and the protective eye of God as guardian. While the *VCM* joins this establishment of elided space with the evangelistic miracle of the massive stones used to construct the physical cell, the sustained emphasis on the elision points to Bede's ruminative expectation. This descriptive episode is meant to call to mind the particular potency of sanctity and its ability to reconcile the fallen and heavenly realms, as well as the eschatological elision of the New Jerusalem and the New Paradise.[130] This reconciliatory emphasis further establishes Cuthbert's sanctity for the various restorations of Creation which follow on Farne.[131]

130 See Rev. 22: 1–5, Cf. Ezk. 47, as well as Bede's *Commentary on Revelation* for 22:2, which interprets the 'Tree of Life' in the New Jerusalem as the same tree that was in the centre of Eden: *Bedae Presbyteri Expositio Apocalypseos*, ed. R. Gryson, CCSL 121A (Turnhout, 2001), p. 565, 'Idem quippe lignum, quod Iohannes in aurea Hierusalem, Moyses in paradiso florere descripserunt' ('Indeed, this tree that John described in the golden Jerusalem is the same one that Moses described flowering in Paradise'). See further D. Mathewson, *A New Heaven and a New Earth: the Meaning and Function of the Old Testament in Revelation 21. 1–22.5* (London, 2003), esp. pp. 186–216.

131 A similar elision of spaces takes place in the *VCM* Ch. XLIV, where the calf-skin covering used as Cuthbert's roof is compared to the roofs of Solomon's temple. Bede declares that while the temple's walls rival the heavens in glory and jewels, the calf-skin covering is more marvellous because it still is efficacious in producing miracles, where the temple has been destroyed by Chaldean flames and the crumbling of stone.

With the establishment of Cuthbert's cell Bede moves into the various restoration miracles. In Ch. XVI Bede relates the well miracle, truncating it and then combining it with another aqueous miracle: the transformation of water into wine. Whereas the *VCA* highlights the theme of obedience in the episode, as well as drawing out the interpretative connection to the biblical precedents for such a miracle,[132] Bede instead focuses on the miracle itself as a way of highlighting the connection between Cuthbert's sanctity and his ability to restore a portion of Creation. The *VCM* here divides the miracle into three small poetic sections. First, the miracle itself is described as occurring through Cuthbert's prayers, rather than the mundane obedient digging of the brothers.[133] The well miracle in the *VCM* has no overt purpose, other than to further depict Cuthbert's virtuous ability to transform the postlapsarian present into a prelapsarian moment. This is evidenced by Bede evoking paradise in his lexis, as well as by connecting the miracle to the *locus amoenus* tradition: 'Fontis inops fuerat locus hic, sed sanctus amoenam / Excutit insolita precibus dulcedine limpham' ('The place was lacking a fountain, but the saint by his prayers produces a pure spring of extraordinary sweetness') (ll. 406–07). The use of the noun *fons* in this context is meant to evoke the spring that watered Eden in Genesis 2:6: 'Sed fons ascendebat e terra' ('But a spring rose out of the earth'). This interpretation is supported by Bede's ekphrastic collocation of *locus* and *amoenus* in line 406,[134] which echoes the use

132 See above, pp. 47–56.

133 *VCM*, ll. 406–07.

134 For example, see Isidore, *Etymologiarum sive Originum Libri XX*, ed. W. M. Lindsey, 2 vols (Oxford, 1911; repr. 1957), vol. 1, XIV.8.33 (I follow the edition here, which has no page numbers, but gives book, chapter, and section): 'Amoena loca Varro dicta ait eo quod solum amorem praestant et ad se amanda adliciant. Verrius Flaccus, quod sine munere sint nec quicquam his officia, quasi amunia, hoc est sine fructu, unde et nullus fructus exsolvitur. Inde etiam nihil praestantes inmunes vocantur' ('According to Varro, "pleasant places" (*locus amoenus*) are so called because they promote love (*amor*) and draw to themselves things that ought to be loved. According to Verrius Flaccus they are so called, because they are without a "public function" (*munus*), nor is anything in them like business'). All references to the text are from this edition. See also E. R. Curtius, *European Literature and the Latin Middle Ages*, trans. W. R. Trask (Princeton, 1990), pp. 190–202, at 192: 'the word *amoenus* ("pleasant, lovely") is used. It is Virgil's constant epithet for "beautiful" nature'.

of the *locus amoenus* tradition to describe Eden, but which is now used to describe the newly consecrated ground on which Cuthbert's celestial/earthly cell sits. This is echoed by Bede's later description of paradise in *In Genesim*: 'esse locum scilicet amoenissimum, fructuosis nemoribus opactum, eundemque magnum et magno fonte fecundum' ('namely as a very delightful place, shaded with fruitful groves, and also fertile with a great spring').[135] Bede retains some of the *VCA*'s emphasis on the fountain's present efficacy, '[p]raebet adhuc cunctis suavem potantibus haustum' ('[i]t still offers a sweet draught to all who drink it') (l. 409),[136] but moves quickly on to specify the meditative focus of the miracle by explication and the addition of a related miracle excised from later in the *VCA*:

> Nec mirum haec domini famulum potuisse mereri,
> Qui quondam, saliente sitim dum pelleret unda,
> In meracum latices valuit convertere nectar.
> (ll. 410–12)

> (Nor is it any wonder that the Lord's servant could have merited this: once, while thirsting he struck [a rock] and a spring welled up; he was able to turn this water into pure wine.)

The springing forth of the well on Farne in the *VCM* is primarily concerned with the role of Cuthbert's merit ('mereri') in bringing about the restoration of Creation, rather than alleviating Cuthbert's thirst which is described in the *VCA* and (later) the *VCP*. Likewise, the attachment of the connected miracle, the poetic transformation of water into 'nectar' or 'wine', further emphasises Cuthbert's efficacy, deriving from his internal sanctity, as he is able ('valuit') to do so. Mosaic typology is still present in the well miracle itself, as is Bede's additional Christological reference to the miracle at Cana, but both are subsumed into bare descriptions, as Bede considered the connection to be so apparent as to be unworthy of explicit presentation. This shifts the focus from typological interpretation to the present exegesis concerning sanctity's power to restore Creation.

135 Bede, *In Genesim*, p. 46. This passage is taken wholesale from Augustine's *De Genesi Ad Litteram*, p. 231; this correspondence is noted by both the CCSL edition, p. 46, and Kendall, *On Genesis*, p. 111.

136 Note Bede retains the adjective *suavis* from the *VCA*.

Bede engages directly with Cuthbert's restoration of animate Creation by expanding the obedient raven miracle from III.4 of the *VCA* into two separate episodes: *VCM* XVII and XVIII. This alteration serves to emphasise both the role of sanctity in its ability to restore Creation, and the exemplary function of these miracles. In Ch. XVII Bede centres on Cuthbert's active life of cultivation, where, like Antony,[137] he seeks to live by means of his own labour. Here Bede inserts a group of thieving birds which, instead of stealing thatch, set out to steal Cuthbert's harvest.[138] The entire episode has its roots in the Antonian narrative, where the hermit seeks to be self-sufficient through farming, after recognising the hardships of those who brought him sustenance, only to have wild beasts from the desert come and graze.[139] Antony verbally admonishes the beasts and banishes them; they not only depart, but also never again enter the garden.[140] Bede closely follows the structure of the Antonian episode, removing any other human agents to focus better on Cuthbert and his deeds, but also alters the miraculous power of the saint over animate Creation from an imperious declaration to a prelapsarian unification. This is brought about not only by the general framing of Cuthbert's eremitic life on Farne discussed above, but also by the alteration of Cuthbert's response to the thievery. Where Antony seizes one of the intruders and banishes the entire group, Cuthbert calmly speaks to them.[141] The extraordinary humility of Cuthbert's speech itself contrasts with Antony's commanding the desert creatures to leave; Cuthbert responds pastorally, not only pointing out the birds' implied error, but also humbly acknowledging that if their actions were permitted by God, he would not forbid them:

137 Colgrave notes this correspondence in the relevant chapter of Bede's *VCP*, and that 'the prototype of this miracle is obviously the story which Bede quotes of Antony driving away the wild asses from his little plot in the desert in which he had planted a few herbs (*Vit. Ant.* c. 25)', p. 350, note to Ch. XIX. This episode is in Ch. 50 of the edition of *Vita Antonii* cited in this work.

138 *VCM*, ll. 416–18.

139 *Vita Antonii*, Ch. 50, p. 177.

140 *Vita Antonii*, Ch. 50, p. 177.

141 *VCM*, l. 419.

Quid precor inlicito messem contingitis ausu,
Quae vestro sulcis non est inserta labore?
Pauperies an vestra meam transcendit, ut istud
Incurvam merito falcem mittatis in aequor?
Quod si forte deus iube his instare rapinis,
Non veto; sin alias, vos finibus indite vestris.
(ll. 420–25)

(Why, pray, do you seize with unlawful attempt the harvest which was not sown in the soil by your own labour? Does your poverty surpass mine, such that you would be worthy to put the curved blade to the earth? But if by chance God commands you to pursue this theft, I do not oppose it; if otherwise, stay within your limits.)

This serves to elaborate further Cuthbert's exemplary sanctity, founded in Bede's depiction of the saint as the epitome of the Gregorian harmonised active and contemplative lives; even in his interaction with animate Creation his pastoral role is evoked. The birds respond in kind, not simply obeying the saint and altering their behaviour, as the wild beasts do with Antony, but also being fully restored into prelapsarian relationship with Cuthbert, who is now their shepherd, even as they are his faithful sheep.[142] While the emphasis on the peaceful coexistence between the saint and animate Creation has its roots in hagiographical tradition, especially the *Vita Antonii*,[143] the imagery of the shepherd and sheep allows for the connection to Cuthbert's pastoral role over his human flock. It is telling that Bede's example, cited in the introduction of this book, for prelapsarian relationship in his *In Genesim* is based on this hagiographical commonplace, where the obedience of birds to saints is a testament to first Creation ('testimonium primae creationis'), that is, of paradise:

Denique, testimonium primae creationis legimus uiris sanctis atque humiliter Deo seruientibus et aues obsequium praebuisse, et rictus cessisse bestiarum et uenenum nocere non potuisse serpentium.

142 *VCM*, ll. 426–30.

143 *Vita Antonii*, Ch. 51, p. 177: 'feras, sicut scriptum est, secum pacificaba' ('the beasts, as it is written, "kept peace with him"'). Biblical reference, Job 5:23.

(Finally, as evidence of the first Creation, we read not only that birds have rendered obedience to saints humbly serving God, but also that they have been removed from the yawning jaws of beasts, and that the poison of serpents has been unable to harm them.)[144]

Ch. XVIII, instead of centring on Cuthbert's ability to restore prelapsarian Creation, focuses on the miracle's potential as an exemplar for righteous behaviour. By dividing this section into two chapters, Bede is able to differentiate the species of bird involved, where Ch. XVII has the general and nondescript birds ('volucres'), Ch. XVIII can focus entirely on the raven ('corvi'), an animal brimming with cultural and theological baggage.[145] In the *VCM*, Bede focuses only on a specific portion of this baggage, supplying an exegesis that includes the fundamental aspect of the ravens' rapacious behaviour and their expected role of general disobedience.[146] The ravens' act of thieving vandalism is not, therefore, unexpected, and does not immediately reflect upon avian kind as a whole. [147] It does, however, heighten the miraculous nature of their eventual obedience. The narrative structure of the miracle follows that of the *VCA*: the ravens stealing the thatch, Cuthbert warning them twice, the ravens' departure, their contrition, Cuthbert's forgiveness, and their return with a gift of hog's lard for the saint. Bede once again does not reproduce the distinction between levels of command in the *VCA*, *deposco* and *praecipio*, but nonetheless shifts the force of Cuthbert's communication with the ravens. In the first instance, where the *VCA* presents Cuthbert driving the birds away with a wave of his hand, in the *VCM* the saint's action is described with the verb *monere*, whose semantic range includes both 'admonishment' as well as the pedagogical notion of 'instruction': 'Hos vatis cessare monet spretisque suadelis' ('The prophet warns/instructs

144 Bede, *In Genesim*, p. 29; see above, p. 61 in relation to *VCA*, and below, p. 158, in relation to *VCP*.

145 The source for this hagiographical trope is the story of Elijah and the crows/ravens. The presentation here is also influenced by the raven story in Gregory's *Dialogues*, vol. 2, pp. 160–70.

146 This is born out by comparison with Bede's *VCP*, which explicitly highlights the ravens as inherently prideful, and thus their ensuing obedience as more miraculous. See below, pp. 161–63.

147 *VCM*, l. 431.

them to cease, and they despise his exhortations') (l. 434). Bede's shift here aligns with the overall focus of the *VCM* on Cuthbert as teacher. Similarly, in the depiction of Cuthbert's second command, Bede gives force to the saint's admonition by an imperative use of the verb *abere*, which he places at the end of line 435 to further the term's emphasis: '"Hospitium fratrum quid" ait "laceratis? Abite / Huius et extorres patriae durate per aevum!"' ('"Why do you tear asunder the monks' guesthouse?" he said, "Go, and be forever exiled from this homeland!"') (ll. 435–36). The ravens obediently flee, and three days later one returns to beg for absolution and forgiveness. In the *VCA*'s account, the specific actions of the penitent raven mirror that found in texts describing monastic behaviour such as Gregory's *Dialogues*,[148] and while Bede retains this general idea, his presentation involves both poetic embellishment as well as a deliberate echo of the similar scene in which Cuthbert forgives and heals the spying monk in *VCM* VIII:

> triduoque peracto
> Alter adest vati pedibusque volutus et alas
> Lugubre sparsus veniam reditumque precatur.
> (ll. 437–39)

> (and after three days one of them comes to the prophet and casts itself at his feet, and stretching out its wings in sorrow, begs for forgiveness and permission to return.)

Bede emphasises the penitent actions, transforming the subtle bowing of the head ('inclinato capite') with wings outstretched in the *VCA* to full prostration ('volutus') and the humble cries for pardon ('humili voce'), into a potent lamentation: 'Lugubre [...] precatur'. He also ties the episode to Ch. VIII, where the penitential actions of the spying brother are given as follows: 'Aeger adest vati, supplex genibusque volutus' ('He stands suffering before the prophet; and casts himself on his knees in supplication') (l. 238). The paralleling is clear, where the first half of each line describes the penitent arriving, both with phrase 'adest vati', and the second their act of prostration before Cuthbert, both with the same word,

[148] See above, pp. 64–65. Also, see Gregory's *Dialogues*, vol. 2, pp. 160–70.

'volutus'. This kind of verbal repetition is rare in the *VCM*, and points to a deliberate echoing that is meant to highlight Cuthbert's sanctity, as his forgiving the sins of both the brother and the ravens is an imitative exemplar in the saintly extreme. This deliberate parallel to the actions of the spying monk can therefore be connected with Bede's primary alteration and focus in the episode, in which he teases out the didactic function of such a miracle. Whereas the *VCA* concludes the episode with another attempt at displaying the present efficacy of Cuthbert's miracle in its attempt to create sites for potential pilgrimage, Bede instead interprets it as another model by which his monastic audience can learn humility and obedience:

> Cerne viam corvi et caecum depone furorem,
> Qui precibus noxam fletuque et munere purgat.
> Nec pudeat vitae volucrum de pectore formam
> Sumere, cum moneat Sapientia: respice calles,
> Quos formica terat, sensumque addisce sagacem.
>
> (ll. 446–50)

> (Behold the ways of the raven, and lay down your blind fury, which redeemed its offence through prayers, weeping, and a gift. Do not be ashamed to take a model [for life] from the lives of birds, since the [Book of] Wisdom encourages us to look at the paths worn by ants, and learn wise behaviour.)

The biblical reference here is Prov. 6:6, which includes a command to observe the actions of ants so that the watcher might learn wisdom: 'Vade ad formicam, o piger, et considera vias ejus, et disce sapientiam' ('Go to the ant, O sluggard, and consider her ways, and learn wisdom'). The interpretative practice evidenced in the verse is equivalent to what Bede suggests here: watch the ravens, and learn obedience and humility. The penitent raven miracle in Ch. XVIII is, therefore, focused on the exemplary actions of both Cuthbert, as pastoral exemplar, and the ravens, as obedient penitents. That this alteration is done through Cuthbert's prelapsarian interaction with animate Creation aids in Bede's transformation of the miracle into material useful for monastic rumination.

Death and Psalmody

Bede's presentation of Cuthbert's final hours in *VCM* XXXVI and XXXVII is likewise adapted to portray Cuthbert observing the Divine Office. In these chapters, Cuthbert's obedience is united with that of the monks of Lindisfarne, presenting them as a group engaged in the Office on the last night of the saint's life. This unification of the community under the Divine Office resonates with Bede's overall transformation of Cuthbert's life, and further highlights his pastoral role. In another significant expansion from the *VCA*, the *VCM* centres Cuthbert's final hours on the keeping of the Night Office and the spiritual potency of the act of psalmody. For example, Ch. XXXVI describes how the collected brothers on Farne, as well as Cuthbert himself, fill the island with psalmody, driving the physical and spiritual shadows away:

> Iamque soporata torpebant omnia nocte,
> Excubiis *vigiles* premeret dum Farne tenebras,
> Quaque *preces resonant* modulis, *fremit insula psalmis*.
> Vatis et ipse suae praegustans *gaudia palmae*
> *Aethereis expendit ovans sub laudibus umbras*.
>
> (ll. 775–79)

(Now during that night all things fell towards sleep, while those on Farne kept away the shadows with *vigils; prayers resounded* throughout with their rhythms; *the island murmurs with psalmody*. The prophet himself, anticipating tasting the *delights of his victory, triumphantly measures out the shadows by means of heavenly praise.*)

Both Cuthbert and the brothers on Farne are depicted here partaking in the Night Office, marking the hours, and keeping away sleep and shadow. Bede describes this practice as providing a foretaste of the victory of heaven that awaits Cuthbert who, despite imminent death, is filled with joy in the singing of psalms. This represents a substantial change from the *VCA*, where the primary focus of Cuthbert's death is his partaking of communion and subsequent relinquishing of his spirit without any struggle, 'sine gemitu obiit in uiam patrum' ('without a sigh [he] went in the way of his fathers').[149] In the *VCM* Bede focuses instead on Cuthbert's partaking of the Divine Office as well as communion, and thereby transforms Cuthbert's last earthly action from a passive relinquishing of his spirit into a release joined with the continuation of praise:

> Pocula degustat vitae Christique supinum
> Sanguine munit iter, vultusque ad sidera et almas
> Sustollit gaudens palmas, animamque supernis
> Laudibus intentam laetantibus indidit astris.
> (ll. 782–85)

> (He tastes the cup of life and safeguards his journey upward with the blood of Christ, and, rejoicing, he lifts his face and his blessed hands to the starry heavens, and sends forth his soul, stretching out toward heavenly praise, to the joyful stars.)

The metrical weight of the final line falls on the wordplay between the alliterative pair 'laudibus' and 'laetantibus', and focuses the end of Cuthbert's earthly life on the subject of praise, here functioning contextually within the keeping of the Night Office.

Bede furthers this focus on monastic duty in the connected, and original, episode in the subsequent chapter, XXXVII, which presents the monks' engagement with the Divine Office as functioning prophetically. Here the *VCM* departs from the narrative of the *VCA*, and describes how the priest who had attended Cuthbert in his cell comes out and tells the awaiting brethren on Farne that the saint has died. They are found chanting Ps. 59, which Bede somewhat disingenuously describes as occurring by chance ('forte'), and which enumerates numerous afflictions that will beset the faithful, as well as the assistance God will eventually provide. The focus on Cuthbert's final earthly action as the singing of the Night Office is furthered by the description of the monks signalling their brethren on Lindisfarne with torches: '*Qua vigil* e speculis pernox *servaverat horam* / Aethereas pandunt vatem penetrasse sub auras' ('they signal that the prophet has entered the heavenly realms of the air, where *awake* in his hermitage, he *had kept the hours of the night*') (ll. 795–96). Bede joins the monks on Lindisfarne to this depiction as they were also, 'by chance', at the same point of *Nocturns*, and were

[149] *VCA*, pp. 130–37.

singing Ps. 59 as well. While Bede highlights this as a wondrous ('mirum') coincidence, the structure of the Office makes the situation believable. Psalmody operates here as the prophetic voice of Cuthbert once did, and the elements fittingly assault Lindisfarne thereafter, until the storm abates and Eadberht restores the glory of the community. The combined effect of Chapters XXXVI and XXXVII is to present a passing of the torch from Cuthbert to Eadberht and the monastic brethren of Lindisfarne. Where Cuthbert has consistently led by virtue of his saintly obedience, both directly and indirectly acting pastorally, here the brethren join together with him in keeping the Divine Office, and continue to do so after the saint's ascension to heaven.

Conclusion

As a Latin poetic *vita*, Bede's *VCM* is unique in the corpus of Anglo-Saxon hagiography. While firmly linked to the *opus geminatum* tradition, the poetic *vita* was utilised by Bede for several distinct purposes: as a combination of *ruminatio* and poetic practice for Bede himself; as a ruminative text for its audience; and as a way of re-fashioning Cuthbert into an idealised Gregorian monk-pastor whose actions illuminate all of Britain. The restoration of Creation via the sanctity of Cuthbert is given its first explicit exegetical framing in the Cuthbertine tradition; and it is Bede's use of this frame here in the *VCM*, stemming ultimately from Augustine, which shapes not only the representation of the relationship between Cuthbert and Creation in Bede's own *VCP*, but also the various *vitae* of St Guthlac which follow. The *VCM* goes to great lengths to tie Cuthbert's restoration of Creation to monastic obedience via the saint's observing of the Divine Office – a transformation from the highly eremitic sanctity of Cuthbert in the *VCA*. Cuthbert emerges in the *VCM* as an ideal Gregorian balance between the active and contemplative lives, and in his final hours passes on this model of behaviour to his successors on Lindisfarne. When Bede returned twenty years later to write the *VCP*, this balanced presentation of Cuthbert is given central focus and, indeed, exaggerated.

3

Bede's Exegesis and Developmental Sanctity: The Prose *Vita Sancti Cuthberti*

More than twenty years after completing the *VCM*, sometime near 721, Bede returned to Cuthbert with his prose *Vita Sancti Cuthberti* (*VCP*).[1] This return seems, at first, paradoxical: Bede's *VCM* was composed as the poetic companion piece to the *VCA*, in the *opus geminatum* tradition, and its completion signalled the conclusion of the unit.[2] Scholars have long debated Bede's intentions in composing another (prose) life of Cuthbert.[3] The most widely

1 Colgrave, *Two Lives of Saint Cuthbert*, p. 16, posits 721, due to the dedication in the text to Bishop Eadfrith, who died that year, as well as Bede's reference to the work in his *De temporum ratione* as having been written 'some time ago'. All references and translations are taken from this edition.

2 It should be noted that Bede's *VCP* supplanted the *VCA* as the primary prose text of Cuthbert's life, and circulated with the *VCM* as an *opus geminatum*. See C. Cubitt, 'Memory and Narrative in the Cult of Early Anglo-Saxon Saints', in *The Uses of the Past in the Early Middle Ages*, ed. Y. Hen and M. Innes (Cambridge, 2000), pp. 29–66, at 47.

3 See Knibbs, 'Exegetical Hagiography', 237; W. Berschin, '*Opus deliberatum ac perfectum*: Why Did the Venerable Bede Write a Second Prose Life of St Cuthbert?', in *Cuthbert and Community*, pp. 95–102, at 96, who believes Bede intended to supplant the *VCA*; K. Lutterkort, 'Beda Hagiographicus: Meaning and Function of Miracle Stories in the *Vita Cuthberti* and the *Historia ecclesiastica*', in *Beda Venerabilis: Historian, Monk and Northumbrian*, ed. L. A. J. R. Houwen and A. A. MacDonald (Groningen, 1996), pp. 81–106; A. Thacker, 'Bede's Ideal of Reform', in *Ideal and Reality in Frankish and Anglo-Saxon Society*, ed. P. Wormald et al. (Oxford, 1983), pp. 130–53, at 138–42, argues that the alterations in and impetus for the *VCP* lie in Bede's desire to reform, and that Cuthbert provided a perfect model in which to do so; Stancliffe, 'Pastor and Solitary', pp. 21–44, at 28, agrees with Thacker that the adaptation is for the purpose of making Cuthbert more Gregorian and Benedictine, in connection with Bede's reformative concerns. For a different perspective see McMullen, 'Rewriting Landscape',

accepted view, synthesised by Thacker, is that Bede's alterations to his source material from the *VCA* stemmed from his desire for ecclesial reform.[4] This argument is supported by Bede's use of Cuthbertine material in the *Historia* which, in Thacker's words, places the Cuthbert of the *VCP* as the conclusion to and 'finest fruit' of 'the conversion initiated by Gregory the Great'.[5] Bede returns a third time to the life of Cuthbert in his *Historia,* inserting select stories as the conclusion to Book IV. This placement of Cuthbertine material highlights Bede's lifelong fascination with the saint, as well as providing evidence for his refashioning of Cuthbert as a saint with wide appeal.[6] Bede's adaptation of Cuthbertine material in the *VCP* and *Historia* transforms the relationship between Cuthbert and Creation, from the model of static sanctity presented in the *VCA* and *VCM* into a developmental sanctity that ultimately leads to the restoration of select portions of Creation.[7] Bede's particular conception of the sanctity towards which Cuthbert is journeying is firmly fixed in a Gregorian form of monasticism, but is augmented with a model of spiritual progression based on the Evagrian *Vita Antonii*. The movement towards an idealised Gregorian monk-pastor as saint is framed as a progression within the Antonian model, whereby Cuthbert's transition from ascetic to cenobite, and eventually bishop, is based upon individual moments of spiritual progression. Bede combines this saintly process in the *VCP* with his clearly didactic purpose for the work, which seeks to interpret and provide exegesis for the events in Cuthbert's life, as well as to highlight Cuthbert's pastoral excellence and subsequent emulative behaviour. Bede presents Cuthbert primarily as a reformer who transforms the already Christian into something more holy. The result of these changes, building on developments already

who posits that part of Bede's changes were an attempt to curb the propagandistic intentions of the author of the *VCA*.

4 Thacker, 'Bede's Ideal of Reform', pp. 138–42.

5 Thacker, 'Bede's Ideal of Reform', p. 145.

6 Thacker, 'Bede's Ideal of Reform', pp. 144–45.

7 While scholars have noted the developmental aspect of the *VCP*, there has as yet been no full exploration of the Antonian modelling for this, or of the specific ways Bede alters Cuthbert's relationship with Creation to assist in this development. See, for example, Newlands, 'Bede and Images of Cuthbert', 81–91.

noted in the *VCM*, is a presentation of Cuthbert and Creation distinct in form and intent from that of the *VCA*. Bede presents the relationship between the saint and Creation in the *VCP* as functioning in two ways: first, to provide the necessary impetus for Cuthbert's progression in sanctity; second, to provide evidence of this progression. In the *VCP*, Bede further develops the *VCM*'s link between his own exegesis concerning the Genesis narrative, based on Augustine, and Cuthbert's restoration of Creation.

Cuthbert's Developmental Sanctity

In the *VCP* Bede, for the most part, mirrors the structure of his own *VCM* rather than the *VCA*, breaking Cuthbert's life up into forty-six chapters arranged chronologically. This structure aids the portrayal of Cuthbert as an idealised English saint through its symbolic resonance. Berschin demonstrates how Bede, following Augustine,[8] considers this number as the 'perfection of the Lord's body';[9] Bede is employing this symbolism to 'describe a saint who achieved perfection through following the Lord'.[10] This perfection is achieved via Bede's transformation of Cuthbert into a developmental figure modelled on St Antony,[11] and displayed by detailing specific points of Cuthbert's progression in sanctity. This framework follows that of the *Vita Antonii*, where the saint must progress to a certain point in the saintly life before certain kinds of miracles can be performed, the height of which, for both the *VCP* and the

8 See Augustine, *De Trinitate Libri XV*, ed. W. J. Mountain and F. Glorie, 2 vols, CCSL 50–50a (Turnhout, 1968), p. 172: 'sed quia ipsa perfectio corporis domini tot diebus ad partum perducta comperitur sicut a maioribus traditum suscipiens ecclesiae custodit auctoritas' ('but because of the perfection itself of the body of the Lord is found to have been brought so many days to the birth, as the authority of the Church maintains upon the tradition of the elders').

9 Berschin, '*Opus deliberatum*', p. 101, in reference to Bede's Homily II.1, Bede, *Opera homiletica. Opera rhythmica*, ed. D. Hurst and J. Fraipont (Turnhout, 1955), I.24, p. 189: 'Qui etiam numerus annorum *perfectioni dominici corporis* aptissime congruit' ('This number [46] of years is also most apt for the perfecting of our Lord's [physical] body').

10 Berschin, '*Opus deliberatum*', p. 101.

11 On Bede's knowledge of the *Vita Antonii*, see Lapidge, *Anglo-Saxon Library*, p. 204.

Vita Antonii, is prophecy. Newlands has noted the developmental nature of Bede's *VCP*, but argues against an Antonian source, as she perceives the *Vita Antonii* to be 'organized around the different locations of the saint, not around character'.[12] However, the influence of Antony's progression towards what Kurtz calls 'spiritual majority'[13] is evidenced not only by similar statements of internal development, but also through the placement of Cuthbert's miracle-working only after he has reached a certain point in his development (echoing Antony): a model distinct from Gregory the Great's Benedict, who miraculously fixes a broken sieve while still a child.[14] This progression furthers Bede's transformation of Cuthbert in terms of the movement towards spiritual majority (restoring Creation and exercising prophecy), which Bede depicts through the Gregorian/Benedictine progression towards eremitism and divine contemplation.

Newlands has thus far provided the most thorough analysis of the developmental transformation of the *VCP*. She explores this feature of the text as a way of arguing that Bede alters his presentation of Cuthbert in the *VCP* to create a unifying saint who is a 'harmonious amalgam of Gaelic and Roman Christianity',[15] building on Thacker's argument concerning Bede's reforming tendencies.[16] In addition to the moments Newlands mentions, however, there are a number of other points of saintly progression within the *VCP*.[17] In order to interrogate this developmental transformation, as well as to argue for an Antonian modelling, I have selected several representative chapters to analyse.

12 Newlands, 'Bede and Images of Cuthbert', at 83.

13 See B. P. Kurtz, 'From St. Antony to St. Guthlac: A Study in Biography', *University of California Publications in Modern Philology* 12 (1926), 103–46, at 116: 'The power to perform miracles like those recorded in Scripture or *in exemplum partum* is indubitable evidence of holiness. Athanasius (c.14, 15; 38; 48, 50, 54, 57 ff.) represents this power in Antony as the stage of spiritual majority resulting from nearly two decades of discipline and solitude'; see also Thacker, 'Social and Continental Background', p. 290.

14 Gregory's *Dialogues*, vol. 2, p. 128.

15 Newlands, 'Bede and Images of Cuthbert', 77.

16 Newlands, 'Bede and Images of Cuthbert', 84–86, 88–90, 91; discussing *VCP*, Chs. I, III, IV, XIV, XV, and XVII.

17 See *VCP*, Chs. IV, VII, IX, X, XI, XVI, XVII, XXVI, and XXXVI.

Bede's transformation of Cuthbert's *vita* from static to developmental sanctity is already evident in Ch. I of the *VCP*, as he offers a short synopsis of the major progression of Cuthbert's life which leads from monastic discipline to eremitism.[18] It is his next addition, however, that sets Bede's transformation of the hagiography in motion; Cuthbert can only become a saint through God's continual assistance in his necessary progression: 'Sed ut haec in maiori aetate posset, superna illum gratia ad uiam ueritatis paulatim a primis iam pueriae incitauerat annis' ('But in order that he might be able to do these things in his later years, the heavenly grace had urged him little by little into the way of truth, from the earliest years of his boyhood').[19] As Newlands notes, this is a development from the equivalent episode in the *VCM*; the use of the adverb *paulatim* and the phrase *via veritatis* 'suggest that Bede now views the life of Cuthbert as a journey towards perfection'.[20] Further still, these changes alter the effect of the child's prophecy concerning Cuthbert's eventual role as bishop from fixed providential sanctity into the first of many moments highlighting Cuthbert's ongoing internal development. This is accomplished by first anchoring the notion of spiritual progression with biblical support through the image in 1 Cor. 13:11: '*Cum* enim *esset paruulus, ut paruulus sapiebat, ut paruulus cogitabat*, qui postmodum *factus vir*, plenissime ea *quae paruuli erant deposuit*' ('For when he was a child he understood as a child, he thought as a child; but after he became a man, he put away childish things entirely').[21] Bede connects this progression to God's dispensation in helping Cuthbert to develop spiritually.[22] Bede also turns Cuthbert's response into a chiding and prophetic statement to foreshadow his pastoral role as bishop, as well as to highlight Cuthbert's spiritual progression in the episode itself:[23]

18 *VCP*, p. 154.
19 *VCP*, pp. 154–55.
20 Newlands, 'Bede and Images of Cuthbert', 85; she discusses this further on p. 86 in relation to her comparison between the Bedan lives of Cuthbert.
21 *VCP*, pp. 156–57. Biblical text in italics.
22 *VCP*, p. 156.
23 See Newlands, 'Bede and Images of Cuthbert', 85–86.

relicta continuo ludendi uanitate domum rediit, ac stabilior iam ex illo tempore animoque adolescentior existere coepit, illo nimirum spiritu interius eius praecordia docente.[24]

(He forthwith gave up the idle games and, returning home, he began from that time to be steadier and more mature in mind. That Spirit assuredly instructed his heart from within.)

In this way, Bede depicts Cuthbert already progressing towards the internal stability of sanctity, with the emphasis on temporal progression ('ex illo tempore'), as well as the pedagogical ('docente') role of the Spirit. This introductory chapter establishes the first half of Bede's structural convention for miracles in the *VCP*, whereby they serve as the impetus for progression in the saintly life; this is connected to the second part, in which the miracles function as evidence for that progression.

Bede's emphasis in the *VCP* on Cuthbert's progression in sanctity is evident in his reordering of the chapters. In the *VCM*, the healing of Cuthbert's knee by the angel is followed by the first triptych, consisting of the raft miracle in Ch. III, the vision of Aidan in Ch. IV, and Aidan's calming-the-sea miracle in Ch. V, which together serve several functions, including the elevation of Cuthbert by comparison with Aidan. In the *VCP*, however, Bede removes Aidan's miracle in Ch. V, and instead focuses on how the first two miracles push Cuthbert forward in his spiritual progression. The raft miracle in Ch. III functions even more clearly as a moment of saintly progression than in the *VCM*, as the healing of Cuthbert's knee in Ch. II leads directly into his complete dedication to God here.[25] The *VCP*'s account of the raft miracle follows that of the *VCM* in general form, with the monks being washed out to sea, the crowd jeering, Cuthbert rebuking and praying, the miraculous turning of the winds occurring, and the crowd being brought to a greater

[24] *VCP*, pp. 158–59. Colgrave is here translating *adolescentior* as a form of *adolesco* ('growth, growing up, or advancing'), which figuratively can be used to refer to maturing rather than physical growth or aging; for example, Virgil's *Georgics* II, l. 362.

[25] *VCP*, p. 160. Cf. Newlands, 'Bede and Images of Cuthbert', 87: 'This chapter's opening sequence reminds us that Cuthbert is only a boy, *puer*; his prayers are presented then as part of his progress in holiness'.

understanding of the power of God in his saints. However, the raft miracle is fundamentally altered in purpose in the *VCP*. This is apparent even in the description of the storm itself; where the *VCM* lavishes imaginative detail for rumination, the *VCP* is less ornate and employs a plainer style, suggesting that Bede was writing with a different, perhaps broader audience in mind. In the first descriptive section the wild river ('feri [...] fluenti'), the raging current and wind ('Fluminis et venti subito feriuntur ab ictu'), and the white-capped expanse of the sea ('Oceani canum pronae') are rendered in the *VCP* simply as 'subito ab occasu tempestiuus assurgens' ('a sudden storm of wind, arising from the west'), and describes how 'ui fluminis ac uentorum uiolentia superati nequaquam ualebant' ('they [the monks] were overcome by the force of the river and the violence of the winds and could do nothing').[26] Though Bede does retain the avian simile from the *VCM*, the poetic force of the storm and the sea's foamy turbulence are now omitted; the description of the sea in the *VCP* is truncated and subordinated to the actions of the human agents. This subordination is evidenced in Cuthbert's admonishment to the jeering crowd: while in the *VCM* the speech includes the declaration of God's power over elemental Creation, in the *VCP* Cuthbert simply proposes that it would be better and kinder to pray for the monks' safety.[27]

This emphasis on human agency is mirrored by Bede's expansion of the purpose of the crowd's actions, both descriptively and in their response. Newlands argues – equating rustics with paganism – that this episode has been 'developed as a confrontation between Cuthbert and paganism' to display further Cuthbert's saintly power.[28] I would argue, however, that the *VCP* does not present a Martinian battle against paganism, but instead presents Cuthbert as a pastoral reformer, foreshadowing the saint's later reforms on Lindisfarne. The pagans of this episode are, in fact, rustic Christians whose syncretistic beliefs are in need of pastoral reform, an

26 *VCP*, pp. 162–63. Compare *VCM*, ll. 99–100, 102.
27 *VCP*, p. 162.
28 Newlands, 'Bede and Images of Cuthbert', 88.

emphasis that aligns with Bede's overall transformations.[29] The depiction of the crowd as a group of Christians in need of reform is evidenced in a number of ways. First, in keeping with the single hint in the *VCM*, Bede highlights the status of the gathered crowd four times as 'common people': in his initial description of them gathered on the riverbank, as 'uulgaris turba' ('common people'); in the description of their hatred of the monks, whom they see as despising a sense of common human law, 'qui *communia mortalium iura* spernentes' ('they spurned the *common law of mortals*'); in the immediate description before the crowd's speech, 'illi *rustico* et *animo* et *ore* stomachantes aduersus eum' ('they fumed against him with *boorish minds* and *boorish words*'); and finally in the moment of the crowd's reform of belief, '[v]identes autem *rustici* erubuerunt de sua infidelitate' ('When *the countryfolk* saw this, they were ashamed of their own unbelief').[30] Bede's *VCM* mentions their rustic nature once, and never ties it directly to their misunderstanding and unbelief. In the *VCP*, by contrast, Bede directly connects the people's response to their status; and it is in this status as unlettered country folk that Cuthbert's pastoral dimension comes into focus: they are in need of orthodox reform, not conversion.

When analysed within the context of the *VCP* as a whole, the crowd's response to the monks' plight, which Newlands interprets as a clash between their paganism and the monks' Christianity,[31] likewise emerges from this place of syncretic Christianity, allowed to proliferate because of pastoral neglect. This designation can be seen earlier in the *VCM*, where the crowd is described as Christian in the final line with the possessive 'suorum' ('[God's] own'),[32] but

29 See J. T. Rosenthal, 'Bede's Use of Miracles in "The Ecclesiastical History"', *Traditio* 31 (1975), 328–35, at 333, who argues that Bede almost never presents miracles as being used to cause conversion: 'Conversion was for Bede a rational or a spiritual process, whereas miracles were wonders to be savored by those who had *already* joined the elite'. He analyses the *HE* and finds that of the twenty-four accounts of conversion, only a few of the earliest conversions involve miracles as 'an instrumental part of the process'.

30 *VCP*, pp. 162–65.

31 See Newlands, 'Bede and Images of Cuthbert', 88.

32 *VCM*, l. 118; Newlands relegates this to a footnote with no further comment, 'Bede and Images of Cuthbert', 88, n. 55: 'The final word of this episode in the metrical life, "suorum" (119), refers to the mob as God's own; thus they are not heathens'.

is more clearly expressed here in the *VCP*, where the impetus for their jeering stems from a distorted Christianity.[33] Their subsequent embarrassment at Cuthbert's display of power and their praising God is therefore a return, not a conversion. The *VCP* expands the depiction of the crowd as unlettered Christians by incorporating it into Bede's transformation of Cuthbert into the paragon of sanctity. When Cuthbert starts his reproach, he calls them brothers ('fratres'), and within their own statement they express their disdain of the monks in Christian terms:

> Nullus [...] hominum pro eis roget, nullius eorum misereatur Deus, qui et ueteres culturas hominibus tulere, et nouas qualiter obseruare debeant nemo nouit.[34]
>
> (Let no man pray for them, and may God have no mercy on any one of them, for they have robbed men of their old ways of worship, and how the new worship is to be conducted, nobody knows.)

Not only is their complaint given in Christian terms, the heart of it rests on the lack of knowledge about the new manner of worship, an implicit critique of the monks' pastoral failing. This can be connected with the initial description of the crowds' dissent, which equally centres on how these new rules are unknown to them: 'noua et ignota [...] statua [...] uiuendi' ('new and unknown rules of life').[35] The conflict, therefore, is not a battle between pagan and Christian, but is centred on the crowd's present understanding of Christian praxis, and the new rules and forms of worship brought in by the monks. This progression towards novel orthopraxy is reflected in Bede's reforming tendencies, and thus in his depiction of Cuthbert in the *VCP*. The tension here rests on the implicit failure of the monks who have not instructed the rustic people in the new ways of worship. When Cuthbert performs his miracle their unfaithfulness to God is transformed into praise of Cuthbert's faith and humble praxis, which by extension leads towards emulative behaviour. Taken together the raft miracle functions as evidence for Cuthbert's progression in the saintly life, as well as foregrounding

33 *VCM*, l. 110. Bede uses the word *reprobis* ('false') here.
34 *VCP*, pp. 164–65.
35 *VCP*, pp. 162–63.

Bede's emphasis of Cuthbert's pastoral role in the *VCP*, in connection with the saint's role as reformer.

Bede's union of Antonian progression and the Gregorian monk-pastor ideal takes a dramatic step forward in Ch. X with the famous service of the sea animals, here specified as sea-otters.[36] Bede also restructures the order of events and breaks up the second triptych of the *VCM*, as well as the order of events in the *VCA*, by separating this miracle from its original mirrored miracle of Cuthbert as guest-master rendering service to the angel. This division of the two episodes serves to highlight Cuthbert's saintly development, as Bede bridges the episodes with two separate chapters that focus on the saint's progression in sanctity: Ch. VIII, where Bede portrays Cuthbert learning from Boisil and receiving a prophecy concerning his eventual rise to bishop; and Ch. IX, where Bede highlights the saint's progression within the monastic hierarchy with Cuthbert attaining the role of prior after Boisil's death. The sea-otter miracle of Ch. X becomes the next step in Cuthbert's progression, as can be seen by Bede's introductory framing of the episode: 'Cum uero sanctus uir in eodem monasterio uirtutibus signisque succresceret, famaque operum eius circunquaque crebresceret' ('Now while the holy man was going from strength to strength in that monastery by his signs and miracles, and the fame of his works had spread everywhere').[37] This manifest progression in sanctity, evidenced by Cuthbert's virtue and signs, is the reason Æbbe sends for him to come to Coldingham, so that he might exercise his pastoral gifts.[38] By contrast, the *VCA* gives the bare description that he arrived and stayed for a few days,[39] while the *VCM* simply launches into an account of Cuthbert setting off for the seashore to pray.[40] These changes in the *VCP* serve to highlight Cuthbert's developmental Antonian sanctity.

Bede further focuses on Cuthbert's pastoral activities in another important addition to the *VCP*: 'uiam iusticiae quam precabatur

36 *VCP*, p. 190, 'lutraeae'.

37 *VCP*, pp. 188–89. The peculiar reading 'circunquaque' rather than 'circumquaque' comes from Colgrave's edition.

38 *VCP*, p. 188.

39 *VCA*, p. 80.

40 *VCM*, ll. 220–23.

omnibus *actu* pariter et *sermone* pandebat' ('[he] opened up to them all the path of righteousness about which he preached, as much by his *deeds* as by his *words*').[41] The emphasis here on the consonance between his works ('acti') and words ('sermones') echoes the Gregorian ideal of the union of the active and contemplative lives.[42] The *VCP*'s presentation of the sea-otters' ministrations as miraculous evidence for Cuthbert's spiritual development provides a direct expression of his pastoral efficacy. Bede's description of the miracle itself follows both the *VCA* and *VCM* in general, but retains the nightlong submersion and psalmody of the *VCM*.[43] This helps to highlight Cuthbert's monastic obedience as Bede surrounds the otters' miraculous service with psalmody. So, for example, where the *VCA* simply states that Cuthbert returned to public prayer with the brethren ('orationem commune cum fratribus'),[44] the *VCP* describes how he joined the other monks in keeping the Divine Office, the reference to psalmody creating an envelope for the miracle: 'Ipse quoque mox domum reuersus, canonicos cum fratribus hymnos hora competente compleuit' ('He forthwith returned home and sang the canonical hymns with the brethren at the appointed hour').[45] Bede's emphasis here unites monastic obedience with Cuthbert's balanced active and contemplative life, as well as his pastoral efficacy, as he joins the brothers for the Office.

Bede returns to his focus on Cuthbert's pastoral side in the *VCP* by retaining the exegetical interpretation he added to the narrative in the *VCM*, emphasising the Christological typology of Cuthbert's declaration to the healed monk to tell no one of the miracle until Cuthbert has died.[46] Yet even here Bede alters the presentation,

41 *VCP*, pp. 188–89.

42 See Coates, 'Pastor and Solitary', at 610–12, for a discussion of Gregory and the varying influences in Bede's conception of the union of the active and contemplative lives; see above, pp. 94, 107.

43 *VCP*, p. 188.

44 *VCA*, p. 80.

45 *VCP*, pp. 190–91. Colgrave notes (p. 347) how 'Cuthbert would return in time for the matin office, or lauds, which in Benedictine monasteries began immediately after dawn ("incipente luce", *Regula Benedicti*, c.8, p. 41). In the *V.A.* (II, 3) he returns home at "cockcrow" to unite in prayer with the brethren.'

46 *VCP*, p. 190.

removing the *VCM*'s reference to Jesus' healing of the blind men in Mtt. 9, and instead referencing Jesus' Transfiguration in Mtt. 17:9:

> In quo nimirum praecepto eius secutus est exemplum, qui discipulis in monte gloriam suae maiestatis ostendens ait, *Nemini dixeritis uisionem, donec filius hominis a mortuis resurgat*.[47]
>
> (In this command he followed the example of Him who, when He showed the glory of His majesty to the disciples on the mount, said: "Tell the vision to no man until the Son of Man be risen again from the dead".)

Bede is here utilising the Transfiguration to depict Cuthbert as representative of the union of the active and contemplative lives. Firstly, Bede highlights Cuthbert's pastoral prowess by implicit Christological comparison, emphasising the pastoral care of Jesus and his disciples, which is a much closer relationship than that between Jesus and the blind men he healed. Just as Peter, James, and John are depicted in Mtt. 17 as falling down in fear, and are then restored from their fear by Jesus,[48] so too is the spying monk in the *VCP*. As Newlands observes, Bede uses miracles in the *VCP* 'not so much for the wondrous events themselves as for what they reveal about Christian conduct and episcopal leadership'.[49] Bede himself follows this tropological exegesis in his own commentaries on the Transfiguration, consistently focusing on the moral lesson rather than an overarching allegorical one, as Joan Hart-Hasler has demonstrated in comparing Bede with his patristic sources.[50] For example, Bede's *Homily* I.24 discusses the Transfiguration as it appears in Matthew in tropological terms:

> ut doceat omnes qui hanc uidere desiderant non in infimis iacere uoluptatibus non illecebris seruire carnalibus non cupiditatibus adhaerescere terrenis sed aeternorum amore ad superna semper erigi angelicae munditiae pietatis pacis dilectionis et iustitiae uitam

47 *VCP*, p. 190. Italics used for biblical text.

48 Mtt. 17:6.

49 Newlands, 'Bede and Images of Cuthbert', 92.

50 J. N. Hart-Hasler, 'Bede's Use of Patristic Sources: The Transfiguration', *Studia Patristica* 28 (1993), 197–204, at 203.

quantum mortalibus possibilie est semper imitari debere iuxta eum qui dixit.[51]

(In order to teach everyone who desires to see this [glory] not to lie in base pleasures, not to serve carnal enticements, not to cleave to earthly lust, but to be always raised by the love of the eternal towards heaven; and [to teach] that as far as is possible for mortals, they should always imitate angelic purity, piety, peace, love, and justice, just as he said.)

The heavy emphasis on the Transfiguration's role as a pastoral example is equivalent to Bede's altered presentation of Cuthbert in the *VCP*. When Bede has Cuthbert order the brother to remain silent about the healing miracle, he is not merely following hagiographical precedent or borrowing directly from the *VCA*, but is instead emphasising his saint's pastoral qualities. Bede argues, in the same homily, that the charge to remain silent in the Transfiguration is fundamentally a pastoral concern, necessary to help clear any potential interpretative confusion when people who had only heard of the vision beheld the cross.[52]

Bede also focuses, however, on Cuthbert's contemplative life in this passage by tapping into the rich tradition of patristic interpretation of the Transfiguration. The Transfiguration on Mount Tabor was frequently used by the Desert Fathers as an anagogical and tropological metaphor by which their eremitic ascents towards a vision of the divine could be achieved.[53] Ward argues that the sea-otter miracle in the *VCP* elides with that of prelapsarian Creation, with Cuthbert functioning typologically as the 'New Adam, once more at peace with all creation, naming the animals, who were the first servant and the first friend'.[54] The *VCP* unites all these traditions to depict the miracle not as a mirror of Cuthbert's obedient service to the angel, but instead as another example of Cuthbert's sanctity

51 Bede, *Homily* I.24, in *Opera homiletica. Opera rhythmica*, p. 172.

52 Bede, *Homily* I.24, in *Opera homiletica. Opera rhythmica*, p. 172: 'aut certe hi qui eandem uisionem audientes credidissent uiso crucis opprobio scandalizarentur' ('or those who had believed when they heard about this vision might stumble by the reproach of the cross').

53 See A. Andreopoulos, *Metamorphosis: The Transfiguration in Byzantine Theology and Iconography* (Crestwood NY, 2005), p. 25.

54 See Ward, 'Spirituality of Cuthbert', 72.

within the Gregorian/Benedictine ideal of the monk-pastor. The restoration of Creation is depicted here as evidence of Cuthbert's spiritual progression, both as pastor, imitating Christ, and fiery hermit, climbing the heights of the spiritual life. The union of these qualities in a moment of theophany in Cuthbert blinds the disobedient monk, and furthers Bede's overall transformation of Cuthbert into an idealised English saint.

Bede transforms the second and third miraculous provisions of sustenance in the *VCP*, the dolphin flesh (Ch. XI) and the eagle-and-fish miracle (Ch. XII) respectively, in order to display Cuthbert's achievement of another milestone on the road to sanctity: prophecy. This development is also tied explicitly with Cuthbert's continuing pastoral and evangelistic exploits, fitting Bede's overarching transformation of the saint into an idealised monk-pastor in the *VCP*. In Ch. XI Bede sets the miracle in the context of Cuthbert's growth by way of a quotation from Gregory's *Dialogues*, again emphasising the saint's development: 'Coepit inter ista uir Dei etiam prophetiae spiritu pollere, uentura praedicere, praesentibus absentia nuntiare' ('Meanwhile the man of God began to grow strong in the spirit of prophecy also, to foretell the future and to describe to those with him events that were happening elsewhere').[55] This connection between sanctity and prophecy echoes the structure of the *Vita Antonii*, in which the saint only begins exercising prophecy in his old age after he has progressed to his most remote eremitic location, beginning with the prophetic healing of a man in Ch. 57.[56] The framing of Cuthbert's progression towards prophecy thus subordinates the ensuing miracle in the service of displaying this growth. This is a clear contrast to the focus on monastic obedience of the *VCA*, or on Cuthbert's internal stability in his obedience amidst the storm in the *VCM*. Once again Bede has restrained his descriptive powers (as displayed in the *VCM*), presenting the storm here not as an expected result of the time of year (6 January), but rather simply as a storm for which the brothers were unprepared.[57] By diminishing the wildness of the storm, Bede is able to focus more

55 *VCP*, pp. 192–93.
56 *Vita Antonii*, Ch. 57, p. 179.
57 *VCP*, p. 192.

on Cuthbert's response, furthering his exemplary behaviour: 'quo tamen tempore uir Dei non marcida luxu otia gerere, nec somnis uacare uolebat inertibus, sed pernox in oratione perstare satagebat' ('but the man of God refused to spend this time of leisure in sluggish sloth or give himself up to idle slumber, choosing instead to spend the night in prayer').[58] This description is an addition to the *VCA*, and is distinct from Bede's focus on Cuthbert's internal stability in the *VCM*: 'Ille, ut erat placidus' ('He, being at peace') (l. 259). In this way, Cuthbert's decisive action serves to display his spiritual progression as a lesson for imitation.

Cuthbert himself focuses on this imitation in the first prophecy of the episode, by causally linking the brothers' faith with his own and God's provision:

> Credo si non fides titubat non uult nos hodierna dies ieiunos permanere [...] Precorque eamus alicubi quaerentes, quid nobis epularum in gaudium suae festiuitatis prestare dignetur.[59]
>
> (I believe that, if our faith does not waver, He will not allow us to remain fasting to-day [...] Let us go somewhere, I beg you, and find out what banquet He will deign to bestow upon us, so that we may keep His festival with joy.)

This exhortation is unique to the *VCP*, and further transforms the episode into a presentation of Cuthbert's continuing pastoral excellence, even now that he has grown into the saintly role of prophet. Thus Cuthbert's subsequent leading of the men to the shore, where they discover the miraculous provision of dolphin flesh, becomes a pastoral link with the second prophecy, that the storm would calm after three days. This second prophecy is similarly transformed in the *VCP* to function didactically:

> Videtis dilectissimi quae sit gratia confidendi et sperandi in Domino? Ecce et cibaria famulis suis preparauit, et ternario quoque numero quot diebus hic residendum sit nobis ostendit.[60]

58 *VCP*, pp. 192–93.
59 *VCP*, pp. 194–95.
60 *VCP*, pp. 194–95.

> (You see, beloved, what divine favour comes from trusting and hoping in the Lord. Look how He has prepared food for His servants, and has also showed us by the fact there are three pieces, how many days we must remain here.)

This is a drastic alteration of the narrative of both the *VCA*, in which Cuthbert orders but does not interpret,[61] and the *VCM*, which has no equivalent passage. The fulfilment of Cuthbert's prophecy, therefore, is presented as further evidence for his spiritual progression as a result of his pastoral excellence. The role of Creation has once again been subsumed within Bede's primary reformative concerns, though the relationship between Cuthbert and Creation is equivalent to the other *vitae*, as it is always founded upon Cuthbert's sanctity; in the *VCP* Bede has altered the manner in which that sanctity is achieved, and refocused it on the pastoral.

In Ch. XII Bede transforms the eagle-and-fish miracle to highlight once again Cuthbert's progression in sanctity evidenced in prophecy, as well as to display how this development is connected to Cuthbert's ongoing active, evangelistic life. While the *VCA* implicitly presents the interaction between Cuthbert and his companion as one of a teacher instructing his pupil, Bede brings this dynamic to the forefront in the *VCP*, introducing their discussion with an explicitly didactic interpretation: 'ait ad puerum temptans eum' ('he said to the boy to test him').[62] It is in the discussion itself that Bede makes his second major alteration to the episode by removing all biblical reference or allusion. Unlike the *VCA*, which includes four biblical quotations,[63] and the *VCM*, which alludes to Elijah being fed by the ravens in 1 Kings 17:6,[64] the *VCP*, significantly, transfers all the authority of interpretation to Cuthbert himself. While the connection to Elijah's ravens is (to a certain degree) an echo of the hagiographical motif, Bede is here highlighting the pastoral excellence of his saint by having Cuthbert, rather than biblical quotation, provide interpretation for the miracle: Cuthbert is acting here as an exegete of Creation, much like Bede himself later would

61 *VCA*, p. 84.

62 *VCP*, pp. 194–95.

63 See *VCA*, p. 86.

64 See *VCM*, ll. 298–300.

in his *In Genesim*. Thus in the conversation, Bede presents Cuthbert dialogically as a teacher employing the characteristic imperative:

> *Disce* [...] filiole fidem semper et spem habere in Domino, quia nunquam fame perit, qui Deo fideliter seruit.[65]
>
> (*Learn*, my dear son, always to have faith and trust in the Lord; for he who serves God faithfully never perishes of hunger.)
>
> *Cernis* [...] aquilam illam porro uolantem? Etiam per huius ministerium possibile est Domino nos hodie reficere.[66]
>
> (*Do you see* that eagle [...] flying far off? It is possible for God to refresh us today even by the ministration of that eagle.)

The miracle itself is described in terms of progression equivalent to the *VCA*, fulfilling Cuthbert's prophecy. Once again Bede downplays here the heavy emphasis on monastic obedience in the *VCA*, removing any use of the term *praecipio*, and instead portraying Cuthbert's orders in less weighted language, including the imperative '*Seca* citius medium' ('*Cut it* quickly in half'), and the verb *iubere*: 'Fecit ut *iusserat*' ('He did as he *was bidden*').[67] This change in emphasis leads directly into Bede's final transformation of the episode, in which he alters the purpose of the miraculous sustenance from implicit aid in Cuthbert's evangelistic pursuits (by feeding them they are sustained to continue preaching in the mountains), into a tool of evangelism. Here Bede has elided the two groups of people mentioned in the *VCA* into one: the unidentified company of men with whom they broil and eat the fish, and the mountain people they continue to evangelise. Cuthbert and his companion are described as continuing onwards after the miracle and only eating the fish when they turn aside into the next village, enter a house, and share the fish while Cuthbert preaches. The fulfilment of the saint's prophecy has now become a direct functional tool for evangelism, and the role of Creation has been subordinated to the dominant narrative of Cuthbert as developing monk-pastor. The

65 *VCP*, pp. 196–97.
66 *VCP*, pp. 196–97.
67 *VCP*, pp. 196–97.

relationship between Cuthbert and Creation once again provides evidence for his progression in the saintly life, and also foreshadows Cuthbert's most miraculous restorations of Creation on Farne. This is most evident in the causal connection between complete faith in service to God and that restoration, which echoes in phrasing Bede's full exegesis in Ch. XXI: Ch. XII's '*fidem semper* [...] habere in Domino' ('*always* to have *faith* [...] in the Lord') and 'qui Deo *fideliter seruit*' ('who *serves* God *faithfully*'),[68] foreshadow Ch. XXI's 'Qui enim auctori omnium creaturarum *fideliter* et integro corde *famulatur*' ('For if a man *faithfully* and wholeheartedly *serves* the maker of all created things').[69]

Cuthbert's advance towards eremitic contemplation and attainment of the necessary spiritual development to fully restore Creation is dramatically expanded by Bede in his account of Cuthbert's transition to Lindisfarne in Ch. XVI, to focus on another crucial step in Cuthbert's progression in sanctity. In an introduction mirroring the structure of Ch. X, Bede frames the ensuing move to Lindisfarne as a direct result of Cuthbert's development, recognised by a higher-ranking member in the monastic community:

> Cum ergo uenerabilis Domini famulus multos in Mailrosensi monasterio degens annos, multis uirtutum spiritualium claresceret signis, transtulit eum reuerentissimus abbas ipsius Eata in monasterium quod in Lindisfarnensium insula situm est, ut ibi quoque regulam monachicae perfectionis et praepositi auctoritate doceret, et exemplo uirtutis ostenderet.[70]
>
> (So when the venerable servant of God had passed many years in the monastery at Melrose and had distinguished himself by the many signs of his spiritual powers, his most reverend abbot Eata transferred him to the monastery which is situated in the island of Lindisfarne, in order that there also he might both teach the rule of monastic perfection by his authority as prior and illustrate it by the example of his virtue.)

Bede's ordering of events here is significant, as the emphasis rests not primarily on Cuthbert's spiritual virtue, but instead on the

68 *VCP*, pp. 196–97.

69 *VCP*, pp. 224–25.

70 *VCP*, pp. 206–07.

many years he has spent developing in a monastic context, for which those miracles ('signis') are the external proof of his spiritual progression. Just as his maturing virtues are the reason for Æbbe's request for him to come to Coldingham to exercise his pastoral gifts, so too here it is his further development, exemplified by the previous chapter's presentation of his prophetic, pastoral, and evangelistic efficacy, that causes Cuthbert's abbot, Eata, to transfer him to Lindisfarne.

In this episode Bede further unites his two hagiographical exemplars in the person of Cuthbert through parallels with the character of Antony on the one hand, and an episode in the life of Benedict from Gregory's *Dialogues* on the other. Bede's use of the *Vita Antonii* as a model is clear, for example, in the expansion of the reason for Cuthbert's move to Lindisfarne. This is evidenced by Cuthbert's humble obedience to his superiors within the ecclesial hierarchy which continues after he has progressed to prophetic miracles, echoing a description given of Antony. Ch. 67 of the *Vita Antonii,* after a lengthy set of chapters describing Antony's spiritual feats (including prophecy, exorcising of demons, and visions, including the passing of souls to heaven), describes how the saint was still obedient within the ecclesial hierarchy.[71] Although Antony is a wonder-worker and prophet, as Ch. 62 highlights,[72] in Ch. 67 he is also humble in spirit within the ecclesial hierarchy:

> Nam omnes clericos usque ad ultimum gradum ante se orare compellens, episcopis quoque atque presbyteris, quasi humilitatis discipulus, ad benedicendum se caput submittebat.[73]
>
> (For he was compelled to pray for all the brothers even to the lowest rank before himself, as well as submit his head, as a disciple of humility, for blessing by bishops and elders.)

The progression here from the enumeration of Antony's wonder-working, only achieved by his life of discipline and spiritual majority, to the description of his obedience within the ecclesial

71 *Vita Antonii,* Chs. 62–66, pp. 181–82.

72 *Vita Antonii,* Ch. 62, p. 181.

73 *Vita Antonii,* Ch. 67, p. 182. This is very similar to Cuthbert's later statement in the *VCP* concerning the elevated position which monks should be granted.

hierarchy provides a model for Bede's presentation of Cuthbert's obedience in this episode.

Cuthbert's pastoral role as a reformative figure is brought to the fore in the rest of this episode, in which Bede expands the *VCA*'s description of Cuthbert helping the monks to write down and fix their first *regula* not simply to echo, but to surpass, the equivalent reformative actions of Benedict. In the *Dialogues* II.3, Gregory describes how Benedict was implored by the monks of a nearby monastery to come in from his eremitic life to be their abbot.[74] He at first refuses, proclaiming that his holy manner of life was unsuited to them, but is eventually persuaded and proceeds to govern the abbey in a fixed and rigid manner.[75] Bede echoes this transition in the *VCP* by Cuthbert's actions when appointed to Lindisfarne, which include not simply ordering and teaching, but crucially, doing so in combination with his exemplary actions: 'mox instituta monachica fratribus uiuendo pariter et docendo tradebat' ('soon equally by his life and by his doctrine taught the monastic rule to the brethren').[76] While Bede explains that the monks of Lindisfarne had generally followed a monastic rule since the monastery's foundation by Aidan,[77] the implication, when compared with the *VCA*'s depiction of their *regula mixta*, is that the actual state of praxis was more syncretic upon Cuthbert's arrival. This is evidenced in the second parallel with Benedict, who encounters fierce opposition to his leadership when he seeks to reform the manner in which the abbey conducted its monastic life.[78] The monks' reaction to Benedict's changes is almost identical to that of some of the Lindisfarne monks to Cuthbert's reforms:

> Erant autem quidam in monasterio fratres, qui priscae suae consuetudini quam regulari mallent obtemperare custodiae.[79]

[74] This is also mirrored in the transition of Cuthbert from eremite, to abbot, and back to eremite.

[75] Gregory's *Dialogues*, vol. 2, p. 140.

[76] *VCP*, pp. 208–09.

[77] *VCP*, p. 208.

[78] Gregory's *Dialogues*, vol. 2, pp. 140, 142.

[79] *VCP*, pp. 210–11. The complaint is also connected to the rustic people's complaint in Ch. III.

(Now there were certain brethren in the monastery who preferred to conform to their older usage rather than to the monastic rule.)

It is here, however, that Cuthbert's trajectory departs from that of Benedict. Where the *Dialogues* relate how the monks refuse to reform, and eventually try to kill Benedict with poisoned wine,[80] the monks of Lindisfarne, though arguing with Cuthbert and assailing him with bitter insults, are eventually converted through the saint's exemplary sanctity: 'Quos tamen ille modesta patientiae suae uirtute superabat, et cotidiano exercitio paulatim ad melioris propositi statum conuertebat' ('Nevertheless he overcame these by his modest virtue and patience, and by daily effort he gradually converted them to a better state of mind').[81]

Bede has lifted Cuthbert above Benedict in the perfections of his active life through the complete reform of the dissenting Lindisfarne monks. Cuthbert's praiseworthy actions while living at Lindisfarne include continuing his nightlong vigils; always weeping when celebrating Mass; rebuking those who sinned while also being kindhearted in his pardons; and wearing ordinary garments so that neither their elegance nor decrepitude would prove problematic.[82] What is unique about this list is Bede's focus on actions that are inherently imitable in pattern, if not in degree, when compared with Cuthbert's fiery prophecies and miracles. Cuthbert's vigils are exemplary as vigils, but are depicted here in their saintly extreme; his rebuking and pardoning work are imitable within a monastic hierarchy; and Cuthbert's dress, the most prosaic of the four, speaks to the humble praxis of monastic life. So in this episode Bede once again combines Cuthbert's pastoral excellence (here via his imitable actions), with further depictions of his sanctity. Cuthbert's sanctity is most evident in the portrayal of his tearful compunction in celebrating the Mass, a portrayal which also retains echoes of eremitic spirituality. The notion of tears of compunction has its roots in not only the writings of the Church Fathers,

80 Gregory's *Dialogues*, vol. 2, p. 142.
81 *VCP*, pp. 210–11.
82 *VCP*, pp. 210–12.

especially Cassian's *Collationes*,[83] but also in the Desert Fathers, who focused on Jesus' exclamation during the Sermon on the Mount, '[b]eati qui lugent: quoniam ipsi consolabuntur' ('Blessed are they that mourn: for they shall be comforted') (Mtt. 5:5), as evidence for what later became known as the *gratia lachrymarum* ('the grace of tears').[84] As both Sandra McEntire and James Palmer discuss, this popular patristic trope was well known in early medieval England, as it appears in numerous foundational texts including Gregory's *Dialogues*, Benedict's *Regula*, and in various texts by Bede himself, including the *Historia*.[85] Katherine Harvey highlights how this trope can also be found in the *Vita Martini* and Augustine's *Confessiones*.[86] What is most relevant to Bede's application of the action to Cuthbert is its theological status as fundamentally a gift of God, an unearned blessing that nonetheless is bestowed upon those implicitly holy enough, most commonly in the earlier texts to those who became saints.[87] The ascription to Cuthbert of this spiritual gift further establishes his sanctity, an external sign of internal holiness, while simultaneously offering an imitable exemplar in the desire for such a gift, which is connected to humility in prayer.[88] Contextually this

83 C. Stewart, 'Benedictine Monasticism and Mysticism', in *The Wiley-Blackwell Companion to Christian Mysticism*, ed. J. A. Lamm (Oxford, 2013), pp. 216–34, at 220–23, argues for the importance of Cassian to early medieval Latin monasticism.

84 S. McEntire, 'The Doctrine of Compunction from Bede to Margery Kempe', in *The Medieval Mystical Tradition in England IV*: The *Exeter Symposium IV, Papers Read at Dartington Hall, July 1987*, ed. M. Glasscoe (Cambridge, 1987), pp. 77–90, at 77; see also S. DeGregorio, 'Affective Spirituality: Theory and Practice in Bede and Alfred the Great', *Essays in Medieval Studies* 22 (2005), 129–39. Bede himself discusses compunction in numerous texts, including the *Commentary on Samuel*, see S. DeGregorio, 'The Venerable Bede on Prayer and Contemplation', *Traditio* 54 (1999), 1–39, at 36.

85 See McEntire, 'The Doctrine of Compunction', pp. 77–79; and J. M. Palmer, '*Compunctio* and the Heart in the Old English Poem *The Wanderer*', *Neophilologus* 88 (2004), 447–60, at 449–50.

86 K. Harvey, 'Episcopal Emotions: Tears in the Life of the Medieval Bishop', *Historical Research* 87 (2014), 591–610, at 593: *Vita Martini*, Ch. 27, pp. 64, 66; Augustine, *Confessionum Libri XIII*, ed. L. Verheijen, CCSL 27 (Turnhout, 1971), pp. 113–14.

87 See McEntire, 'The Doctrine of Compunction', p. 78; Palmer, '*Compunctio* and the Heart', 449–50; and Harvey, 'Episcopal Emotions', 593.

88 See A. G. P. Van der Walt, 'Reflections of the Benedictine Rule in Bede's Homiliary', *Journal of Ecclesiastical History* 37 (1986), 367–76, for the connection between compunction and humility and prayer.

imitability of the gift of tears also aligns with Bede's depiction of Cuthbert's progression in the saintly life, a pattern that allows for imitability in a way that the predestined sanctity of the *VCP* and the *VCM* does not; a saintly pastor's actions can be followed, while a numinous prophet's cannot (except in the most extraordinary circumstances). Bede concludes the chapter with a statement highlighting how Cuthbert's exemplary pastoral behaviour allows the reforms Cuthbert wanted to take place:

> His et huiusmodi spiritualibus exercitiis uir uenerabilis et bonorum quorunque ad se *imitandum* prouocabat affectum, et improbos quosque ac rebelles uitae regularis a pertinatia sui reuocabat erroris.[89]
>
> (The venerable man, by these and other spiritual exercises of the same kind, aroused in all good men the desire *to imitate him* and recalled the wicked and those who rebelled against the rule from the obstinacy of their error.)

It is only from this point of Cuthbert's progression, in which he has fully joined the roles of Antonian prophet and wonder-worker with the monk-pastor ideal presented (and exemplified) by Gregory, that the saint can progress towards divine contemplation as a hermit, and the connected restoration of Creation.

Cuthbert as Hermit: St Cuthbert's Isle and Farne

Bede makes his clearest statement concerning the correct progression of the holy life in Ch. XVII, how only after many years of development within an active monastic life might a person be worthy of divine contemplation as a hermit:

> At postquam in eodem monasterio multa annorum curricula expleuit, tandem diu concupita, quaesita, ac petita solitudinis secreta, comitante praefati abbatis sui simul et fratrum gratia multum laetabundus adiit. Gaudebat namque quia de longa perfectione conuersationis actiuae, ad otium diuinae speculationis iam mereretur ascendere. Laetabatur ad eorum sortem se pertingere de quibus canitur in

89 *VCP*, pp. 212–13.

psalmo, *Ambulabunt de uirtute in uirtutem, uidebitur Deus deorum in Syon.*[90]

(Now after he had completed many years in that same monastery, he joyfully entered into the remote solitudes which he had long desired, sought, and prayed for, with the good will of that same abbot and also of the brethren. For he rejoiced because, after a long and blameless active life, he was now held worthy to rise to the repose of divine contemplation. He rejoiced to attain to the lot of those concerning whom the Psalmist sings: "The saints shall go from strength to strength; the God of gods shall be seen in Zion".)

Within this outline Bede once again highlights Cuthbert's continuing obedience and humility within the monastic hierarchy, in that he is allowed to progress into an eremitic life with the permission of the abbot. This transition is connected with another presentation of Cuthbert's continuing pastoral excellence, where he departs from Lindisfarne with the 'good will' of the monks he had been teaching. Bede frames his statement by way of quoting Ps. 83:8, which describes a kind of spiritual progression ('Ambulabunt de uirtute in uirtutem'). Augustine interprets this movement in Ps. 83:8 from virtue to virtue as a progression from individual virtues given by grace, including *temperantia, fortitudo, prudentia, justitia,* to the final virtue, which is divine contemplation:

Et audi quia ab huius actionis uirtutibus, in illam contemplationem ibimus. Sequitur ibi: *Ibunt a uirtutibus in uirtutem*: quam uirtutem? Contemplandi.[91]

(And that you might understand from this behaviour the virtues, we will advance in that contemplation. It therefore follows, *they shall go from virtue to virtue*. What virtue? That of contemplation.)

Bede's interpretation of Cuthbert through the lens of this psalm reinforces with biblical and exegetical support his progression from active virtue, as exemplified in the *VCP* thus far, to the (higher) virtue of eremitic divine contemplation.[92]

90 *VCP*, pp. 214–15.

91 Augustine, *Enarrationes in Psalmos*, p. 1158. Italics for biblical text.

92 It should be noted that Bede uses Ps. 83:8 in only one other text, to describe St Fursey in the *HE*. The use in the *HE* is related to that of the *VCP* only insomuch as it presents a correlation between Fursey's worthiness and the granting of his

Bede echoes the structure of the *Vita Antonii* once more,[93] emphasising the movement deeper and deeper into seclusion as the correct progression of the contemplative life, by adding an intermediary location for Cuthbert's seclusion, commonly identified as the present day St Cuthbert's Isle:[94]

> Et quidem in primis uitae solitariae rudimentis, secessit ad locum quendam qui in exterioribus eius cellae partibus secretior apparet.[95]
>
> (Now indeed at the first beginning of his solitary life, he retired to a certain place in the outer precincts of the monastery which seemed more secluded.)

While the identification of the location as St Cuthbert's Isle remains uncertain, the functional purpose of an intermediary site as a way to highlight Cuthbert's progression remains. The saint moves physically from the communal space of Lindisfarne to the solitary, but nonetheless intimately connected, space on Cuthbert's Isle, and in so doing progresses significantly in his spiritual, eremitic career, while still remaining in contact with the communal life of the monastery. Bede highlights this use of the intermediary site for progression in his word choice, as his employment of the connective comparative 'secretior' binds the isle to Lindisfarne. In this way the isle echoes Antony's first cell, which was located just outside the bounds of his village.[96] It is from here that Antony must begin his

---

visions, just as Cuthbert's worthiness allows him entry into divine contemplation. See Bede, *HE*, III.19, p. 270.

93 See *Vita Antonii*, Ch. 3, in which he begins eremitic life right outside his own village (p. 161); Ch. 8, in which Antony moves further away, into the tombs (p. 164); Ch. 11, in which Antony sets off for mountains in the desert (p. 164); Ch. 49, in which Antony retreats further into the upper Thebaid to escape his escalating fame (p. 176); and Ch. 50, in which Antony settles on a mountain near a clear spring surrounded by date palm trees (pp. 176–77). See also Gittos, *Liturgy, Architecture, and Sacred Places*, p. 32.

94 See *VCP*, p. 349, note to Ch. XVII: 'The first scene of his hermit life is supposed to be the rocky islet called St Cuthbert's Isle, about one hundred yards from Lindisfarne, and south-west of the priory. It is surrounded by water at high tide, but can be approached over a slippery ridge of stones at low tide.'

95 *VCP*, pp. 214–15.

96 *Vita Antonii*, Ch. 3, p. 163: 'quicumque in Christi seruitate sibimet prodesse cupiebat, haud procul a sua uillula separatus instituebatur' ('but whomever desired to benefit from service to Christ, would practice this discipline in solitude near their own village').

contemplative life, but also must leave to seek ever-deeper solitude and its corollary, greater contemplation. Cuthbert, likewise, dwells on the isle a while at the beginning of his eremitic life, and then moves further away from Lindisfarne.[97] It is Cuthbert's desire for 'greater things', his desire for progression in the saintly life, that leads to his eventual settling of Farne, where after his victory over the inhabiting demons he is able to reach spiritual majority, and thus bring about moments of Creation's restoration.

Bede utilises the physicality of this scene to further his depiction of Cuthbert's saintly progression, here exemplified through his lengthy and literal description of the distinction between Lindisfarne as a tidal island (*semiinsula*),[98] and Farne as a true island. Lindisfarne represents, as St Cuthbert's Isle does, an intimate connection with the active life and the secular world joined by the causeway; thus when Bede describes the sea-bound nature of Farne, he is also describing the necessary seclusion by which divine contemplation is achieved: 'sed aliquot milibus passuum ab hac semiinsula ad eurum secreta, et hinc altissimo, et inde infinito clauditur oceano' ('but it is some miles away to the south-east of this half-island, and is shut in on the landward side by very deep water and on the seaward side by the boundless ocean').[99] Bede retains his echo of the Antonian narrative in Cuthbert's move to Farne from the *VCM*, with its elaborate battle with the demonic inhabitants preceding his founding of a dwelling there, with substantial expansion.[100] The demons which Antony conquers are the final test in his progression in the saintly life, only after which is he capable of performing miracles, as is displayed in chapters 12–15 of the *Vita Antonii*: in Ch. 12, Antony moves even further into the desert, setting up his cell in an abandoned fort; then in Ch. 13 he is assaulted by demons which he conquers; and then in Ch. 15 Antony performs his first real miracle, where by his prayer he and his companions cross the

97 *VCP*, p. 214.

98 *VCP*, p. 214.

99 *VCP*, pp. 214–15. See also K. M. Wickham-Crowley, 'Living on the *Ecg*: the Mutable Boundaries of Land and Water in Anglo-Saxon Contexts', in *A Place to Believe In: Locating Medieval Landscapes*, ed. C. A. Lees and G. R. Overing (University Park, PA, 2006), pp. 85–110, at 94–5.

100 *VCP*, p. 214.

Arsenoitic Canal unharmed by the vast number of crocodiles. The demonic confrontation fulfils a similar role in the *VCP*, where it is only upon Cuthbert's successful banishing of the demonic inhabitants of Farne that he achieves spiritual majority, and thereby has progressed far enough in the saintly life for the restoration miracles that follow.

The battle between the demons and Cuthbert is altered by Bede to unite the saint with the hagiographical tradition of the *miles Christi* in a much more explicit manner in the *VCP*, by way of quoting sections from Eph. 6 concerning the Armour of God:

> Verum intrante eam milite Christi, armato *galea salutis, scuto fidei, et gladio spiritus quod est uerbum Dei, omnia tela nequissimi ignea extin*cta et ipse nequissimus cum omni satellitum suorum turba porro fugatus est hostis.[101]
>
> ([B]ut when the soldier of Christ entered, armed with *the helmet of salvation, the shield of faith, and the sword of the spirit which is the word of God, all the fiery darts of the wicked one were quenched*, and the wicked foe himself was driven far away together with the whole crowd of his satellites.)

Bede retains the causal connection from the *VCM* between this spiritual victory and the physical and spiritual building of a cell in which to engage in *contemplatio*, as he calls Cuthbert both a soldier of Christ ('milite Christi') and a founder: 'monarcha terrae' [...] 'condidit ciuitatem' ('monarch of the land [...] built a city').[102] This elision, in combination with his expulsion of the demons, once again highlights Cuthbert's success where Adam failed, and connects the episode with the wider exegesis as an Edenic motif. In keeping with his earlier account in the *VCM*, Bede removes all other human agents in this construction narrative to elide further the physical and spiritual edifices.[103] Where the *VCA* goes to great

101 *VCP*, pp. 214–15. Italics signify biblical text from Eph. 6:16–17.

102 *VCP*, pp. 216–17. The term *monarcha* has as its primary meaning 'monarch', and is applied not only to kings and emperors, but also anchorites, abbots, and bishops, as well as Christ and God, though application to non-royal figures has no earlier attestation than Bede's *VCP*; see *DMLBS*.

103 He does, however, relocate this to Ch. XVIII, which mentions the brother's help in the construction of his hermitage, *VCP*, p. 218.

lengths to connect Cuthbert's foundation to his brethren at Lindisfarne, the *VCP* focuses instead on Cuthbert's founding of his retreat by his actions alone in order to reveal better the sanctity he has achieved. The entire founding episode centres on the cell's function as a focusing device for Cuthbert to pursue divine contemplation, with its hollowed-out floor and surrounding walls binding the saint's vision to the heavens alone.[104] Gone is the relational focus on obedience, as well as the evangelistic nature of the miracle and its relationship to future pilgrimage from the *VCA*;[105] instead, in keeping with the *VCM*, the external construction is meant to mirror Cuthbert's internal spiritual state, retaining Bede's consistent insistence on this process in the *VCM*. Helen Gittos, in her enumeration of potential interpretations for this episode in the *VCP*, describes how the foundation scene here is also 'a prefiguration of the heavenly city, a way of contrasting what can be achieved in this world with what Cuthbert will experience in the next'.[106] Her assessment, however, better describes the *VCM*, which reveals how Bede has sublimated this complex elision within his depiction of Cuthbert's progression in the saintly life in an earthly context in the *VCP*. In this act of exiling the demons, therefore, and constructing his cell, Cuthbert attains the requisite sanctity to be worthy (*merere*) of eremitic divine contemplation.

Bede's alteration of the well miracle is likewise connected to his overarching concern with Cuthbert's progression, especially with his portrayal of the saint as an idealised Gregorian monk-pastor. Instead of following the *VCA* and *VCM* in rendering the miracle in a sustained and singularly focused episode, Bede now connects it with a novel exposition of Cuthbert's manner of eremitic life up to his movement into full seclusion. This is immediately apparent in the framing title of Ch. XVIII, 'Quomodo precibus aquam de arida produxerit, uel qualiter ipse in anachoresi uixerit' ('How he produced water from dry land by his prayers and how he lived as a

104 *VCP*, p. 216.

105 While Bede does mention the size of the great stones and the perceived assistance of angels, the placement, after the description, and function of the miraculous portion of the construction simply serve the overarching purpose of showing the worthiness which Cuthbert has attained here. See *VCP*, p. 216.

106 Gittos, *Liturgy, Architecture, and Sacred Places*, p. 32.

hermit'),[107] and in the progression of the chapter itself, which places the well miracle as part of the process of constructing the eremitic retreat. This is signalled by the first sentence after the depiction of the miraculous well: 'Facta ergo iuuantibus se fratribus mansione ac domibus praefatis, incipit habitare solus uir Domini Cuthbertus' ('Now when this same dwelling-place and these chambers had been built with the help of his brethren, Cuthbert the man of God began to dwell alone').[108] Bede not only alters the narrative depiction of the episode itself, but also includes exegesis to illustrate better these aspects of his transformed saint. The general structure of the narrative resembles the *VCA*, where Cuthbert calls the monks to assist him because there is no water on Farne, instructs them to dig in the middle of his dwelling, and by the next day the pit is filled with water.[109] Bede's primary alteration in the *VCP* comes in the form of excisions and expansions to this depiction. As numerous scholars have noted, Bede removes all of the *VCA*'s biblical references to Moses and Samson, and instead supplies verses from two psalms in Cuthbert's speech to the brothers to provide biblical support for the miracle:[110]

> sed rogemus obsecro illum *qui conuertit* solidam *petram in stagn*um *aq*uae *et rupes in fontes aquarum*, ut *non nobis sed nomini* suo *dans gloriam* de hac quoque rupe saxosa nobis ueniam fontis aperire dignetur [...] credo *torrente uoluptatis* su*ae potabi*t nos.[111]
>
> (but let us, I beseech you, pray Him "*who turns* the solid *rock into a standing water and the flint into fountains of waters*", that, giving "*glory not unto* us *but unto his name*", He may vouchsafe to open to us also a spring of water from the stony rock [...] I believe that He will "*make* us *drink from the river of His pleasures*".)

107 *VCP*, pp. 216–17. The titles for individual chapters are represented in most manuscripts and will be considered authorial.

108 *VCP*, pp. 218–19.

109 *VCP*, p. 216–18.

110 See Shockro, 'Bede and the Rewriting of Sanctity', at 4 and 14, where she argues that Bede is here simplifying by removing the potential confusion in equating Samson and Cuthbert, as analogous connections break down quickly, and instead simply focusing on the passages in the Psalms: 'in which the story of Moses receiving water from a rock was an example of the limitless possibility of God's power', 14.

111 *VCP*, pp. 218–19. Italics signify biblical text.

While at first glance the removal of the biblical precedent for all well miracles in hagiography appears an odd choice, the application of the new verses, from Ps. 113 and 39 respectively, helps to bring into focus Bede's intention for this episode. Whereas the *VCA*, via the biblical references, and the *VCM*, via its singular focus, centre upon the miraculous power of the saint to draw forth water from dry stone, the *VCP* is intent on once again uniting Cuthbert's active pastoral and eremitic lives. As Sandra Duncan has noted, the verse from Ps. 113 not only echoes the familiar passage in Exodus concerning Moses, but also is interpreted by Augustine as representing 'the gifts of the Holy Spirit and the post-Resurrection teaching of the Church as fountains of water'.[112] Duncan also argues for a connection to the Eucharist by interpreting the reference to Maundy Thursday ('coena Domini') later in Ch. XVIII as suggesting the sacraments.[113] While the sacramental overtones are to some degree apparent, the specific exegesis she refers to in Augustine's *Enarrationes in Psalmos* points rather to Christ as pastor, where He, like Cuthbert here, is interpreting and making accessible the mysteries of Scripture. In the first section of Augustine's exegesis on Ps. 113:9 he connects the standing water with the exegetically rich metaphor of water as eternal life from the New Testament, a connection implicitly echoed by the 'aquae uiue' of the *VCA* and parallel hagiographical well miracles: '*Qui conuertit petram in stagna aquarum, et rupem in fontes aquarum.* Seipsum enim [...] liquefecit [...] ut fieret in eis fons aquae salientis in uitam aeternam' ('*Who turned the hard rock into pools of water, and the stone into fountains*. For He melted Himself [...] that He might be a *fountain springing unto everlasting life* in them').[114] Augustine emphasises the pastoral dimension here by interpreting the hard rock as the teachings of Jesus which could not be understood before his resurrection: 'quia prius, cum ignoraretur, durus uidebatur' ('before, when He was unknown, He seemed hard').[115] The rock, that is the teachings, become accessible (standing water) only after Jesus' resurrection when he pastorally

112 Duncan, '*Signa de Caelo*', 406.

113 Duncan, '*Signa de Caelo*', 406–7.

114 Augustine, *Enarrationes in Psalmos*, p. 1640.

115 Augustine, *Enarrationes in Psalmos*, p. 1640.

explains the teachings. Likewise Jesus sends the Holy Spirit, which itself is once again connected back to water to be imbibed:

> Ista petra, ista duritia conuersa est in stagna aquarum, et ista rupes in fontes aquarum, cum resurgens exposuit eis, incipiens a Moyse per omnes prophetas, quia sic oportebat Christum pati; et misit Spiritum sanctum, de quo dicebat: *Si quis sitit, ueniat et bibat.*[116]
>
> (The rock, that hardness was converted into pools of water, that stone into fountains of water, when on His resurrection, He expounded to them, beginning with Moses and all the prophets, how Christ must suffer this (Luke 24:26–27); and sent the Holy Spirit, of whom He said, *If any man thirst, let him come and drink* (John 7:37).)

The emphasis here lies on Jesus' interpretative pastoral teaching ('exposuit') which allows the living water or Holy Spirit to be imbibed. Bede's depiction of Cuthbert as the person whose prayers cause this physical and spiritual blessing to occur therefore has a distinctly pastoral overtone: 'Unde dubium non erat hanc orationibus uiri Dei [...] elicitiam fuisse aquam' ('Hence there was no doubt that this water had been drawn [...] through the prayers of the man of God').[117] In this context, the image of the prosaic digging by the brothers, though derived from the *VCA*, points to the metaphor of digging as a process of Scriptural study, as Augustine discusses in his *In Joannis Evangelium Tractatus*:

> Putemus scripturam Dei tamquam agrum esse, ubi uolumus aliquid aedificare. Non simus pigri, nec superficie contenti; fodiamus altius, *donec perueniamus ad petram*: *Petra autem erat Christus* (1 Cor. 10:4).[118]
>
> (Let us consider the Scripture of God to be like a field where we wish to build something. Let us not be lazy or content with the surface, *but let us dig until we come to the rock*: *and that rock was Christ*.)

While the digging metaphors are not equivalent, the process of digging until the seeker finds Christ is consonant. The second biblical reference added by Bede, Ps. 35:9, further connects to this interpretation, as this psalm and its 'torrents of pleasure' from which

116 Augustine, *Enarrationes in Psalmos*, p. 1640.

117 *VCP*, pp. 218–19.

118 Augustine, *In Iohannis Evangelium Tractatus CXXIV*, ed. R. Willems, CCSL 36 (Turnhout, 1954; repr. 1990), p. 232.

Cuthbert declares they shall drink are interpreted by Augustine as symbolising Christ.[119] The miraculous water, therefore, can be read as a continuation of Bede's presentation of Cuthbert's pastoral excellence, whereby he is the mediator between the brothers and the waters of life, whether the Holy Spirit or Christ, which can be accessed by delving through Scripture, but only under the direction of a wise and holy teacher such as Cuthbert.

Bede transforms the well miracle further with a detail that occurs nowhere else in the Cuthbertine tradition, or in hagiography more generally, the water's containment:

> Quae uidelicet aqua mirum in modum primis contenta ripis, nec foras ebulliendo pauimentum inuadere, nec hauriendo nouit deficere, ita moderata gratia largitoris, ut nec necessitati accipientis superflueret, nec sustentandae necessitati copia deesset.[120]
>
> (And this water was wonderfully kept within its first limits so that it never bubbled over and covered the floor, nor failed through exhaustion of its supply; but the grace of the Giver so controlled it that it did not exceed the necessities of the receiver, nor was the supply for those necessities ever lacking.)

Bede uses the description of the well miracle here for three united purposes: first, to provide accurate physical description of the scene as a foundation for exegesis; second, to display Cuthbert further as the idealised monk-pastor, balancing even his asceticism; and third, as an implicit gesture towards Cuthbert's ability to restore Creation via echoes of the Genesis narrative. For the first purpose, Bede is continuing his careful delineation of physicality to provide potential for interpretation and application. Thus, Bede has taken the *VCA*'s description of a fountain bursting forth from the floor of his dwelling and specified that instead of flooding everywhere and rendering the dwelling uninhabitable, as it physically should, it is contained and does not bubble over and cover the floor. This is akin to the process of description concerning Lindisfarne's status as a part-island ('semiinsula') in Ch. XVII, a detail unnecessary but added for both physical accuracy and potential interpretation. As

[119] Augustine, *Enarrationes in Psalmos*, pp. 332–33.
[120] *VCP*, pp. 218–19.

scholars have noted, the second purpose of this depiction involves connecting the miracle with the overarching emphasis on Cuthbert's asceticism, as founded in the balanced life.[121] Cuthbert's particularly balanced form of asceticism is emphasised by Bede throughout the *VCP*, and is clear in each of the moments touching upon the ascetic experience: for example, in Ch. V, when Cuthbert receives just enough food in the horse-and-loaf miracle to sustain himself and his horse;[122] in Ch. VI, where Cuthbert abstains from alcohol but eats enough food to keep himself hardy for work;[123] in Ch. X, where he spends the night immersed in the ocean in prayer and returns in time for matins;[124] and in Ch. XVI, where Cuthbert follows the middle path ('ita temperanter agens') while living on Lindisfarne by wearing neither elaborate garments nor an ascetic hair-shirt.[125] Bede's version of the well miracle reveals this same balance within Cuthbert's progression towards the ascetic life. He is given water, but only enough for his own need to support his eremitic divine contemplation.[126] The third function of this description is a gesture towards Cuthbert's restoring of Creation through its echo of the Genesis narrative. This is evidenced by the emphasis in the miracle on the binding of the water, its being kept within its first limits and being controlled by the grace of the Giver, which echoes the form of Bede's exegesis concerning the separation of the waters from the firmament in Gen. 1:6–8. In his *In Genesim* Bede addresses the apparent paradox of the description of the waters being placed above the firmament, as the nature of water is to flow and sink to the lowest point, by expounding God's binding and containing of water:[127]

121 See Gittos, *Liturgy, Architecture, and Sacred Places*, pp. 32–33.

122 *VCP*, p. 170.

123 *VCP*, p. 174.

124 *VCP*, pp. 188, 190.

125 *VCP*, p. 212.

126 J. O'Reilly, 'Reading the Scriptures in the Life of Columba', in *Studies in the Cult of Saint Columba*, ed. C. Bourke (Dublin, 1997), pp. 80–106, at 85–89, suggests, in relation to the well miracle in the *Vita Columbae*, that this focus on provision for actual need could be read 'as prefiguring the eucharist', even though Adomnán gives the 'common baptismal interpretation' for the miracle.

127 Bede, *In Genesim*, p. 10.

> ligat aquas in nubibus suis ut non erumpant pariter deorsum [...] infra caelum ligat aquas ad tempus cum uult ut non pariter decidant.[128]
>
> ([God] binds up the waters in his clouds, so that they break not out and fall together [...] binds the waters under heaven when he wishes and as occasion warrants, so that they do not fall down together.)

This notion of binding the waters is connected to the idea of specific limits set by God on them, as Bede says a few lines later: 'factam caeli aquarumque creationem ac dispositionem non excessisse praescriptos sibi in uerbo Dei terminus' ('when the creation and arrangement of heaven and the waters were done, they did not go beyond the boundary prescribed for them in the Word of God').[129] This divine ordering, where water is bound against its inherent nature and placed in specific limits by God in the Genesis narrative, is echoed in Cuthbert's miracle, and also foreshadows the upcoming miracle of the sea's obedience to Cuthbert (in connection with Bede's explicit exegesis in the beam miracle). Taken together the well miracle displays (as the framing title delineates) how Cuthbert conducted his life as a hermit; in the physical accuracy, the balanced asceticism, and the echo of prelapsarian Creation, Bede continues to highlight the development of his monk-pastor, hermit saint.

The uniqueness of the *VCP*'s depiction becomes clearer when compared with Bede's final reiteration of the well miracle in the later *Historia*. Unlike the *VCM*, which connects the well miracle with Christ's miracle of turning water into wine, or the *VCP*'s layered presentation of the bounded water, the *Historia* gives an account almost identical to that of the *VCA*, which focuses on the provision of water as a physical necessity and as a continuingly efficacious miracle for pilgrimage.[130] This depiction can best be understood within the context of the *Historia* as a whole, where Cuthbert is presented as essentially a pastoral reformer, not a miracle worker. Bede's focus in the *Historia* on Cuthbert as reformer is evident in his highly selective presentation of the saint, which removes all the

128 Bede, *In Genesim*, p. 11.
129 Bede, *In Genesim*, p. 11.
130 Bede, *HE*, IV.28 [26] p. 436.

ante-mortem miracles except the well and barley miracles, which are kept only to provide evidence for God's support in his eremitic endeavours. This use of the miracles as evidence for divine support can be seen in Bede's expansion and repositioning of Cuthbert's short Antonian speech in the barley miracle (in which Cuthbert states that he will not continue to pursue the eremitic life if God does not support him) into the beginning of Cuthbert's decision to pursue the contemplative life:

> quod aditurus insulam protestatus est fratribus, dicens: 'Si mihi diuina gratia in loco illo donauerit, ut de opera manuum mearum uiuere queam, libens ibi morabor; sin alias, ad uos citissime Deo uolente reuertar'.[131]
>
> (When he was about to go to the island, he made this declaration to the brothers: 'If by divine grace it is permitted to me to be able to support myself in this place by the work of my hands, I will gladly stay there; but if matters turn out otherwise, I intend, God willing, to return to you forthwith'.)

Comparatively the sections on Cuthbert's monastic progression up to his assignment to Lindisfarne, his bishopric, and his pastoral excellence have been given far greater space and even at some points expanded from the *VCP*.[132] These excisions and expansions serve to present Cuthbert as a pastoral reformer heavily involved with the ecclesiastical establishment, rather than as an eremitic saint.

Cuthbert and Restored Creation

Bede's transformation of Cuthbert into the perfect union of Antonian and Gregorian sanctity is completed in his adaptation of the final three miraculous interactions between Cuthbert and Creation:

[131] Bede, *HE*, IV.28 [26], pp. 434–35.

[132] See Bede, *HE*, IV.27 [25], pp. 431–35, where Bede summarises Cuthbert's life in terms of his transitions within the monastic hierarchy, from tutelage under Boisil, to Melrose, to Abbot Eata transferring him to Lindisfarne; Ch. 28 [26], pp. 435–39, where Bede briefly describes Cuthbert's eremitic life on Farne before jumping to his appointment as bishop; and Ch. 29 [27], pp. 439–43, where Bede describes Cuthbert's death, but spends most of his description portraying a conversation between Cuthbert and Brother Herbert, again highlighting the pastoral even in Cuthbert's eremitic endings.

the thieving birds in Ch. XIX, the penitent ravens in Ch. XX, and the sea-and-beam miracle in Ch. XXI. The exploration of Cuthbert's progression in the saintly life in these chapters is concluded by Bede's full exegesis on Creation, and unites these final miracles as evidence for the saint's ability to restore Creation. Cuthbert's toil on Farne recalls Adam, and his curse (Gen. 3: 17–19), which Cuthbert lifts by his holiness depicted in his eventual prelapsarian relationship with the birds. Bede's presentation of Cuthbert as an exemplary monk-pastor is evident in the introductory framing of the first avian miracle, the barley crop of Ch. XIX, which provides an alternative impetus for Cuthbert's decision to support himself by his own efforts. In the *VCA*, Cuthbert is said to engage in subsistence farming by reflecting on the biblical principle of 2 Thes. 3:10: 'quoniam si quis non vult operari, nec manducet' ('He that will not work, neither shall he eat').[133] In the *VCP*, however, Bede has once again removed the *VCA*'s biblical reference, instead adding that Cuthbert's impetus stems from reflecting on the deeds of the hagiographical fathers, and implicitly upon the deeds of Antony on which this episode is based: 'Post modum uero proprio manuum labore iuxta exempla partum uiuere magis aptum ducebat' ('But afterwards, in accordance with the example of the fathers, he considered it more fitting to live by the labour of his own hands').[134] Even the description of Cuthbert's request to the brethren mirrors that of Antony: Cuthbert asks for tools to work the land and wheat to sow,[135] just as Antony asks for a spade, an axe, and some corn.[136] After the initial failure of the crop, Bede expands the Antonian concern for the burden placed upon the monks[137] into a lengthy speech by Cuthbert when asking for barley, which includes a declaration that he will return to Lindisfarne and give up his eremitic life if God does not provide for him from the labour of his own hands.[138] This entire expansion not only emphasises Cuthbert's thoroughly Antonian nature, but also frames the ensuing miraculous growth

133 See *VCA*, pp. 100–01.
134 *VCP*, pp. 220–21.
135 *VCP*, p. 220.
136 *Vita Antonii*, Ch. 50, p. 176.
137 *Vita Antonii*, Ch. 50, p. 176.
138 *VCP*, p. 220.

of the barley as another example of Cuthbert's extraordinary pastoral concerns; even as a committed hermit, he is concerned with the well-being of the monks previously under his charge.

The presentation of the birds' attack on Cuthbert's miraculous barley crop in the *VCP* follows the *VCM* in general, but is again altered to highlight further the saint's pastoral excellence, which provides theological support for his authority over a portion of animate Creation. This can be connected, once again, to Bede's conception of the wider symbolic function of the miraculous interaction of saints and birds as evidence of prelapsarian Creation.[139] Thus, after describing how the birds come to eat Cuthbert's crops, Bede breaks off the narrative to interject an explanatory note outlining how Cuthbert related this and other tales for the purpose of edifying his listeners:

> sicut post ipse referebat, solebat enum sepe quia laeti uultus et affabilis erat ad confirmandam fidem audientium, aliquia etiam de eis quae ipse credendo optinuerit in medium proferre.[140]

> (As he afterwards related – for, being of a happy disposition and very friendly, he was often in the habit of disclosing some of the blessings which he had gained by his own faith, in order to confirm the faith of his listeners –)

Before the miracle has even been presented, Bede is already directing interpretative attention to the pastoral utility of this miracle; its location in the narrative progression of the episode further highlights the role of Cuthbert's delineated Bedan sanctity in the upcoming restoration miracle. Cuthbert's speech to the birds is nearly equivalent to that of the *VCM*, with the exception being the final command and the description of the birds' ensuing obedient response. Where the *VCM* presents a prelapsarian moment of subservience in which Cuthbert becomes the birds' keeper, via his polite command to stay within their bounds and the following description of their relationship,[141] Bede transforms this episode in the *VCP* to reflect the authority and emphasis on obedience to commands

139 See above, pp. 13, 49, 115, 163.

140 *VCP*, pp. 220–23.

141 *VCM*, ll. 424–25.

found in the *Vita Antonii*. The episode therefore reflects its Antonian parallel to a much greater degree (where Antony commands the invading animals to leave, and is immediately obeyed).[142] The connection to the Antonian narrative is made explicit as Bede, in another unique addition, explains that, in enacting these miracles on Farne, Cuthbert is following the deeds of his eremitic forebears; for the well miracle, he is following Benedict, and for the present bird miracle, Cuthbert is following Antony:

> Et hic quoque uenerabilis Christi famulus in duobus miraculis duorum patrum est facta secutus. In aqua uidelicet elicita de rupe, factum beati patris Benedicti [...] Porro in arcessitis a messe uolatilibus reuerentissimi et sanctissimi patris Antonii sequebatur exemplum, qui a lesione hortuli quem ipse plantauerat uno onagros sermone compescuit.[143]
>
> (And on this occasion also, the venerable servant of Christ followed in these two miracles the deeds of two of the fathers; namely, in obtaining water from the rock, a deed of the blessed father Benedict [...] in driving away the birds from the crops, he followed the example of the most reverend and holy father Antony, who with one exhortation restrained the wild asses from injuring the little garden that he himself had planted.)

In specifying the miracle's hagiographical sources Bede provides unambiguous evidence for Cuthbert's saintly development; he has progressed to the point where direct comparison with two of the men whom Bede felt best expressed holy living is possible. The first bird miracle demonstrates that Cuthbert has achieved the requisite sanctity, specifically delineated in the monk-pastor model of Gregory, that he can now emulate the restoration miracles of the holy fathers.

Bede connects his additional avian miracle with the thieving and penitent ravens from the *VCA* in *VCP* Ch. XX, by further comparing Cuthbert's miracle with that of Benedict: 'in exemplum praefati

142 *Vita Antonii*, Ch. 50, p. 177: 'Abite, et in nomine Domini ne huc appropietis ulterius' ('Depart, and in the name of the Lord come not nigh to this spot'); 'post hanc denuntiationem quasi timentes, numquam illuc bestias appropinquasse' ('And from that time forward, as though fearful of his command, they no more came near to the place').

143 *VCP*, pp. 222–23.

patris Benedicti' ('after the example of the above-mentioned father Benedict').[144] Here too Bede's exegetical impulse takes over and he doubles up his didactic interpretation of the miracle by not only retaining the ending from the *VCM*, in which he quotes Proverbs to give biblical precedent for learning righteous behaviour from animate Creation, but also by adding a new interpretation of Benedict's miracle as a parallel for Cuthbert's: 'in quo auium oboedientia et humilitate palam contumatia et superbia condempnatur humana' ('in which human pride and contumacy are openly condemned by the obedience and humility of birds').[145] Thus the miracle is encapsulated in exegesis and transformed to fit the Bedan mould of the *VCP*, where Cuthbert's distinct kind of sanctity, exemplified in the monk-pastor model, provides the necessary support for the restoration of Creation that occurs. The pastoral didacticism in the exegesis above hints at the overall change Bede has made to the miracle, evidenced by the alteration he makes to the gift of hog's lard which the ravens bring Cuthbert.[146] While in the *VCA* and the *VCM* Cuthbert uses the lard to treat his own shoes, in the *VCP* Bede depicts Cuthbert using the miracle as another tool for his ministry. This is accomplished by Cuthbert greasing the shoes of those visiting him, before launching into a didactic interpretation of the miracle as exemplar for obedient behaviour: 'atque ad unguendas caligas praebere solebat, contestans eis quanta hominibus oboedientiae, quanta sit cura humilitatis habenda' ('and to offer it [lard] to grease their shoes; declaring how carefully men should seek after obedience and humility').[147] While the miracle as a whole is transformed by these changes into an episode of much clearer didactic application, the emphasis on the restoration miracle itself is nonetheless maintained, as is borne out by the changes Bede makes to the miracle itself.

The *VCP*'s description of the penitent raven miracle follows the *VCA* and *VCM* in most respects: the ravens steal thatch from the

144 *VCP*, pp. 222–23. Cf. Gregory's *Dialogues*, vol. 2, pp. 160–64.

145 *VCP*, pp. 222–23.

146 *VCP*, p. 224; see Newlands, 'Bede and Images of Cuthbert', 96–98, who notes this alteration and its relationship to didacticism.

147 *VCP*, pp. 224–25.

guest-houses for their own nests; Cuthbert's initial warning which they ignore; Cuthbert's order for them to depart which they obey; the single raven's return and contrition; and Cuthbert's forgiveness for the birds who both return to live on Farne. Newlands suggests that Bede, in his didactic transformation of the episode, seeks to 'humanize' the birds to provide an easier example for emulation, by removing 'the onomatopoeic "raven" sound of the anonymous vita, "crocitare" (*VCA*, 102)', and also substituting 'feathers soiled and ruffled with human emotion' for the 'raven's outstretched wings' of the *VCA*.[148] She likewise sees the lexical choice in both the *VCM* and *VCP* for the bird's active pursuit of forgiveness, 'precatur' and 'precabatur' respectively, as furthering the humanising process.[149] While it is clear that Bede is emphasising the emulative potential in the episode, especially in connection with the image of Cuthbert's pastoral excellence, the depiction in the *VCP* is much closer to the *VCA* than Newlands suggests, and the 'humanizing' is perhaps better understood as echoing the *VCA*'s focus on the wider implications of the raven's contrition in relation to the restoration of Creation. First of all, Bede's removal of the verb 'crocitare' is balanced by his additional descriptive detail of the ravens' bills ('rostro')[150] with which they carried off the thatch, as well as his detailed description of their distinction from humanity, where the single raven's various actions in begging forgiveness are described as that of an animal using the only motions it could.[151] Bede likewise invokes a part of the raven's natural character, their status as the proudest of birds (auis superbissima),[152] in his concluding interpretation of the miracle to add weight to his didactic message. The physical actions taken by the raven, the ruffling of its feathers and the drooping of its head, are no more human than those ascribed to the raven in the *VCA*, which, as discussed above, follows to a greater extent the description of this process in the *Regula Benedicti*.[153] Just as in the *VCA*, then, for the ensuing miracle to retain its

148 Newlands, 'Bede and Images of Cuthbert', 97.

149 Newlands, 'Bede and Images of Cuthbert', 97; *VCM*, l. 439; *VCP*, p. 224.

150 *VCP*, p. 224.

151 *VCP*, p. 224.

152 *VCP*, p. 224.

153 See above, pp. 64–65.

didactic efficacy its initial miraculous nature must be maintained, as the emulative interpretation can only occur after the physical miracle has occurred.[154]

It must be remembered that in his *In Genesim* Bede presents the image of the miraculous obedience of birds to saints as evidence of prelapsarian Creation in the postlapsarian world,[155] and in *VCP* XXI, he gives his exegesis in which the ability of saints to restore Creation is justified. This emphasis on the physical relationship between Cuthbert and the ravens, however interpreted, rests on the overarching view of the miracle presenting a prelapsarian moment. The heavy emphasis on Cuthbert's pastoral excellence in his understanding ('intelligens') of the bird's attempt at contrition, as well as in his immediate forgiveness, displays Cuthbert's progression in sanctity: the ravens, the 'proudest' of birds, behave in a prelapsarian manner. While Bede does remove the shepherd and sheep analogy from the bird miracle in XIX of the *VCP*, in his conclusion to this chapter gestures in that direction through an adaptation of the *VCA*'s focus on the continuing efficacy of the Farne miracles:

> Denique ad dandum hominibus exemplum correctionis, multos deinceps annos ipsa insula manebant et nidificabant, neque aliquid molestiae cuiquam irrogare audebant.[156]
>
> (And in order to give mankind an example of [moral correction], they remained for many years in that island and built their nests, not daring to inflict any injury on anyone.)[157]

Other than the continuing emphasis on the didactic nature of the miracle in the *VCP*, Bede's alteration here unites it with his overarching emphasis on Cuthbert's role as reformer and moral exemplar; the continuing efficacy of the raven miracle is not presented in order to display Cuthbert's power per se, but is instead meant to correct

154 Shockro, 'Saints and Holy Beasts', likewise highlights the necessary animality of creatures in saints' *vitae*, in reference to a moment in the *Vita S. Columbae*: 'The animal is able to attest to the piety of the brother in question because he remains thoroughly animal and uncontrolled by humanity', p. 55.

155 See Bede, *In Genesim*, p. 29; see above, pp. 13, 49, 115, 163.

156 *VCP*, pp. 224–25.

157 While 'correctionis' can mean 'reformation', its use here speaks more specifically to moral correction, so I have thus emended above.

the disobedience of the faithful. That this is done through the obedience of birds, a Bedan echo of 'first Creation', highlights the foundational nature of the restoration of Creation in this episode. Cuthbert's pastoral excellence, exemplified here by his consistent didactic interpretative use of the miracle, declares his progression in the saintly life, which is likewise presented through his ability to perform the restoration miracle.

The explicit exegesis in full concerning the restoration of Creation and its relationship with sanctity enters hagiography in Ch. XXI, the sea-and-beam miracle. This addition by Bede to Cuthbertine tradition represents his most fundamental change concerning the depiction of the relationship between saints and Creation in Anglo-Saxon hagiography, and, as we shall see, was subsequently developed in the Guthlac tradition. The exegesis is employed by Bede to cover all of the previous miracles involving Creation, including the two avian miracles, so that every depiction of Cuthbert and Creation is drawn within its delineating bounds:

> Non sola autem aeris sed et maris animalia, immo et ipsum mare sicut et aer et ignis iuxta quod in superioribus exposuimus, uiro uenerabili praebuere obsequium.[158]
>
> (Moreover not only the creatures of the air but also of the sea, yes, and even the sea itself, as well as the air and fire as we have shown above, did honour to the venerable man.)

Bede is clear that Creation's honouring or obedience ('obsequium') to Cuthbert is conditional upon his perfect sanctity, towards which Cuthbert has progressed throughout the *VCP*:

> Qui enim auctori omnium creaturarum fideliter et integro corde famulatur, non est mirandum si eius imperiis ac uotis omnis creatura deseruiat. At nos plerunque iccirco subiectae nobis creaturae dominium perdimus, quia Domino et creatori omnium ipsi seruire negligimus.[159]
>
> (For if a man faithfully and wholeheartedly serves the maker of all created things, it is no wonder though all Creation should minister

[158] *VCP*, pp. 224–25.
[159] *VCP*, pp. 224–25.

to his commands and wishes. But for the most part we lose dominion over the Creation which was made subject to us, because we ourselves neglect to serve the Lord and Creator of all things.)

While Bede initially relocates the sea-and-beam episode to the end of the Farne creation miracles in the *VCM*, the retention of that move serves a distinct function here in the *VCP*. Here Bede presents the obedience of the sea to Cuthbert as the height of his ability to restore Creation and the greatest evidence for Cuthbert's progression in sanctity. This is demonstrated by the heavy emphasis on the wonder of the sea as an elemental force serving the saint, as the *VCP* explicitly highlights twice. The first occurs in the initial line of the chapter where the phrase 'immo et ipsum mare' ('and even the sea itself') distinguishes the sea as the most surprising portion of animate and inanimate Creation which served Cuthbert.[160] Bede then reiterates this emphatic statement after the exegesis to begin the episode proper: 'Et ipsum inquam mare promptum famulo Christi ubi opus habuit' ('The very sea, I say, was ready to do service to the servant of Christ when he needed it').[161] The emphasis on the sea seems at first arbitrary, as it appears no more miraculous than the sea-otters drying Cuthbert's feet. It is only when considered in the context of the complex Anglo-Saxon imaginative interaction with the sea as a whole that Bede's point gains force. As numerous studies have shown, the literary engagement with the sea by the Anglo-Saxons, in both Old English and Anglo-Latin, displays a profound polyvalence,[162] which is based upon the fundamental characteristic of the ocean's unimaginable vastness,

[160] *VCP*, pp. 224–25

[161] *VCP*, pp. 226–27.

[162] See for example, W. Rudolf, 'The Spiritual Islescape of the Anglo-Saxons', in *The Sea and Englishness in the Middle Ages: Maritime Narratives, Identity and Culture*, ed. S. I. Sobecki (Cambridge, 2011), pp. 31–57; Appleton, 'Anglo-Saxon Landscapes', pp. 198–99, argues for a conceptual difference between sea and ocean, where the waters surrounding Britain and Scandinavia in the accounts of Ohthere and Wulfstan in the Old English *Orosius* are labelled *sæ* rather than *garsecg*, and are thus centralising the sea as a linking place between lands; Clarke, *Literary Landscapes*, pp. 10, 19–21, argues for both the surrounding ocean's limiting and connecting roles in Gildas and Bede; A. A. Lee, *The Guest-Hall of Eden: Four Essays on the Design of Old English Poetry* (New Haven, 1972), sees water as both the turbulence of corporeal life in *Andreas*, and functioning as a descent into hell in *Beowulf*, pp. 92–93 and 205–7 respectively.

allowing it to function as a plane of interpretative projection.[163] Neville interprets this and similar miracles as highlighting what she calls the 'semi-autonomous' agency of the sea, which in Old English poetry in general is depicted as 'normally undirected (or perhaps even self-directed) and apparently uncontrollable'.[164] Thus, she concludes, the sea's subservience would have been truly miraculous, as the sea 'normally maintains an impersonal to hostile disregard for human affairs'.[165] The native conception of the sea's inherent vastness and untamed wildness is clearly at play in Bede's depiction.

Bede himself incorporates this vision into the beginning of the *Historia,* following Classical precedent, by locating Britain and the Orkney islands within and bounded on the west by what he calls the 'Oceano infinito'.[166] Helen Appleton has highlighted how Bede's use of *oceanus* in his *Historia* reflects his perception of Britain as conceptually located at the edge of the map.[167] Bede is not, however, consistent in his terminology, as evidenced by his variation of sea lexis in the *Historia,* where he utilises *oceanus* and *mare* to cover both the sea between Britain and mainland Europe, traversable and connective, and the outer sea to the west, boundless and isolating. In general, he is predisposed to use *oceanus* when describing the aqueous field in which islands reside, as he does with Britain as a whole, Farne, the Orkneys, and the Isle of Wight. Bede also tends

[163] Rudolf, 'Spiritual Islescape', p. 46: 'Without doubt the sea served as a major plane of projection for the Anglo-Saxons' ideas of their earthly and transcendental existence'; see also D. Anlezark, 'The Anglo-Saxon World View', in *The Cambridge Companion to Old English Literature,* ed. M. Godden and M. Lapidge, 2nd edn (Cambridge, 2013), pp. 66–81, at 68–71, for a general discussion of this insular mentality; and see T. O'Loughlin, 'Living in the Ocean', in *Studies in the Cult of Saint Columba,* ed. C. Bourke (Dublin, 1997), pp. 11–23, at 13–14, 25, for a discussion of the ocean surrounding Ireland and Britain as conceptually representing the limits of the world: 'This expanse of water was qualitatively different from other waters in that it was at the very limit of inhabitable reality and its shore marked the point of the separation between the waters and the dry land (Gen. 1.9)', p. 13.

[164] Neville, *Representations,* pp. 170, 175.

[165] Neville, *Representations,* p. 172.

[166] Bede, *HE,* I.1, p. 14. Also, see F. Michelet, *Creation, Migration, and Conquest: Imaginary Geography and Sense of Space in Old English Literature* (Oxford, 2006), p. 128.

[167] Appleton, 'Anglo-Saxon Landscapes', p. 178.

to use *oceanus* when focusing on the ocean's wildness, particularly when portraying the saints subduing it. Bede's use of the term *mare* often, though by no means always, refers to the sea between Britain and Europe, and is used to highlight the navigable and therefore connective sea-paths that unite the two. Thus when describing the water between Britain and the Orkneys in I.1 he employs *oceanus*, whereas he terms the water between Britain and the shores of Gaul *mare*: '[a] tergo autem, unde *Oceano infinito* patet, Orcadas insulas habet' ('Behind the island, where it lies open to the *boundless ocean*, are the Orkney islands').[168] Later in Book I he again refers to the sea beyond the west of Britain, in which the Orkneys lie, as *oceanus*, 'Orcadas etiam insulas ultra Brittaniam in *oceano* positas' ('the Orkneys, some islands which lie in the *ocean* beyond Britain'),[169] but also uses the term to refer to the sea between Britain and Europe, as in his description of the Arian heresy crossing to infect Britain: 'et hac quasi uia pestilentiae trans *Oceanum* patefacta' ('and this quickly opened the way for every foul heresy from across the *Ocean*').[170] It is important to note, however, that when Bede describes miracles in which the sea is calmed, including Cuthbert's in the *VCP*, he prefers to use the term *oceanus*, as he does when depicting the sea miracle of Bishop Germanus:

> Qui cum prompta deuotione preces et iussa sanctae ecclesiae suscepissent, intrant *oceanum* et usque ad medium itineris, quo a Gallico sinu Brittanias usque tenditur, secundis flatibus nauis tuta uolabat.[171]
>
> (These, ready with zeal, complied with the requests and commands of the holy Church and embarked on the *Ocean*. The ship sped along safely with favouring winds and had reached half-way across the channel between Britain and Gaul.)

The varying use of these terms at first impairs analysis, but when they are all taken together, especially in reference to Farne itself, a consistent vision appears: that Farne is meant to be viewed as truly insular, as set amidst the infinite ocean (*Oceano infinito*), and that the

168 Bede, *HE*, I.1, pp. 14–15.
169 Bede, *HE*, I.3, pp. 22–23.
170 Bede, *HE*, I.8, pp. 34–35.
171 Bede, *HE*, I.17, pp. 54–55.

miraculous service of the ocean to Cuthbert is meant to display the extraordinary circumstance of that boundless expanse of water's obedience to the hermit. This is most evident in both of Bede's prose descriptions of Farne where it is conceptually set amidst the boundless ocean, as seen in the *VCP*, 'et inde *infinito* clauditur *oceano*' ('and on the seaward side [encompassed] by the *boundless ocean*'),[172] and in the *Historia*, 'appellatur Farne, et ab eadem ecclesia nouem ferme milibus passuum in *Oceano*' ('called Farne which is in the Ocean, about nine miles away from [that same church]').[173] Even though in the beam miracle itself Bede uses *mare* to refer to the subservient sea, his penchant for lexical overlap, combined with his repeated emphasis on the location of Farne amidst *Oceanus*, connects the water with the wild and untamable *garsecg* which surrounded the world. Bede is utilising the conceptual function of the sea as a limiter of humanity, dynamic and beyond human control, boundless except for those limits imposed by God himself at creation. Through the combination of Bede's double emphasis, and the relocation to the end of the Farne Creation miracles, he is presenting this final restoration miracle as evidence of Cuthbert's extraordinary spiritual development. Cuthbert has progressed in sanctity to a state where not only the winds change with his prayers, and the proud ravens obey his orders, but also the sea, the wildest and most powerful of all elemental forces, recognises his authority.

Bede concludes the sea-and-beam miracle in the *VCP* with an expansion of the interpretative addition placed in the *VCM*, which narratively describes how the miracle functions as an exemplar from which humanity might learn holy behaviour:

> Uidentes autem, mox et uiri uenerabilis sanctitatem mirabantur, cui etiam elementa seruirent, et suae mentis tarditatem debito cum pudore culpabant, quos etiam insensibile elementum quam sit sanctis obtemperandum doceret.[174]

[172] *VCP*, pp. 214–15, emendation for reason of sense and context.

[173] Bede, *HE*, IV.27 [25], pp. 430–31. Colgrave and Mynors elaborate the expression 'that same church' as 'the Church of Lindisfarne', which is not in the Latin. While understood in context, for the purpose of quoting the passage here, I have emended the translation.

[174] *VCP*, pp. 227–28.

(As soon as they saw this, they marvelled at the holiness of the venerable man for whom even the elements did service; and with fitting shame they blamed their slothful minds, for even the insensible elements taught them what obedience ought to be shown to saints.)

In this expansion Bede highlights the causal connection between Cuthbert's sanctity and the restoration of Creation; it is the saint's holiness ('sanctitatem'), marvelled at by the brothers, which provides the necessary reason for the subsequent service of the elements. Further still, Bede once again highlights the distinct nature of Cuthbert's sanctity, as this interpretation is given from the perspective of his spiritual charges, the forgetful brothers, who are pushed towards reformation of their behaviour and spiritual development. Thus, even in this most wondrous of all restoration miracles, the service of the infinite ocean, Bede's Cuthbert is still a unified and exemplary monk-pastor, whose activities always push towards reforms of one kind or another, external or internal. Bede's emphasis on this interaction, whereby the sea's subservient position to the saint functions as a pastoral check on disobedient behaviour, is evidenced further by an original miracle in the *VCP*, where Cuthbert calms another storm. In Ch. XXXVI, Bede adapts the description given in the *VCA* of Cuthbert's return to the eremitic life on Farne after his two years as bishop by attaching a lengthy additional sea miracle. In the miracle, the sea detains a few brethren who ignore a casual command of Cuthbert to eat a goose the saint gives them before they depart: 'primo sumite cibus, et aucam illam [...] coquite et comedite, et sic in nomine Domini nauem ascendite, ac domum redite' ('first take some food; and that goose [...] cook and eat it, and so in the name of the Lord go on board your vessal and return home').[175] As is evident in the lexis here, Cuthbert's command is given via the weak imperative 'sumite', which carries none of the semantic force of words like *praecipio*[176] or *iubens*.[177] Bede mentions the weak nature of this command in the introductory material before

[175] *VCP*, pp. 268–69.

[176] As in the *VCA*, or rarely in the *VCP*; see *VCP* p. 190 where *praeceptum* is used to describe Cuthbert's order to the spying monk: 'in quo nimirum *praecepto*'.

[177] As given by Bede for Cuthbert's order to the birds in the first avian miracle, *VCP*, p. 222.

his narrative of the episode begins: 'quo clarius elucescat quantum uiris sanctis obtemperandum sit etiam in his quae *negligentius imperare uidentur*' ('so that it may be more evident how obedient we ought to be to holy men, even in those matters about which they seem *to give very casual commands*').[178] It is, therefore, unsurprising that the brothers neither cook nor eat the goose, especially as the other food they took was enough to sustain them.[179] The result of this mild disobedience is a fierce storm that grounds them for seven days on Farne.[180] Unlike the miracle in Ch. XXI, the brothers do not make the connection between their disobedience and the actions of the sea, and Cuthbert must step even further into his pastoral role and convict them with his characteristically joyful words: 'placido uultu et laeto potius sermone redarguit eorum inoboedientiam' ('with unruffled mien and even with joyful words, he convicted them of disobedience').[181] The episode concludes with a reiteration of the restored relationship between Cuthbert and the sea which is altered by Bede to function as a didactic tool, in which the sea's service to the saint is utilised pastorally: 'Gaudebant quia intellexere tantam fuisse Deo curam de fideli suo famulo, ut contemptum eius etiam per elementa uindicaret' ('They rejoiced because they saw that God took such care of His faithful servant that He even punished, by means of the elements, those who esteemed His servant lightly').[182] As the episode is a complete addition by Bede, he is here freer to focus on the pastoral and didactic function of the miracle, but it nonetheless echoes and further supports the exegesis and restoration of Creation based on Cuthbert's distinct kind of sanctity from Ch. XXI.

178 *VCP*, pp. 266–67.
179 See *VCP*, pp. 268–69.
180 *VCP*, p. 268.
181 *VCP*, pp. 268–69.
182 *VCP*, pp. 270–71.

Conclusion

Bede's *VCP* is distinct in the Cuthbertine tradition not only for its lasting influence, but also for the transformation of Cuthbert's life into a reformative model that seeks to highlight the imitative virtues of the Gregorian pastor-monk, of which Cuthbert is the saintly epitome. Bede transforms the nature of Cuthbert's sanctity from the static model of the *VCA* to a developmental one by adapting the progression towards spiritual majority found in the *Vita Antonii*. The relationship between Cuthbert and Creation is utilised within this model for two primary reasons: to provide the necessary impetus for Cuthbert's progression in sanctity, and to provide evidence of that progression. The exegetical framework in which this relationship operates is that of Bede's own interpretation of the Genesis narrative, ultimately based on Augustine, in which the truly saintly are able, temporarily, to restore a portion of Creation into its prelapsarian state.

4

Enargaeic Landscapes and Spiritual Progression: Felix's *Vita Sancti Guthlaci*[*]

FELIX'S *Vita Sancti Guthlaci* (*VSG*)[1] is more indebted to hagiographical convention than most antecedent and contemporary saints' lives,[2] to the extent that Cubitt argues 'it is almost entirely a textual confection, created from a modicum of personal recollection and much hagiographical borrowing'.[3] While little is known about Felix himself, the breadth of his education, evidenced in these wide-ranging sources, is clear and strongly suggests he was a monk, or a cleric familiar with monastic life.[4] The *VSG* is also one of the few Anglo-Saxon hagiographies written not at the request of a specific monastic house or ecclesial figure, but instead for a royal patron, the East Anglian King Ælfwald. This focus suggests that Felix was likely writing in East Anglia. While Felix's intended audience for the *VSG* beyond Ælfwald is uncertain, the complexity of its depictions and hagiographical references points to a specialised and educated audience, whether monastic or aristocratic. The most recent editor of the text, Bertram Colgrave, persuasively argues for

* Portions of this chapter are taken from my article, 'Felix's Construction of the English Fenlands: Literal Landscape, Authorizing Allusion, and Lexical Echo', in *Guthlac of Crowland: Celebrating 1300 Years*, ed. J. Roberts and A. Thacker (forthcoming).

1 *Felix's Life of Saint Guthlac*, ed. B. Colgrave (Cambridge, 1956). All quotations and translations are from this edition.

2 See *VSG*, pp. 16–19; Thacker, 'Social and Continental Background', pp. 279–89; Cubitt, 'Memory and Narrative', pp. 29–66, at 51–57.

3 Cubitt, 'Memory and Narrative', p. 51. H. Foxhall Forbes, *Heaven and Earth in Anglo-Saxon England: Theology and Society in an Age of Faith* (Farnham, 2013), argues against this, that not only were these places 'genuinely feared' and 'believed to be inhabited by wicked spirits', but that Felix's lexis includes toponymically accurate detail, pp. 91–92.

4 *VSG*, p. 16.

a date between 730 and 740 for the composition of the *VSG*.[5] While the *VSG* itself is dense with neologisms, allusions to hagiographical precedent, and lexical echoes of classical authors like Virgil, one of its most striking features is the level of descriptive detail in its depiction of the East Anglian fens and the eremitic site of the island of Crowland. Scholarly engagement with these depictions has been wide ranging, most recently in Estes's *Anglo-Saxon Literary Landscapes*. Estes argues that the depictions of fens in Anglo-Saxon texts more broadly, usually described as 'wildernesses', are cultural constructions of those areas 'as seen by people who lived elsewhere'.[6] Felix's *VSG* presents a divergent case, however, as both he and his patron would likely have had first-hand knowledge of the East Anglian fens. The *VSG*'s descriptive detail, therefore, requires a different kind of analysis. I would argue that Felix's technique can profitably be discussed in terms of the classical rhetorical techniques of 'ekphrasis', the 'elaborate "delineation" [...] of people, places, buildings, works of art',[7] and more specifically 'enargeia', which Graham Zanker, summarising Dionysius, argues is a much more precise term for the kind of description which seeks to depict what I call 'experiential' reality:

> [Enargeia] is the stylistic effect in which appeal is made to the senses of the listener and attendant circumstances are described in such a way that the listener will be turned into an eyewitness.[8]

Felix's consistent focus on depicting the physical world in a vivid and realistic manner for the purpose of allowing the reader to experience it as an 'eyewitness' certainly can be thought of in similar terms. A possible source for this technique in Felix's prose is his deep knowledge of Virgil, who is often cited by Latin authors as utilising enargeia.[9] The appeal to such a delineated physical world via the senses in the *VSG* primarily serves to describe the saint's eventual restoration of Creation. It is true that Felix provides few

5 *VSG*, p. 19.
6 Estes, *Anglo-Saxon Literary Landscapes*, p. 16.
7 Curtius, *European Literature*, p. 69.
8 G. Zanker, 'Enargeia in the Ancient Criticism of Poetry', *Rheinisches Museum für Philologie* 124 (1981), 291–311, at 291.
9 See Zanker, 'Enargeia', at 300.

topographical details in the form of named places, despite the echoes of what Nicholas Howe calls 'narrative cartography',[10] yet his depiction of the landscape is carefully crafted to reflect the experience of entering the physical fens.

Unlike most of his hagiographical models, in particular the *VCA*, Felix is not primarily depicting physical space for the purpose of creating and/or expanding a tomb cult, but is instead more interested in the active role of Creation in the spiritual development of the saint. Thus Felix sets out to describe the fens and Crowland as real, physical places:

> Est in meditullaneis Brittanniae patribus inmensae magnitudinis aterrima palus, quae, a Grontae fluminis ripis incipiens, haud procul a castello quem dicunt nomine Gronte nunc stagnis, nunc flactris, interdum nigris fusi vaporis laticibus, necnon et crebris insularum nemorumque intervenientibus flexuosis rivigarum anfractibus, ab austro in aquilonem mare tenus longissimo tractu protenditur.[11]
>
> (There is in the midland district of Britain a most dismal fen of immense size, which begins at the banks of the river Granta not far from the camp which is called Cambridge, and stretches from the south as far north as the sea. It is a very long tract, now consisting of marshes, now of bogs, sometimes of black waters overhung by fog, sometimes studded with wooded islands and traversed by the windings of tortuous streams.)

What is most readily apparent here is the specificity of the locality: the fen is in the middle of Britain near the banks of the 'Grontae' (later the Cam), located near the camp called 'Gronte' (later Cambridge). With these details, the *VSG* is affixed to a physical location

10 N. Howe, *Writing the Map of Anglo-Saxon England: Essays in Cultural Geography* (New Haven, 2008), pp. 6–8, discusses the idea of narrative cartography and the role of texts like Sigeric's itinerary which provides 'a sequence of stages for a journey that is at once a cross of natural and cultural terrain' by the use of specifically named locations along the journey, p. 8. See also N. Howe, *Across an Inland Sea: Writing in Place from Buffalo to Berlin* (Princeton, N.J., 2003), pp. 64–70, for a discussion on the general structure of a journey narrative, which moves 'between the known and unknown in ways that simple linearity or the overly precise grid cannot render', p. 65.

11 *VSG*, pp. 86–87.

which the audience of the text could have experienced.[12] Felix not only locates the narrative in a specific and identifiable place in the English landscape, but also describes the fens experientially: the subsequent employment of the temporal referent *nunc* collocated with *interdum* and *necnon* pulls the reader forward and reveals the fenland landscape as if one were walking through it.[13] The reader is drawn forward through the fens, meeting first the stagnant ponds ('stagnis'), followed by the related marshes ('flactris'), which are often overhung with fog. The lexical range of these terms creates a sensual reality rare in hagiographical accounts. As Hugh Magennis notes, 'despite its conventional aspects [...] Felix's description presents an account of the "Black Fens" of East Anglia which is both topographically accurate and highly atmospheric'.[14] The term *stagnum* carries not only the visual imagery of a bog or pool of standing water, but also the olfactory displeasure of rotting matter,[15] while the term *flactra*[16] expands the landscape by making what could be a single pool of stagnant water into an entire marsh. These images and smells are followed by water made black ('nigris') with peat, overhung by what Felix calls 'vapor', another unusual

12 K. A. Kilpatrick, 'The Place-Names in Felix's *Vita sancti Guthlaci*', *Nottingham Medieval Studies* 58 (2014), 1–56, at 47, points out that the river Welland, which served as one of the traversable paths between the Mercian and East Anglian kingdoms, ran near Crowland, and therefore posits that while Felix was writing under the patronage of the East Anglian King Ælfwald, he must have assumed a Mercian as well as East-Anglian audience. She further argues that Felix's descriptive detail and the specific location of Guthlac's hermitage in a barrow likewise stem from what would have been a recongisable landscape to his audience.

13 It should also be noted that the collocation of *nunc* used to depict a chronological sequence of events all occurring in the present, each subsequent to but also overlapping with the previous event, has echoes in Virgil. See Virgil, *Georgics*, I. ll. 266–67: '*nunc* facilis rubea texatur fiscina uirga, / *nunc* torrete igni fruges, *nunc* frangite saxo' ('*Now* weave the graceful basket of reddish twigs, *now* parch grain by the fire, *now* grind it on the stone').

14 H. Magennis, *Images of Community in Old English Poetry* (Cambridge, 1996), p. 180.

15 *Stagnum*: 'pool, pond, or standing / stagnant water'; as a verb can mean 'to stagnate'. There is an instance *of stagnum* c. 1125 in reference to an island in the Fens, 'insula stagnensis': *Revised Medieval Latin Word-List from British and Irish Sources*, ed. R. E. Latham (Oxford, 1965), p. 450.

16 *DMLBS*, 1: 'marsh, reed-bed'. The only usages of this term are in Felix and Orderic Vitalis (who is copying Felix).

word that speaks not just of a mist, but more specifically of a warm fog composed of unpleasant gases.[17] Felix expands his enargaeic description of this dismal landscape with further detail, wooded islands appearing as it were out of the fog, between which flow tortuous streams ('flexuosis rivigarum'). The reader's vision is also drawn outwards by Felix to see the full scope of the landscape, and how it stretches from the south to the sea in the north. Throughout this description, Felix's consistently enargaeic appeal to the senses, primarily sight and smell, invites the reader to enter into and to experience the landscape which Guthlac will sanctify and restore.

This vivid depiction of the fens owes much of its potency not only to careful lexical deployment, but also to the necessarily sequential representation of geographical knowledge exhibited by the Anglo-Saxons. Howe argues that such knowledge is contained in a form of narrative wherein places are not located by 'pointing to them on an illustrative map or by setting their coordinates on a grid, but by writing them in an ordered sequence that typically begins with a well-defined and isolatable site'.[18] While Howe is speaking of longer journeys here, his argument remains applicable to the opening of *VSG*: Felix's description of the fens begins with an identifiable site, the banks of 'Grontae' not far from 'Gronte', which extends south. The following narrative does not, however, employ further identifiable markers, but Felix's conception of the landscape, 'not as a vista to be contemplated but a sequence of signs to be walked',[19] is nonetheless evident throughout. The signs are not specific in the way most journey narratives are, but instead focus on lavish enargaeic details which cumulatively serve to evoke a place fundamentally bereft of significant landmarks. Helen Foxhall Forbes has argued that Felix's depiction is not simply a confection, as Cubitt argues, but that its descriptive details, the 'dark and stagnant water or boggy places [...] are matched in other contexts, such as the descriptions of features and landmarks described

17 *DMLBS*, 1: a. 'Gaseous emanation (esp. from something heated), vapour; b. (as forming cloud, mist, or sim.)'.

18 Howe, *Writing the Map*, pp. 5–6.

19 Howe, *Writing the Map*, p. 39. He is referencing boundary clauses here.

in boundaries of estates recorded in charters'.[20] Felix is attempting, therefore, in combination with his conception of travel in a landscape described narratively, to present as close an approximation to the act of traversing the fens as possible; thereby affixing his saint's subsequent spiritual progression and eventual restoration of Creation to the specific postlapsarian world of the fens and Crowland.

Felix's initial description of Guthlac's hermitage site, the island of Crowland, is created through an equivalent process of interweaving derivative hagiographical precedent with original elements to depict an enargaeic location. For example, in a departure from the Antonian tradition, Guthlac is not led to his desert retreat by any direct intervention of God through prophetic vision or miniature miracle,[21] but instead he has to seek the knowledge of local people. This inclusion of local knowledge as the vessel through which Guthlac's eremitic location is revealed further grounds the *vita* in the physical landscape. By specifying the named guide, Tatwine, who later takes the saint in his own boat to the island, Felix lends not only an air of veracity, in the generic manner of hagiography, but also further delineates the fens by contrast. The location of the hermitage is carefully constructed as an idealised English wilderness, an insular version of the Egyptian desert, by both the local inhabitants in general and Tatwine specifically, as well as through various lexical choices which echo hagiographical deserts.[22] Both the local people and Tatwine refer to the fenland area with the loaded hagiographical term *heremus*: in the former, 'plurima ipsius spatiosi *heremi* inculta' ('many wild places in this far-stretching *desert*'), and in the latter, 'insulam in abditis remotioris *heremi* partibus'

20 Cubitt, 'Memory and Narrative', p. 51; Foxhall Forbes, *Heaven and Earth*, p. 92: 'Sometimes features such as meres or pools were designated with animal or bird names as qualifiers (such as the "titmouse mere" in the boundary of an estate given to Abingdon in 956), but watery places in particular might alternatively be designated "black", "foul", "noxious", or "dark", probably reflecting their actual natural appearance, but perhaps also implying negative connotations'.

21 Compare *Vita Antonii*, Ch. 49, p. 176.

22 See C. Lee, 'Reluctant Appetites: Anglo-Saxon Attitudes towards Fasting', in *Saints and Scholars: New Perspectives on Anglo-Saxon Literature and Culture in Honour of Hugh Magennis*, ed. S. McWilliams (Cambridge, 2012), pp. 164–86, at 181–82, for a discussion of the use of an Antonian desert in relation to practices of fasting in the Guthlac legends.

('island in the more remote and hidden parts of that *desert*').[23] Felix further constructs this English desert with another use of *heremus* in Tatwine's description of the inability of others to abide Crowland due to the demons inhabiting it: 'incognita *heremi* monstra et diversarum formarum terrores' ('uknown portents of the *desert* and its terrors of various shapes').[24] The location is not only one of the wild places of the desert, but is even more remote, superlatively wild, and saturated with the necessary demonic presence for a hagiographical battlefield. Felix is calling upon numerous sources here, from Gregory's *Dialogues* to Jerome's *Vita Pauli*, to the Evagrian *Vita Antonii*.[25] He is not, however, merely copying set pieces without reflection, intent, and alteration; verisimilitude is the point, connecting his highly conventional *VSG* to the physical landscape, and creating the necessary physicality for the saint's development in which he is able to restore Creation. This connection between demons and the desert is in keeping with the Anglo-Saxon perception of the fens as the potential landscape of inimical spirits, whether specifically Christian demons or folkloric beings.[26] It also resonates with the historical dangers of landscapes like the fens, in which bands of thieves wait or, as Lindy Brady notes, foreshadowing the later episode with the phantasmal host, 'that of the wild

23 *VSG*, pp. 88–89.

24 *VSG*, pp. 88–89.

25 See Thacker, 'Social and Continental Background', pp. 256, 283, and esp. 324; see also *VSG*, pp. 15–20; Foxhall Forbes, *Heaven and Earth*, pp. 90–92; Kilpatrick, 'Place-Names', 24–25.

26 See Thacker, 'Social and Continental Background', p. 317. Clarke, *Literary Landscapes*, p. 32, connects the demonic presence to Felix's use of the term *tumulus*. Kilpatrick, 'Place-Names', p. 55, argues that there is substantial archeological evidence for the existence of barrows in and around Crowland, and connects these with the general scholarly consensus that such monuments 'were often regarded by Christian contemporaries as sites where evil spirits and demons dwelled'. N. Whyte, 'The After-Life of Barrows: Prehistoric Monuments in the Norfolk Landscape', *Landscape History* 25 (2003), 5–16, at 6, argues that the assignation of inimical spirits to barrows was well underway by the seventh century. S. Semple, 'A Fear of the Past: The Place of the Prehistoric Burial Mound in the Ideology of Middle and Later Anglo-Saxon England', *World Archaeology* 30 (1998), 109–26, at 117–20, argues for the role of Christianisation in this process of placing inimical spirits in barrow contexts based on Old English literature.

Briton concealed within poised to attack'.[27] It is a place defined by its unapproachability, its wildness and separateness from the ordered world of humanity, exemplified by the community with which it is contrasted, and potentially filled with powers inimical to humankind. Despite these demonic echoes and cultural associations the fens remain a terrestrial landscape, part of the fallen but redeemable world, and thus not hostile itself towards the saint.

Few aspects of the landscape in Felix's *VSG* have attracted as much critical attention as his description, again highly physical, of the spot Guthlac chooses to build his eremitic retreat, the barrow of Crowland:[28]

> Erat itaque in praedicta insula tumulus agrestibus glaebis coacervatus, quem olim avari solitudinis frequentatores lucri ergo illic adquirendi defodientes scindebant, in cuius latere velut cisterna inesse videbatur; in qua vir beatae memoriae Guthlac desuper inposito tugurio habitare coepit.[29]
>
> (Now there was in the said island a mound built of clods of earth which greedy comers to the waste had dug open, in the hope of finding treasure there; in the side of this there seemed to be a sort of cistern, and in this Guthlac the man of blessed memory began to dwell, after building a hut over it.)

This depiction once again incorporates elements from various sources, most notably Ch. 6 of Jerome's *Vita Pauli*. As Thacker notes,[30] Jerome describes hermits living in desert spaces, utilising the noun *heremus*, along with a particular hermit who lived in a cistern and survived on five figs a day, using the same noun Felix later employs, *cisterna*.[31] This echoing of hagiographic texts such

27 L. Brady, 'Echoes of Britons on a Fenland Frontier in the Old English *Andreas*', *RES* 61 (2010), 669–89, at 675–76.

28 For example see Kilpatrick, 'Place-Names', 24–25, 32; S. Semple, 'Anglo-Saxon Attitudes to the Past: A Landscape Perspective: a Study of the Secondary Uses and Perceptions of Prehistoric Monuments in Anglo-Saxon Society', Unpub. D.Phil. (Oxford University, 2002), pp. 252–53; and A. Hall, 'Constructing Anglo-Saxon Sanctity: Tradition, Innovation and Saint Guthlac', in *Images of Medieval Sanctity: Essays in Honour of Gary Dickson*, ed. D. Higgs Strickland (Leiden, 2007), pp. 207–35, at 216–18.

29 *VSG*, pp. 92–95.

30 Thacker, 'Social and Continental Background', pp. 295–96.

31 Jerome, *Vita Pauli*, p. 154.

as the *Vita Antonii* further reflects Felix's desire to create his *vita* in keeping with traditional models, while also expanding the enargaeic description to depict a more vivid physicality.[32]

The term that has attracted the largest amount of scholarly attention of any single portion of Felix's constructed landscape is, however, the noun *tumulus* ('barrow'), a term often used in reference to a barrow or other types of earthen tombs inhabiting the English landscape.[33] Sarah Semple argues that Felix's use of *tumulus*, combined with the description of it being broken in the search for treasure, suggests a barrow, commenting that '[r]ecent analysis of air photographs suggests that the possible site of Guthlac's cell – the remains of a chapel on Anchor Hill – may overlie a round barrow within an irregular ditched enclosure'.[34] Semple sees in Felix's lexical choice an awareness of what she calls 'inverse syncretism',[35] whereby entities of the pre-Christian world have been absorbed into the Christian world view as demonic forces, a feature in harmony with other eighth-century sources.[36] The *tumulus* is, nonetheless, a physical and fixed part of the landscape, whose familiarity as a landscape feature furthers the *vita*'s grounding in enargaeic Creation. The *tumulus* also represents a location potentially inimical to humanity, whether with spirits of the pre-Christian world or their syncretic demon counterparts, and thus its function as an English wilderness is clear: it is the location furthest from the world of order and light, and therefore the place in greatest need of a *miles Christi*, an eremitic soldier of Christ. Felix has employed the noun *tumulus* as an enargaeic tool with full awareness of its connotations. The combination, then, of the present-tense narrative

32 See Thacker, 'Social and Continental Background', p. 324, who argues that the 'frequency of Felix's quotations, and their location at significant points in his narrative, indicate that he intended them to be noticed. They were a product not of incompetence but of art, and Felix wished his audience to recognize in Guthlac an authentic hermit-saint following in the footsteps of Antony, Cuthbert, Martin, Fursey and Columbanus.'

33 Semple, 'Anglo-Saxon Attitudes', p. 252.

34 Semple, 'Anglo-Saxon Attitudes', p. 252.

35 Semple, 'Anglo-Saxon Attitudes', pp. 266–67.

36 See Semple, 'Anglo-Saxon Attitudes', pp. 243–62, where she discusses the connection between barrows and inimical forces in *Beowulf, Maxims II,* and on the Franks Casket.

description of the dismal fen, with the physical landscape feature of the *tumulus*, serves to create a landscape full of detail, but also overflowing with the necessary associations for its use as an eremitic Antonian battleground. Guthlac's progression in sanctity begins with this located and delineated physical space.

Spiritual Progression in the Landscape

Felix's situating of Guthlac's eremitic vocation in the enargaeic and physically delineated location of Crowland in the English fens allows the hagiographer to breathe life into the Antonian model on which much of the *VSG* is based. The saint's spiritual achievement in his temptations and persecutions is cast within the Antonian tradition of progression in the saintly life, though with less emphasis on the purely psychological reflections that the demonic battles represent in the *Vita Antonii*.[37] This progression towards spiritual majority[38] is depicted in the development of Guthlac's response to demonic temptation and persecution, both in his internal and external reactions, which are fundamentally rooted in the physical landscape. This emphasis on demonic temptation is, like all temptations, a reflex to that of Adam.[39] In the saint's successful defence

37 Thacker, 'Social and Continental Background', p. 286: 'Kurtz was wrong to interpret Guthlac's conflicts primarily in psychological terms. The saint was engaged in a heroic combat with the troops of the enemy [...] Even struggles which might be regarded as psychological were described in terms of physical conflict'; and P. Dendle, *Satan Unbound: The Devil in Old English Narrative Literature* (Toronto, 2001), p. 105: 'the same symbols and scenes were undergoing a transformation in the early Middle Ages, and that psychology was no longer of principal interest, as it was for Athanasius, Cassian, and Augustine [...] Felix's *Vita* and the vernacular works are primarily interested in encoding the delicately shifting balances of power between the saintly and the demonic.'

38 See Kurtz, 'Antony to Guthlac', at 116: 'Athanasius (c.14, 15; 38; 48, 50, 54, 57 ff.) represents this power in Antony as the stage of spiritual majority resulting from nearly two decades of discipline and solitude'; also, see Thacker, 'Social and Continental Background', p. 290. See also above, p. 126.

39 The Apostle Paul, for example, interprets the Fall allegorically: 2 Cor. 11:3, 'timeo autem ne sicut serpens Evam seduxit astutia sua ita corrumpantur sensus vestri et excidant a simplicitate quae est in Christo' ('But I fear lest, as the serpent seduced Eve by his subtlety, so your minds should be corrupted, and fall from the simplicity that is Christ'). The allegorical interpretation of the

against the temptations, by which he attains spiritual majority, Guthlac, like Cuthbert, succeeds where Adam fell, allowing Creation to be brought back into prelapsarian relationship.

Guthlac's progression towards sanctity begins with the first temptation, in which Satan appears as a roaring lion shooting poisoned arrows of despair at the young hermit.[40] Felix reveals Guthlac's novice status here through the internal wounding and overwhelming effects of the attack: 'totis sensibus turbatus [...] desperare coepit [...] turbulentum animum [...] ut infinitum et inportabile opus se incoepisse putasset' ('every feeling was disturbed [...] he began to despair [...] his troubled mind [...] that he thought he had undertaken an infinite and insupportable labour').[41] Guthlac's response is likewise structured to further this depiction, as his crying out a portion of the Psalter (Ps. 17:7 here), though a common hagiographical response, is one of desperation, a call for aid: 'In tribulatione invocavi Dominum' ('In my distress I called upon the Lord').[42] The despair which inflicts him derives its theological and narrative power from the physical and psychological reality of a hermit living in the very real difficulties of the fens;[43] it is not a landscape purely of the mind, a desert whose name invokes for Felix's intended audience vague notions of sand and palm trees, but a landscape of the feet, traversable and knowable. As Colgrave notes, this temptation also represents 'the trouble which haunts all beginners in the monastic and mystical life', with a hint of 'Antony's sermon and his warning against allowing the devil to overwhelm the hermit with memories of past sins',[44] but again it is the physical and familiar landscape that creates the power of this struggle.

historical Fall in Gen. 3 was central to the exegetical traditions of the medieval world. For a detailed discussion of patristic and medieval ideas of the Fall and its consequences, particularly relating to typological and tropological readings, see E. Jager, *The Tempter's Voice: Language and the Fall in Medieval Literature* (Ithaca, NY, 1993).

40 *VSG*, p. 94.

41 *VSG*, pp. 96–97.

42 *VSG*, pp. 96–97. See M. J. Toswell, *The Anglo-Saxon Psalter* (Turnhout, 2014), pp. 291–94, for a discussion on Felix's emphasis on the use of the Psalms as weapons in the *VSG*.

43 *VSG*, pp. 94, 96 respectively.

44 *VSG*, p. 184.

His resisting the venom for three days not withstanding,[45] Guthlac's external reaction involves no direct confrontation with Satan, and thus declares his callow status; his response is one of retreat, anguish, and defence. St Bartholomew's arrival into this state serves evocatively to dispel the predominant literal landscape feature, darkness, as Felix describes it, with an explosion of terms for light: 'splendentis' ('resplendent'), 'caeleste adiutorium' ('heavenly aid'), 'angelicae lucis' ('angelic light'), and 'inluminato' ('enlightened').[46] When Bartholomew brings illumination, banishing both the physical night and the metaphorical darkness of despair, the contrast is grounded in the delineated reality of Guthlac's eremitic habitation, fixed within the recognisable physical nature of the fens. Guthlac is able to conquer this despair, via his reliance on the grace and efficacious mercy of God within this physical foundation.[47] The indifferent and shadowed landscape has been used, implicitly by God, to move Guthlac towards greater holiness, just as Augustine and Bede declared it should in their exegeses.[48]

With the conquering of despair Guthlac has taken a step in his progression in the saintly life, and in the subsequent temptation, excessive fasting, Felix depicts another step in his saint's progression through Guthlac's lack of internal turmoil. The episode is, however, once again anchored in the enargaeic landscape of Crowland, this time through the defeated demon's lamentation. The temptation of excessive fasting also relates back to Adam's curse, as the Fall was interpreted by some patristic authors as resulting from Adam's gluttony;[49] when saints and others fast to excess, their unmeasured striving can lead them to gluttony. Guthlac's eventual resisting of this temptation, therefore, is to be understood as the undoing of Adam's curse. The temptation of excessive fasting has

45 *VSG*, p. 96.

46 *VSG*, pp. 96–97.

47 *VSG*, p. 98.

48 See above, Introduction, pp. 9–15.

49 For example, see Cassian, *Collationes XXIIII*, ed. M. Petschenig, suppl. G. Kreuz, CSEL 13 (Vienna, 1886; repr. 2004), p. 122: 'nec primus Adam per gastrimargiam decipi potuit, nisi escae materiam habens in promptu abusus ea fuisset inlicite' ('For the first Adam could not have fallen a victim to gluttony unless he had had material food at hand, and had used it wrongly').

wide currency in medieval texts, especially hagiography, and Felix is here drawing on several sources.[50] Thacker argues that while Felix echoes Antony's warning against this evil, he also draws from Cassian's illustrative story of Abbot John in his *Collationes* Ch. 21.[51] In the *Vita Antonii*, amidst one of Antony's extended sermons, he mentions the temptation to fast to excess as a common demonic approach.[52] It is divorced from the main narrative, and unconnected to any specific place. Likewise, Cassian's depiction of Abbot John is meant as a warning, a short narrative where his excessive fasting leads to the appearance of a demon, from which John discerns that he had been deceived into endangering his body and spirit.[53] Felix has taken these cautionary sources and used them to focus instead on Guthlac's response as evidence for his progression, a structure distinct from his sources. Guthlac once again does not directly engage his tempters, but sings out another psalm. The similarity with his response to the first temptation ends here, however, as the content of this declarative quotation contrasts with the first; where the first singing of the Psalter involved a plea for divine help, a passive reaction, this quotation from Ps. 55:10 focuses the invocation for divine aid as a weapon against the tempter, a decidedly active response: 'Convertantur inimici mei retrorsum' ('Let mine enemies be turned back').[54] His internal state receives no direct comment, though his external response of eating his daily food, including a scrap of barley bread, hints at the internal development that has taken place.[55] While there is no explicit reference to the landscape

[50] Colgrave posits that this temptation to fast in excess was particularly prominent among the Celtic churches, and also that Felix was probably thinking of Antony's warning against this very evil, in Antony's life, *VSG*, p. 184. See *Vita Antonii*, Ch. 25, p. 169; see also S. Downey, 'Too much of too little: Guthlac and the Temptation of Excessive Fasting', *Traditio* 63 (2008), 89–127, at 94–97; Lee, 'Reluctant Appetites', pp. 164–86, at 180–86.

[51] See Thacker, 'Social and Continental Background', pp. 299–300; Cassian, *Collationes*, pp. 32–33.

[52] *Vita Antonii*, Ch. 25, p. 169.

[53] Cassian, *Collationes*, p. 33: 'diaboli circumuentum talique distentum ieiunio, ut lassitudinem non necessariam, immo etiam spiritui nocituram fatigato corpori superponeret' ('Deceived by the wiles of the devil, that such a long fast brought unnecessary weariness, and even spiritual injury, to his fragile body').

[54] *VSG*, pp. 100–01.

[55] *VSG*, p. 100.

here, the scene once again derives its potency from the physical fens; for Guthlac, alone in the fens, the pangs of hunger and the question of appropriate fasting become powerfully relevant. This focus on the fens is anchored by the response of the defeated demons whose lamentations echo throughout the delineated landscape: '*late loca* maestis questibus inpleverunt' ('filling the *land far and wide* with their sad complaints').[56] Felix here has once again drawn the temptation back to his enargaeic fenland landscape. The potency of the demons' loss, and thus Guthlac's victory, is declared in their sorrowing which fills the whole of Crowland and the fens ('late loca'), the landscape central to their battle and Guthlac's progression in sanctity.

Guthlac's Spiritual Majority

Felix's third scene of demonic persecution, in which Guthlac is taken to the door of hell itself, has rightly been identified as central to the *VSG*. Alaric Hall, for example, comments that this is 'the most dramatic of the struggles' with demons in the text, and posits that the subsequent two assaults by demons are therefore 'inevitably anticlimactic'.[57] Hall also notes the progression towards Guthlac's miraculous works, how all the 'struggles with demons are merely a prelude to the real substance of the *Vita*: his miracles'.[58] Similarly, Kurtz locates Guthlac's spiritual majority after the final demonic attack, the bestial horde in Ch. XXXVI.[59] I would argue that this dramatic episode, however, is the precise moment when Guthlac attains spiritual majority. It is no surprise, then, that it is the longest single episode in the *VSG*, as well as the most richly descriptive. In this transformational moment Felix once again ties Guthlac's spiritual progression to the physical landscape. The episode can be divided into two sections with individual parts. In the first section Felix deals with the physical landscape, including the sublunary

56 *VSG*, pp. 100–01.
57 Hall, 'Anglo-Saxon Sanctity', p. 214.
58 Hall, 'Anglo-Saxon Sanctity', p. 214.
59 Kurtz, 'Antony to Guthlac', 116.

air and the earthly ground, thickets, and water. In the second, however, from this physical foundation and in marked contrast with it, he deals with a novel depiction of the traditional and metaphoric landscape of hell.

Felix grounds the final encounter between Guthlac and the demons in the physical landscape, with the demons bursting into Guthlac's cell from earth and sky ('caelo terraque erumpentes'), and covering all the heavens in their vast numbers: 'spatium totius aeris fuscis nubibus tegebant' ('they covered the whole space beneath the heavens with their dusky clouds').[60] Not only do they fill the sky visually, and thus separate the hermit from the sight of heaven and the stars (which even in *Genesis A* were left as a solace for the fallen couple),[61] but also fill that same pregnant liminal space with their raucous din.[62] While the demons appear to represent an overwhelming force, it is important to note the constraint of their dominion here; in another aspect of a theologically defined, physical landscape within postlapsarian Creation, they are relegated to the sublunary world, beneath and forever set apart from the heaven to which Guthlac journeys. The image of Satan and his fallen angels as imprisoned in the air of this world, distinct from the higher realms of ether and the heavenly realms, stemmed primarily from depictions in the New Testament, though it was enriched with late antique classical and Christian philosophy.[63] Peter Dendle provides a comprehensive list of classical, late antique, and patristic sources where this motif appears,[64] but for Felix the most likely

60 *VSG*, pp. 102–03.

61 *Genesis A*, ll. 955–56: 'Ac he him to frofre let hwæðere forð wesan / hyrstedne hrof halgum tunglum' ('But for their comfort He [God] nevertheless let the sky continue, the roof adorned with holy stars'). Doane, *Genesis A*, also comments, with reference to Augustine's *De Civitate Dei* XXII.22–24, that the poet is here making the theological point that 'goodness and beauty are inherent in the creation', with the 'hard realities resulting from' the corruption of the Fall, p. 310.

62 *VSG*, p. 102.

63 See Dendle, *Satan Unbound*, p. 71. See also Discenza, *Inhabited Spaces*, pp. 23–30 for an in-depth discussion of Anglo-Saxon perceptions of the air, the various heavens, and the heavenly bodies.

64 See Dendle, *Satan Unbound*, p. 149. He also highlights Augustine, *De Civitate Dei*, ed. B. Dombard and A. Kalb, 2 vols, CCSL 47–48 (Turnhout, 1955), p. 239: 'qui in hoc quidem aere habitant, quia de caeli superioris sublimitate

sources would be Scripture itself, the *Vita Antonii*, and the relevant works of Augustine. Scripturally this notion is based on two verses in Ephesians, the first of which, 6:12, includes wicked spirits in the sky as part of the powers with which the Christian must wrestle: 'contra spiritalia nequitiae in caelestibus' ('against the spirits of wickedness in the high places'). The term *caelestis*, though it can be used to refer to heaven or the heavens,[65] was understood exegetically here to mean the air of the world, sublunary and part of fallen Creation. This interpretation was firmly established by the time of Athanasius, as Antony quotes Eph. 6:12 in Ch. 21 of the *Vita Antonii*, followed immediately by an exegesis which envisages a scene very much like the one that Guthlac encounters: 'Ingens eorum turba per istum uolitat aerem' ('Vast is their host flying in the air around us').[66] The second verse, Eph. 2:2, joins these wicked spirits to Satan, whose title declares his power over the air: 'principem potestatis aeris huius' ('the prince of the power of this air'). The lexis here is less problematic, with *aer* referring to the lower air of this world.[67] The *Vita Antonii* also utilises the image of sublunary demons, including the noun *aer*, to grant authority to Antony's experience of being spiritually lifted upwards while being hindered in the heavenly journey by demons in the air: '[p]rohibentibus transitum *aeris* daemonibus' ('prohibiting the passage through the *air* by demons').[68] Antony remembers and quotes the verse to explain their presence. Augustine discusses this sublunary boundary in his homily on Ps. 148:9, and concludes that the devil was bound to this air as if within a prison: 'Propterea ad ista caliginosa, id est ad hunc *aerem*, tamquam ad *carcerem*, damnatus est diabolus' ('Therefore, the devil is damned to this darkness, that is, to this *air*, just like a

deiecti merito inregressibilis transgressionis in hoc sibi congruo uelut carcere praedamnati sunt' ('They inhabit that air, because they were thrown down from the heights of heaven for their irredeemable transgression, and condemned to that place as a suitable prison'), p. 149.

65 *DMLBS*, 4: 'celestial, heavenly, belonging to (Christian) heaven; 5: a. heavenly things; b. heavenly regions'.

66 *Vita Antonii*, Ch. 21, pp. 167–68.

67 *DMLBS*, 1: 'air, atmosphere, weather'; see Eph. 2:2.

68 *Vita Antonii*, Ch. 65, p. 181. See Dendle, *Satan Unbound*, p. 149, for a discussion of demons obstructing souls' ascent to heaven in the air.

prison').[69] Thus the gathering of the demons in the air of the physical world of Crowland is both theologically supported and part of Felix's enargaeic detail.

Felix most clearly unites the eventual restoration of Creation and the progression of Guthlac towards spiritual majority within the English landscape through the demons' subsequent immersion of him within it: 'adductum in atrae paludis coenosis laticibus inmerserunt' ('they plunged him into the muddy waters of the black marsh').[70] Felix's use of the phrase 'atrae paludis' connects this scene with his initial description of Crowland in Ch. XXV, where the same descriptive pairing is used ('atrae paludis'), thereby anchoring the trials Guthlac faces in the dark severities of his delineated landscape. This epitome of the desert in England is then given its sharpest definition, within which Felix utilises loaded terminology for the thickets used to tear Guthlac's flesh: 'asportantes illum per paludis asperrima loca inter *densissima veprium vimina* dilaceratis membrorum conpaginibus trahebant' ('they carried him through the wildest parts of the fen, and dragged him through *the dense thickets of brambles*, tearing his limbs and all his body').[71] Not only does the repetition of the phrase 'atrae paludis' in the previous passage invoke all the descriptive detail Felix has laid out, but here Guthlac is taken even further into the desert, through the roughest and harshest part ('asperrimus').[72] Felix's addition of the telling detail of the thickets themselves ('veprium vimina') unites the primary narrative and Guthlac's torturous journey in sanctity with the Augustinian/Bedan exegesis. The image is one that any inhabitant of the fens would have been familiar with, knotted and tangled shoots of bramble growing together in dense structures of stem and thorn, but its use here echoes – and is likely directly influenced by

69 Dendle, *Satan Unbound*, p. 71, quoting Augustine, *Enarrationes in Psalmos*, p. 2171.

70 *VSG*, pp. 102–03.

71 *VSG*, pp. 102–03. Kurtz, 'Antony to Guthlac', p. 113, connects this to Gregory's *Dialogues* where the saint self-flagellates for lustful thoughts by tossing himself into a briar. Other than the depictions including a saint wounded by thorns, however, there is little similarity between the episodes. See Gregory's *Dialogues*, vol. 2, pp. 136–40.

72 *DMLBS*, '1: a. rough, rugged; 2: a. harsh, disagreeable, troublesome'.

– both Augustine and Bede's exegesis concerning the emergence of thorns in Gen. 3:18 discussed earlier.[73] The thorns of these brambles of fallen Creation are being used by God to spur Guthlac towards his spiritual majority. Felix is careful in his depiction of this episode, as the brambles are never inimical towards Guthlac, and only wound him through the agency of the demons who are allowed to do so within the providence of God.

Felix employs these deprivations to portray Guthlac's progression, once again through his responses. Despite being torn and bleeding, Guthlac is unmoved, 'stabilita mente' ('with steadfast mind'), and again responds with the Psalms.[74] This response appears equivalent in kind to the one he gave during the second temptation, yet after further persecutions by the demons, including whipping, Guthlac's firm faith remains: 'illum *inmota mente, robusta fide* in eo quod incoeperat, perstare viderent' ('they saw him persist [with *unmoved mind*] and showing a *robust* [*faith*] in the enterprise he had undertaken').[75] This is a direct developmental parallel with the first temptation; despite continuous physical persecution Guthlac no longer wavers in his eremitic commitment or sees it as impossible work, but instead displays a robust confidence in his calling. Guthlac still has not responded directly to the demons in any active way, but his now unwavering internal response leads almost inevitably, and here grammatically,[76] into the second section of this central battle: the taking of Guthlac to the airy regions and then to the gates of hell.

In the second part of this episode, for the first and only time, Felix describes a landscape other than his enargaeic fenland. In Felix's most monochromatic scene he describes the airy regions, which, as discussed above, form not only the realm for birds to fly in but also part of the sublunary prison for the outcast demons, as an almost tactile and perceptible anti-light, a region where darkness has taken form: 'nubifera gelidi aeris spatia [...] ardua aeris

[73] See above, Introduction, pp. 9–15.

[74] *VSG*, pp. 102–03.

[75] *VSG*, pp. 104–05. Emendations for clarity and consistency.

[76] The sentence begins 'Cum autem', and follows directly into the demons taking him into the freezing skies, *VSG*, p. 104.

culmina [...] septentrionalis caeli plaga fuscis atrarum nubium caliginibus nigrescere videbatur' ('cloudy stretches of the freezing skies [...] the lofty summit of the sky [...] the northern heavens seemed to grow dark with gloomy mists and black clouds').[77] The connection between hellish imagery and the cardinal direction of north here is highly conventional, but once again Felix alters it to create a unique scene.[78] The airy region with its dual potential as a realm both heavenly and infernal is said to darken with an overwhelming collection of synonyms for darkness (*nigrescere, atrarum nubium*, and *fuscis calignibus*). Lexically Felix here recalls not only the dark clouds ('fuscis nubibus') earlier in this temptation, but also the consistent use of terms for cloud, fog, and darkness with which he has constructed the physical fenland landscape. While Felix makes clear that this physical darkness is really a swirling horde of foul spirits,[79] the potency of the image relies on its connection to the shadowed, clouded, and dark landscape of Crowland.

This terrestrial air-scape is meant to conjure terror in both Guthlac and Felix's audience, and it is from this that Felix sets up the ensuing contrast, where Guthlac's earthly temptations and persecutions are compared with hell itself.[80] This distinction is founded on the comparison between the enargaeic landscape of the fens, and the chaotic conflation of metaphor and imagination with which Felix depicts hell. Felix highlights this further through Guthlac's initial reaction to hell, which banishes all thought of the previous temptations.[81] Guthlac's physically-based earthly torments are swept away by the mere sight of hell; thus the persecutions ahead are of greater magnitude altogether. Felix's use of the term *tartarus* for hell here,[82] though a common enough appellative,[83] hints at how Felix constructs the landscape of hell. It is a landscape of the mind

77 *VSG*, pp. 104–05.

78 C. D. Wright, *The Irish Tradition in Old English Literature* (Cambridge, 1993), p. 129: 'The location of hell in the north is commonplace. In classical tradition Oceanus is commonly located in the West, but already in I Enoch the confluence of rivers appears to be located in the north'.

79 *VSG*, p. 104.

80 *VSG*, p. 104.

81 *VSG*, p. 104.

82 *VSG*, p. 104.

83 For example, Bede, *HE*, V.14, p. 504, uses *Tartarus* twice for hell.

rather than the feet, constructed not as an enargaeic depiction of a physical place, but instead as a literary confection from a combination of patristic and hagiographical conventions, as well as the lexis of the classical world:

> Non solum enim fluctuantium flammarum ignivomos gurgites illic turgescere cerneres immo etiam sulphurei glaciali grandine mixti vortices, globosis sparginibus sidera paene tangentes videbantur; maligni ergo spiritus inter favillantium voraginum atras cavernas discurrentes, miserabili fatu animas impiorum diversis cruciatuum generibus torquebant.[84]
>
> (For not only could one see there the fiery abyss swelling with surging flames, but even the sulphurous eddies of flame mixed with icy hail seemed almost to touch the stars with drops of spray; and evil spirits running about amid the black caverns and gloomy abysses tortured the souls of the wicked, victims of a wretched fate, with various kinds of torments.)

The landscape Felix creates is fundamentally different from, though intimately connected to, the terrestrial fens and Crowland. While the English desert is constructed as one would walk through it, with consistent physical detail, colour, scent, tactile features, and boundaries, hell is constructed from an amalgamation of fire, ice, and absence. This warring of paradoxical elements is reflected in the more widely received and contemporary traditions of hell,[85] including the famous description in *Genesis B*:

> þær hæbbað heo on æfyn   ungemet lange,
> ealra feonda gehwilc   fyr edneowe,
> þonne cymð on uhtan   easterne wind,
> forst fyrnum cald,   symble fyr oððe gar.
> sum heard geswinc   habban sceoldon.[86]
>
> (ll. 313–17)

84 *VSG*, pp. 104–05.

85 See C. A. Jones, 'Early Medieval Chaos', in *Verbal Encounters: Anglo-Saxon and Old Norse Studies for Roberta Frank*, ed. A. Harbus and R. Poole (Toronto, 2005), pp. 15–38, at 24, where he highlights the use of chaotic elements in Lucan's *Pharsalia*, II.72–82, at 74, 5.625–636, at 634; Caelius Sedulius' *Carmen Paschale*, II.121–22; Gregory of Tours *Libri miraculorum* I.17; and *Genesis B*, ll. 320–25.

86 *The Saxon Genesis: an Edition of the West Saxon Genesis B and the Old Saxon Vatican Genesis*, ed. A. N. Doane (Madison, WI, 1991). All quotations of the text are from this edition, hereafter referred to as *Genesis B*.

> (There each and every kind of fiend has an eternally burning flame, in a boundlessly long evening. Then at dawn comes the east wind, fiercely cold with frost. Fire or spear-frost, they must always endure hard affliction.)

Felix's use of these elements, however, is distinct in the emphasis on the boundless and chaotic nature of hell. The image of the fiery abyss surging with flames is potent, and yet unstable, its evocative qualities almost lost to its boundless chaos. Felix's description includes neither spatial clue nor width to give the image visual purchase; there is no indication of spatial relationship between the various elements, where the whirlpool lies in respect to the gates of hell or the caverns mentioned earlier. The description instead focuses on the chaotic, like the immensity of the paradoxically linked fire and ice which almost touch the stars. Thacker has suggested that Felix's depiction echoes not only the *Visio Fursei*,[87] but also Bede's account of Dryhthelm's vision of hell in the *Historia*.[88] Dryhthelm's vision, however, provides a helpful contrast, as it displays a similar but ultimately more spatially defined vision of hell. In Ch. 12 of the *Historia*, Dryhthelm describes being led to a deep valley which is given a fixed position: 'leuam nobis sita' ('it lay on our left').[89] Already the vision presents a more descriptively bounded landscape. For example, it locates the alternating torments of hot and cold on either side of the valley: '*unum latus* flammis [...] *alterum* [...] frigore niuium' (*one side* with fire [...] while the *other* [...] icy snow').[90] While the vision's glimpse of hell resembles Felix's emphasis on darkness, it retains a greater sense of delineated spatial relationship. Most often these relationships are defined with respect to Dryhthelm himself, as in the case of the single pit with flames rising up and falling back into it, which appears before him. Likewise the din of lamentation

[87] Thacker, 'Social and Continental Background', p. 306, though the correspondences he discusses are between Felix's depiction of the sky and the gathering black clouds in the northern heavens, how the '*nigra nubis* of Fursey's hell resembles the *nubifera* [...] *spatia* which seemed to Guthlac *fuscis atrarum nubium caliginibus nigrescere*'.

[88] Thacker, 'Social and Continental Background', p. 306.

[89] Bede, *HE*, V.12, pp. 488–89.

[90] Bede, *HE*, V.12, pp. 488–91. Colgrave also notes the widespread nature of this paradoxical combination of extreme hot and cold, p. 490.

is located behind Dryhthelm, 'audio subitum *post terga* sonitum inmanissimi fletus ac miserrimi' ('I suddenly heard *behind my back* the sound of wild and desperate lamentation'), [91] once again providing a greater sense of spatial relationship. Even in the heart of this vision of hell there remains, therefore, some sense of spatial relationship, with Dryhthelm as the anchor-point: the single pit is in front of him and the lamenting mob is behind him. In Felix's *VSG*, there is instead formless turmoil.

Felix furthers the chaotic depiction of hell with the evil spirits that are tormenting the wicked souls. While more in keeping with depictions like that in the *Historia*,[92] his lexis once again serves to present hell as negation and chaos, with the demons running about the black caverns and the gloomy abysses.[93] The features Felix describes are not distinct, as they are in essence the physical presence of absence; the adjective *ater*, for example, adds colour and weight to the negation of space that a cavern is, just as the noun *vorago* ('pit', or more commonly, 'abyss') focuses on absence.[94] These lexical choices add to the depiction of hell as chaos, a place without even the distinctive ordering of spatial relationships and solidity of form.

Felix also adapts the conventional 'waves-touching-the-sky' motif in his depiction of hell. Charles Wright demonstrates that this image was common currency in Anglo-Saxon literature by highlighting its use in both *Beowulf* and the *Visio Pauli*.[95] The poetic image is also used in various apocryphal texts, and thus has a wide range of conventional use.[96] Felix adapts this motif – to the spray from flames

91 Bede, *HE*, V.12, pp. 490–91.

92 See Bede, *HE*, V.12, p. 492, in which Dryhthelm describes being surrounded by demons threatening him.

93 See P. E. Szarmach, 'The Vercelli Prose and Anglo-Saxon Literary History', in *New Readings in the Vercelli Book*, ed. A. Orchard and S. Zacher (Toronto, 2009), pp. 12–40, at 22–23, for discussion of Felix's dense collections of neologisms in this section, including *favillantium*, as well as their Aldhelmian influence.

94 *DMLBS*, 1: a. 'chasm, abyss, gulf; b. whirlpool, eddy'; *Oxford Latin Dictionary*, ed. P. G. W. Glare (Oxford, 2012): 'a deep hole or chasm', p. 2103.

95 Wright, *The Irish Tradition*, p. 135. The image appears in the description of Grendel's *mere*, *Beowulf*, ll. 1373–74a; and the *Visio Pauli*, which Wright notes from redaction VIII of the *Visio*.

96 Wright, *The Irish Tradition*, p. 135: 'The image appears to have had a wider currency in apocalyptic and visionary texts. The Latin Seven Heavens apocryphon

meeting icy hail – as one more portion of his chaotic landscape of hell, and contrasts it purposefully with the use of the noun *siderus* ('star, heaven, or constellation'),[97] the ordered heavens with their luminous constellations and perfect movement. The adaptation, therefore, is in keeping not only with homiletic, poetic, apocryphal, and exegetical convention, but is also in line with Felix's strictly delineated Creation. This is a landscape without eschatological hope, truly inimical and possessing singular potential. It is therefore implicitly contrasted with the terrestrial English wilderness, part of God's initial Creation and capable of being restored to its prelapsarian state, however distorted through humanity's fall.

There is, however, a second traditional strand that Felix interweaves into his construction of hell: the widely attested 'mouth of hell' motif.[98] Felix combines the motif with what amounts to a list of classical, and occasionally patristic, appellatives for hell: 'en tibi patulis hiatibus igniflua *Herebi* hostia patescunt; nunc *Stigiae* fibrae te vorare malunt, tibi quoque aestivi *Acherontis* voragines horrendis faucibus hiscunt' ('The fiery entrances of *Erebus* gape for you with yawning mouths. Behold! the bowels of *Styx* long to devour you and the hot [abysses] of *Acheron* open with dreadful jaws.').[99] While the first appellative for hell, *Erebus*, might seem an apt choice because of its connection to the idea of primordial darkness and chaotic disorder, given what follows it is more likely that Felix is invoking it metonymically along with the mouth of hell motif.[100] While the infernal fires are mentioned, Felix adds the gaping mouths ('patulis hiatibus') of hell ready to devour the wicked soul, even as the demons claimed the fires would consume Guthlac a minute

---

in the *Apocrypha Priscillianistica* describes a river in similar terms: "fluctus eius exaltatur usque ad quintum celum".'

97 See Isidore, *Etymologies*, III.60, in which he differentiates between *stellae, sidera,* and *astra,* with *sidera* defined distinctly as constellation. Augustine also notes the connection between the ordered constellations and the divine ordering of the cosmos as a whole in *De Doctrina Christiana,* ed. K. D. Daur and J. Martin, CCSL 32 (Turnhout, 1962; repr. 1996), pp. 55–56, as well as in *De Genesi ad Litteram*, pp. 39–42.

98 See G. D. Schmidt, *The Iconography of the Mouth of Hell: Eighth-Century Britain to the Fifteenth Century* (Selinsgrove, PA, 1995), pp. 33–83.

99 *VSG*, pp. 106–07. The emendation to 'abysses' is explained further below.

100 Virgil uses the term Erebus metonymically for hell five times in the *Aeneid*: IV, l. 26; VI, ll. 247, 404, and 671; and VII, l. 140.

earlier. Felix's elision here of the pit/cavern image with the almost bestial disembodied mouth furthers his depiction of the landscape of hell as nearly formless chaos. The mouth imagery is heightened by the bowels of Styx that long to consume Guthlac, here again uniting the metonymic usage of Styx as the Christian hell with the idea of devouring. The metonymic chaos is expanded by the depiction of the demons' declaration that the hot gulfs of Acheron yawn with dreadful jaws. Colgrave tries to anchor this description by translating *vorago* as 'gulfs', whose definitive form includes an inlet or opening, but given the metonymic use of Acheron here and the consistent repetition of the image of formless vacuity, it more likely means 'abyss' or 'chasm'. The gaping abyss here is elided with the mouth of hell motif, and yet Felix is not departing too far from his lexical sources, as Virgil himself hints at a similar construction: 'horrendis faucibus' echoes Book VI of the *Aeneid*: 'ipsum [...] faucibus Orci' ('the very jaws of Orcus').[101] This Virgilian echo is, I would argue, a descriptive convenience for Felix, rather than a meaningful allusion.[102]

The sudden burst of classical references here and the overflowing of metonymic hell terms conflated with the mouth of hell motif are striking for a number of reasons, one of which is the apparent incongruity with the rest of the *VSG*. While Felix often employs poetic phrases of two to three words from Virgil, this concentration of classical names is unparalleled throughout the rest of the text. Its aberrant nature was apparent to the scribe of Colgrave's posited original manuscript, which he calls X^1, as there is a marginal notation to aid in the interpretation of this passage in eight of the thirteen extant manuscripts:[103]

101 Virgil, *Aeneid*, VI, l. 273.

102 For an in-depth discussion see my article, 'Felix's Construction of the English Fenlands'.

103 *VSG*, p. 105, where his note describes the following manuscripts: C1, AHDE2G, C2, and N (though he confuses this with the two CCCC ones, as there is no visible notation in C2, which I assume is the binder's cut he refers to, whereas the notation is complete in C1).

Tria sunt flumina inferni, Acheros quod sine gaudio interpretatur ex quo nascitur secundum Stix, id est tristitia, rursus ex ista oritur Cocitus quod luctus interpretatur.[104]

(There are three rivers of hell, Acheron, which is interpreted as without joy, from which springs the second, Styx, which is sorrow, again from this source is Cocitus, which is interpreted as grief.)

The notation is incongruous in many ways, primarily in its focus on the allegorical function of the rivers, rather than their use as descriptive pieces of the hellish landscape. It is unlikely that Felix was the author of the marginal notation, as the interpretation is discordant with the descriptive method of the *VSG*. The Styx and the Acheron are not described as rivers in the text, and Felix uses them as part of the landscape of hell and torment, to further the imagery of fire, teeth, and chaos.

Guthlac takes the last step in his progression towards spiritual majority here, when the description of the hellish landscape is at its most extreme: '*inmotis sensibus, stabili animo,* sobria mente, respondens aiebat' ('with *unshaken nerves,* with *steadfast heart* and sober mind he answered them').[105] Guthlac, for the first time, directly engages his demonic tempters. Felix's echo of the previous temptations and Guthlac's responses are clear, 'inmotis sensibus' echoing 'sensibus turbatus' of the first temptation, as well as 'stabili animo' echoing 'turbulentum animum'. Guthlac has resisted even the sight of hell itself, a victory which, akin to Cuthbert's, highlights the saint's triumph over Adam's failure. The appearance of St Bartholomew thus represents a clear obverse of the shadowed formless hell:

inmenso caelestis lucis splendore medias furvae noctis infuso lumine interrumpens tenebras, sese ab aethereis sedibus radiantis Olimpi coram illis aureo fulgore amictus obtulit.[106]

([I]n boundless splendour of heavenly light [he] broke into the midst of the swarthy darkness of night with outpoured radiance, and

104 *VSG*, p. 105, n. 40 to *sulpheros*; though also on p. 106, n. 14 for *Acherontis*. Translation my own as none provided by Colgrave.

105 *VSG*, pp. 106–07.

106 *VSG*, pp. 106–07.

presented himself before them girt with golden brilliance from the heavenly dwellings of glorious Olympus.)

The intense illumination, 'inmenso splendore', 'caelestis lucis', 'infuso lumine', 'radiantis Olimpi', and 'aureo', is in direct contrast to the landscape of hell, not only physically, but also theologically in that darkness and evil are non-things, whereas light is the purest and most solid of things, the most ordered.[107] The demons, and the landscape of hell, have been defeated, shown to be powerless before the efficacy of heaven. The eventual transformation of Creation back into its prelapsarian relationship with the saint is prefigured here as the air, which before had been flooded with demons and shadow, is filled with angelic singing: 'Cum vero ad medii aeris spatia devenissent, sonus psallentium convenienter audiebatur dicens: Ibunt sancti de virtute in virtutem' ('And indeed when they had reached the spaces of mid-air the sound of voices was heard singing in unison and saying: "The saints shall go from strength to strength"').[108] This moment, with its invocation of Ps. 83:8, shows how the landscape of the air, full of dual potential, is restored to its proper state, the saint directly conquering the immeasurable shrieking that filled the skies at the beginning of this episode. For the remainder of the text the physical landscape of the fens and Crowland are no longer simply indifferent to Guthlac, as he has achieved spiritual majority, and the landscape and all the elements are now ready to be restored into prelapsarian relationship.

Creation and Spiritual Majority

Ch. XXXVI of the *VSG* presents Guthlac's final direct confrontation with the demons, and one of only two after he has attained his spiritual majority. The bestial horde episode has also been considered an example of Guthlac's confrontation with inimical

[107] See Augustine, *De Genesi ad litteram*, pp. 472–73, and *De Genesi contra Manichaeos*, pp. 72–73.

[108] *VSG*, pp. 108–09.

Creation,[109] but it is better understood as part of the episodes which serve to showcase Guthlac's post-spiritual majority. While this includes the more conventional moments of prophetic power,[110] there are two specific instances of what I will call Guthlac's blessed perception, which serve to highlight a unique facet of Felix's portrayal of his saint. The two episodes in question are the appearance of the British host in Ch. XXXIV, and the bestial horde here in XXXVI.[111] Estes likewise considers these chapters linked but concludes, in connection to her wider ecocritical approach, that the bestial horde episode depicts the physical animate world in opposition to Guthlac:

> The intertwining of human, monstrous and animal in depictions of the demonic, all opposed to Guthlac's privileged position in gender, class, and ethnic hierarchies, functions to establish a small class of people – 'englisc,' Christian, male, upper-class – in opposition to all others, with power over land, animals, and other people analogously legitimized through the idea of devotion to God as the highest of values.[112]

The primary purpose of both episodes is, however, to depict Guthlac's novel ability to see through the empty threats of the enemy, as shown by their attempts to threaten him with various guises; in the first, the flesh of Britons is worn, and in the second, the flesh of animals. While the animals here are based upon factual detail, they are no more real than the band of Britons, and therefore are not in any way to be interpreted as part of postlapsarian animate Creation. This distinction is important: if they were real animals,

109 Neville, *Representations*, pp. 126–27, considers this episode, in relation to the later iterations of the Guthlac legend, to be an example of the connection between the inherent hostility of the natural world and the expected presence of demons in the wilderness.

110 See for example *VSG*, Chs. XXXV, XL, XLII, XLIV, and XLVI; Kurtz, 'Antony to Guthlac', p. 116, notes that prophecy is one of the necessary hallmarks of sanctity.

111 See Kurtz, 'Antony to Guthlac', p. 100, where he discusses the connection of these chapters, calling them part of the tradition of demonic intimidation. He includes them along with Chs. VIII, IX, XXIII, XXIV, XXXVI, XXXIX, and XL of the *Vita Guthlaci* as all taking part in the tradition.

112 Estes, *Anglo-Saxon Literary Landscapes*, p. 111.

their aggression towards the saint in his spiritual majority would be highly out of place, especially considering the bird miracles that follow.

The bestial horde episode is framed by Felix as evidence of Guthlac's blessed perception, by which he is able to see through devilish seeming, and by which Felix again highlights Guthlac's success in contrast to Adam's failure, who did not see through the devil's disguise in Eden:[113]

> Verum quia superius, quantum isdem vir venerabilis Guthlac adversus veras apertasque diabolicas insidias valuit, explicavimus, nunc quoque, quid adversus simulaticias malignorum spirituum fraudes praevaluit, exponemus.[114]

> (Since we have explained above how strong this same venerable man Guthlac was against those snares of the devil which were real and open, we will now also show how he prevailed against the feigned deceits of the evil spirits.)

The phrase 'adversus veras apertasque' clearly distinguishes the nature of this devilish attack from the highly physical deprivations which led up to Guthlac's attainment of spiritual majority; what follows, therefore, cannot be interpreted in the same manner. Guthlac is said to prevail against their feigned deceits ('simulaticias [...] fraudes'), and thus the bestial horde can be aligned in both lexis and form with the devils disguised as a marauding band of Britons. In both, Guthlac is confronted with an apparently inimical host while in prayer at night, 'quadam nocte [...] orationum', and 'quadam nocte [...] orationibus' (a certain night [...] in prayers'),[115] who move to attack him, and yet are revealed through Guthlac's perception to be merely devilish seeming: 'hostis pellacis millenis artibus millenas formas persentiens' ('perceiving the thousand-fold forms of the insidious foe and his thousand-fold tricks') in the British host episode, and explicitly 'fantasma' and 'fantasmatum'

113 See Gen. 3:1; see also Rev. 12:9 and 20:2.
114 *VSG*, pp. 114–15.
115 *VSG*, pp. 108–9, 114–15.

('apparition')[116] in the bestial horde episode. Guthlac combats the British host with a complex amalgamation of Ps. 67: 2–3 and the model of Chs. 13, 23, and 39 of the *Vita Antonii*. The form of the episode is taken primarily from Ch. 13 of the *Vita Antonii*, where the saint is often heard combating the demons by singing the same psalm as Guthlac, Ps. 67: 2–3.[117] Felix adapts the Antonian material by having Guthlac quote the first verse, and then enacts verse 3 in the actual narrative: 'velut fumus a facie eius evanuerunt' ('[they] vanished like smoke from his presence').[118] The idea of a warlike host could also have stemmed from the *Vita Antonii* Chs. 23 and 39, which describe how the demons often come as military bands, for example, in Ch. 39:

> Quoties minitantes *ut milites armati*, scorpionibus, equis, belluis, et uariis serpentibus circumdederunt me, et domum in qua eram repleuerunt![119]
>
> (Often they threatening *like soldiers in full armour* surround me as scorpions, horses, wild beasts, and various serpents, and fill up the house in which I was.)

The connection between illusory military bands and bestial hordes has clear precedent in the *Vita Antonii*.

A similar pattern occurs in Ch. XXXVI with the bestial horde. The episode itself is an adaptation of Ch. 9 of the *Vita Antonii*, with a number of important alterations and expansions, most notably the addition of native enargaeic elements to the list of the demons' bestial guises, and Guthlac's response in his victory over them:

116 *VSG*, pp. 110–11, 114–15.

117 *Vita Antonii*, Ch. 13, p. 165. See also S. Downey, 'Intertextuality in the Lives of St. Guthlac', Unpub. Ph.D. thesis (University of Toronto, 2004), pp. 38–39.

118 *VSG*, pp. 110–11. Cf. *Vita Antonii*, Ch. 13, p. 165. Cf. Ps. 67: 2–3: 'Exsurgat Deus et dissipentur inimici eius et fugiant qui oderunt eum a facie eius sicut deficit fumus deficiant sicut fluit cera a facie ignis sic pereant peccatores a facie Dei' ('Let God arise, and let his enemies be scattered: and let them that hate him flee before his face. As smoke vanishes, so let them vanish away: as wax melts before the fire, so let the wicked perish at the presence of God').

119 *Vita Antonii*, Ch. 39, p. 173; see also Ch. 23, p. 168.

Antony:
Sonitus igitur repentinus increpuit, ita ut loco funditus agitato, et parietibus patefactis, multifaria daemonum exinde turba se effunderet. Nam et bestiarum et serpentium formas induentes, omnem protinus locum repleuere phantasiis leonum, taurorum, luporum, aspidum, serpentium, scorpionum, necnon et pardorum atque ursorum. Et haec singula secundum suam fremebant naturam: *rugiebat leo accedere uolens, taurus mugitu et cornibus minabatur*, serpens sibilo personabat, luporum impetus inhiabantur, pardus discolori tergo auctoris sui calliditates uarias indicabat. Truces omnium uultus et uocis horridae dirus auditus.[120]

(Then suddenly sound thundered, shaking the place from its foundation, and manifold demons then poured in through the open walls. They put on the form of beasts and serpents, and filled that place with phantoms of lions, bulls, wolves, bulls, asps, serpents, scorpions, leopards, and bears; each of them raged according to its nature: the lion was roaring, wanting to attack, the bull bellowing and thrusting its horn, the serpent hissing loudly, the wolf eager to attack, and the dappled leopard revealed their various crafts. Altogether the savage forms and horrible voices was a fearful sound.)

Guthlac:
ingenti sonitu totam insulam qua sederet tremere circum putabat; deinde, parvi temporis intervallo succedente, ecce subito velut concurrentium armentorum crepitum cum magno terrae tremore domui succedere exaudiebat. Nec mora, domum ab undique inrumpentes variorum monstrorum diversas figuras introire prospicit. Nam *leo rugiens dentibus sanguineis morsus rabidos inminebat; taurus vero mugitans, unguibus terram defodiens, cornu cruentum solo defigebat;* ursus denique infrendens, validis ictibus brachia commutans, verbera promittebat; coluber quoque, squamea colla porrigens, indicia atri veneni monstrabat, et ut brevi sermone concludam, aper grunnitum, lupus ululatum, equus hinnitum, cervus axatum, serpens sibilum, bos balatum, corvus crocitum ad turbandum veri Dei verum militem horrisonis vocibus stridebant.[121]

(He thought that the whole island in which he was dwelling trembled all around with a tremendous clamour. Then after a short interval, lo! suddenly he heard the noise as of a herd of beasts rushing together and approaching his dwelling with a mighty shaking of the earth. Straightway he saw manifold shapes of various monsters bursting into his house from all sides. Thus a roaring lion fiercely

[120] *Vita Antonii*, Ch. 9, p. 164. Italics used here to show shared text.
[121] *VSG*, pp. 114–15. Italics used here to show borrowed text.

threatened to tear him with its bloody teeth: then a bellowing bull dug up the earth with its hoofs and drove its gory horn into the ground; or a bear, gnashing its teeth and striking violently with either paw alternately, threatened him with blows: a serpent too, rearing its scaly neck, disclosed the threat of its black poison: to conclude briefly – the grunting of the boar, the howling of the wolf, the whinnying of the horse, the belling of the stag, the hissing of the serpent, the lowing of the ox, the croaking of the raven, made harsh and horrible noises to trouble the true soldier of the true God.)

It is immediately apparent that Felix's adaptation expands the depiction with the addition of what Kurtz calls 'sensational'[122] elements. These elements are not, however, purely narrative padding, but function in a similar manner to Felix's enargaeic landscape. They are added to anchor the *VSG* in the physical world, providing relatable ground upon which the saint's miraculous powers can be displayed. Felix transforms what could be a simple and conventional display of saintly power into an easily imaginable, and localised English, narrative. Firstly, Felix expands the actions of the phantasmal beasts with descriptive details: where the lion in Antony merely roars and wants to attack, the lion in Felix not only roars but also threatens Guthlac with its bloody teeth.[123] Similarly, the bull does not just bellow and threaten with its horns but instead is depicted by Felix as digging the ground with its hoofs and driving its horns into the earth.[124] Felix likewise adds descriptive detail to the bear and the serpent.

This expansion creates a scene with enargaeic weight; it is unlikely that Felix's audience, or Felix himself, would ever have seen a lion,

[122] Kurtz, 'Antony to Guthlac', p. 115.
[123] *VSG*, p. 114.
[124] *VSG*, p. 114.

or in all likelihood a bear in England, and the details grant experiential physicality to the scene.[125] Felix anchors the scene further in his adapted summary of the various animal forms and their distinct aural signatures.[126] As can be seen above, Felix has removed the mention of non-native animals from the list – lions, leopards, scorpions, and asps – while retaining the native and familiar wolves, serpents, and oxen. In addition to this he has added the horse, stag, boar, and raven. The application of this English list as the auditory complement to the expanded visual description is striking; Felix has effectively created a cacophony from the familiar, a moment his audience could construct from their everyday lives.

It should be noted, however, that even here Felix's lexical choices are taken from traditional sources. Four of Felix's sonic terms are likely drawn directly from Aldhelm's list of animal noises in his *De Pedum Regulis*: *ululatus*, *hinnire*, *sibilus*, and *crocitus*.[127] In addition to Aldhelm's list, Felix utilises rarer forms of common words from various sources, including *balatus* from *balare* ('to bleat'), and *grunnitus* from *grunnire* ('[of a pig] to grunt').[128] The hagiographer also includes a single neologism, *axatus* ('belling [cry of a stag]').[129] While Felix's proclivity for lexical variety and word creation are a feature of the *VSG* as a whole, his choices here serve to further

125 N. Sykes and T. P. O'Connor, eds, *Extinctions and Invasions: A Social History of British Fauna* (Oxford, 2010), at p. 100, gives an overview of contemporary evidence, which posits that 'the indigenous bear population [of Britain] had become extinct by the early medieval period', and that later archaeological evidence 'derive[s] from European imports, brought into the country for bear-baiting or to satisfy the growing demand for exotica'. See also D. W. Yalden, *The History of British Mammals* (London, 1999), pp. 113–15. While the lion as an image in literature and material culture is evident throughout the medieval period, the only species of lion in Britain, the cave lion (*Panthera spelaea*), was extinct before the end of the Pleistocene (earliest evidence more than 40,000 years old); see A. J. Stuart and A. M. Lister, 'Extinction Chronology of the Cave Lion *Panthera spelaea*', *Quaternary Science Reviews* 30 (2011), 2329–40, at 2336–38.

126 Note that Felix has inverted the order here and placed the list after the description of specific animals.

127 Aldhelm, *De Pedum Regulis*, in *Aldhelmi Opera*, pp. 179–80.

128 *DMLBS*. Felix likely encountered *balatus* in the works of Virgil: for example, *Aeneid* IX, l. 62. A direct source for *grunnitus* is less certain, but the word is found once in St Augustine's *De Natura et Gratia ad Timasium et Jacobum contra Pelagium*, PL 44, col. 273.

129 *DMLBS*.

the enargaeic dimension of the scene. Each word of the sonic vocabulary employed in this scene to describe his native animals is onomatopoeic, and as such helps further to ground this scene experientially, even when drawing from word lists like Aldhelm's *De Pedum Regulis*. These creatures were ubiquitous for the Anglo-Saxons, and the addition of them to the scene supports Felix's vision of English hagiography, embedding the narrative in the sights and sounds of English life.

The primary function of the episode is not, therefore, to describe animate Creation, nor is it to cast animals, in Estes's terms, as 'demonic others',[130] but is instead meant to display Guthlac's novel ability to see through devilish seeming now that he has achieved spiritual majority. This is supported by Felix's imaginative emphasis on the lion in the episode which, given the late hour of the attack, evokes the image of the devil as a roaring lion from 1 Peter 5:8 (a *Capitula* for Compline), much as the dusky lions image does in the *VCA*.[131] The image would, therefore, connect directly to spiritual assault rather than physical attack. Guthlac's victory over the demons without any internal turmoil, then, speaks primarily to his progression in sanctity. The British host and the bestial horde are evocative and highly physical depictions formed to serve this specific end from within the Antonian tradition. Felix is very clear about this, as after the native list of animals the intention of the demonic din is given explicit interpretation: 'ad *turbandum* veri Dei verum militem horrisonis vocibus stridebant' ('[They] made harsh and horrible noises *to trouble* the true soldier of the true God').[132] The Latin word *turba* and its derivatives, most notably *turbatus*, play an important part in Felix's lexis, as the usage here not only unites the scene with Guthlac's first demonic encounter in the *VSG*, but also connects the demonic attacks in general to the consistent imagery of chaos. For example, Felix uses the term to present Guthlac's initial weaknesses during his first temptation of despair: 'miles Christi totis sensibus *turbatus* [...] *turbulentum* animum' ('every feeling of the soldier of Christ was *disturbed* [...] *troubled* mind'), as well as his

130 Estes, *Anglo-Saxon Literary Landscapes*, p. 111.

131 See above, p. 90 n. 84.

132 *VSG*, pp. 114–15.

freedom from that confusion 'inluminato *turbulenti* pectoris gremio' ('his *troubled* heart was enlightened').[133] As discussed earlier, it is this chaotic state of mind that Guthlac must overcome to attain his spiritual majority. It appears to be a particularly dangerous mental state, as Felix never applies the term to any lesser state of distress; for example, in describing the mental state of Wilfrid in his concern for his second lost glove, Felix uses 'egrota mente' ('sick mind').[134]

Felix's use of the noun *turbatus* in the *VSG* also unites the turbulence of Guthlac's mind and soul with both the chaotic imagery of hell, as discussed earlier, and the tumultuous depictions of life in the fallen world, which can be seen in Felix's summary of the *vita* in Ch. XXVII:

> sic et sanctae memoriae virum Guthlac de tumido aestuantis saeculi gurgite, de obliquis mortalis aevi anfractibus, de atris vergentis mundi faucibus ad perpetuae beatitudinis militiam, ad directi itineris callem, ad veri luminis prospectum perduxit.[135]
>
> (So also He led Guthlac [,] a man of saintly memory [,] from the eddying whirlpool of these turbid times, from the tortuous paths of this mortal age, from the black jaws of this declining world to the struggle for eternal bliss, to the straight path and to the vision of the true light.)

The demonic attacks are intent on destroying the saint's stable mind, seeking to drag him back into his pre-spiritual-majority turbulence. Guthlac's response is untroubled and addressed directly to the phantom horde. His despising ('despeciens') of them, as the Antonian precedent suggests, reveals Satan's failure to be twofold: first, Guthlac is unmoved and untroubled, and second, he mocks Satan's fallen position in the hierarchy of being, as the previously honoured angel has now stooped to clothing himself in the flesh and sounds of beasts, another Edenic echo to the devil's temptation of Adam and Eve as a snake.[136] Thus the end of the bestial horde's attack is a comparison with Guthlac's previous state. While the devilish intention has remained the same throughout

133 *VSG*, pp. 96–97.
134 *VSG*, pp. 126–27.
135 *VSG*, pp. 92–93.
136 *VSG*, pp. 114, 116. See above, p. 200.

the text, Guthlac has progressed to the point where he is now able to combat the evocative terror of the British host's spears and fire, as well as the bestial horde's visual and auditory cacophony, with a firm and stable mind. He is unmoved, and responds directly to his tormentors.

The bestial horde episode, while tied to Felix's general founding of the *vita* in the enargaeic Creation of the fens and the recognisable animals of English life, is therefore distinct from his other descriptions of Creation. The beasts are in truth the cloaks of demons, who are using the very real fear that could be generated by actual human interaction with such beasts in the postlapsarian world for the sake of troubling the soul. They are revealed to be powerless, depicted in a similar manner to the *Vita Antonii* Ch. 28.[137] This episode, therefore, does not relate to Guthlac's interaction with actual Creation, with real animals within a soteriological universe. For now that he has attained spiritual majority, the animate world can be brought into prelapsarian relationship with him.

The Restoration of Creation

Directly following the bestial horde episode, Felix presents a series of bird miracles, three involving thieving ravens, and one involving obedient swallows. Felix once again creates these from common hagiographical material, though most influential for this section are the avian miracles of the various *vitae* of Cuthbert. The placement of these miracles, after Guthlac has achieved spiritual majority, is in keeping with the *VCP* and Antonian precedent, as they are the next step on Guthlac's spiritual progression. Guthlac has achieved his spiritual majority, has conquered and driven away the demonic inhabitants of the fens, and thus is able to move on to the restoration of a representative portion of Creation.

Felix titles Ch. XXXVII 'Qualiter corvus cartulam inter undas stagni dimisit; nec illam aquae laedere valuerunt' ('How a [raven] dropped a document into the midst of the pool and how the waters

[137] *Vita Antonii*, Ch. 28, p. 170.

could not harm it').[138] This seemingly mundane framing provides a useful interpretative key to Felix's iteration of the hagiographical commonplace of the thieving raven. While modelled primarily on the various thieving raven miracles of the three Cuthbert iterations, this initial miracle concerning the stolen document is unique to the *VSG*. The episode focuses upon the actions of both the animal and elemental portions of Creation in relation to Guthlac. The important actions are as follows: the raven's theft of the document and subsequent flight into the fens; Guthlac's comforting of and directions to the unnamed monk; the monk's obedience and faith in following Guthlac's commands; and the combination of raven, wind, reed, and water that allow the document to be found intact. While each of these events is innocuous in itself, what appears when the narrative progression is taken as a whole is a vision of Creation restored, where the relationship between God, humanity, and Creation is as it should be. The initial action performed by the raven, the theft of the newly finished document, grounds the episode in physical Creation via its mirroring of actual corvine behaviour, as crows and ravens are often 'curious about objects to which people have recently paid attention, and [...] often pick up such items'.[139] This mundane theft is given no interpretative framing, and is meant to represent a common occurrence; in keeping with Felix's construction of a specifically English hagiography, the familiar enargaeic aspect is heightened here to focus the narrative on the physical world of eighth-century England.

The depiction of Guthlac's restoration of Creation comes about initially through Felix's careful lexical choice of the term *fors* ('chance, fate, providence'),[140] which he uses twice: first, when the

138 *VSG*, pp. 116–17. The manuscript evidence indicates the titles to be authorial, see *VSG*, pp. 26–44, 93 n. 39. Colgrave, for reasons of accurate ornithology, translates *corvi* as 'jackdaws'; see *VSG*, p. 187 for his explanation; I have emended to 'raven' as the hagiographical account is clearly uniting these *corvi* to the ravens of previous exemplars (esp. the *vitae* of Cuthbert).

139 R. Waugh, 'The Blindness Curse and Nonmiracles in the Old English Prose Life of Saint Guthlac', *Modern Philology* 106 (2009), 399–426, at 409. Waugh also notes how the potential sheen of the just finished manuscript page would have added to the raven's potential interest: 'a piece of fresh writing might be even more attractive, since the ink would still be wet and shining', p. 409.

140 *Fors*: 'Chance, fortune, fate; (abl.) by chance, perhaps', *DMLBS*.

unknown monk who had written the document turns, by chance, to see the raven carrying it off: 'cum visus suos *forte* foris divertisset, volantem alitem cartulam in ore suo portantem prospicit' ('looking *by chance* out of doors, [he] saw a bird flying off and carrying a document in its beak'); second, Guthlac leaves the oratory at the right time to see the bird as well: '*Forte* denique eadem hora sanctus Guthlac extra oratorium egrediebatur' ('It happened *by chance* that at the same time St Guthlac was leaving the oratory').[141] Felix highlights the fortuitous timing of the episode, how these two apparently accidental actions come together to create the miracle. This *chance* is immediately redefined as God's providence by Guthlac's consolation of the monk:

> consolari illum coepit, pollicens eit cartulam suam cum Dei omnipotentis auxilio sibi recuperari posse, *sine cuius potestate nec folium arboris defluit nec unus passerum ad terram cadit.*[142]
>
> ([He] began to console him, promising him that his document could be recovered with the help of Almighty God, *without whose power not a leaf of a tree is blown down and no single sparrow falls to the earth.*)

The italicised text here is from Jerome's *Vita Pauli*, itself taken from Mtt. 10:29, which Felix adapts with the change of the subtle word *nutus* ('nod, slight movement', as well as 'power, authority'), for the more direct *potestas* ('possession of control or command, power or authority').[143] The episode in the *Vita Pauli* sheds some light on this application: in it two wild lions appear from the desert to help bury St Paul. In the following description Antony's response highlights what Felix subsumes into his miracle, namely the transformational power of a saint to restore Creation, revealing how the lions recognise Christ's divinity in Antony: 'quod muta quoque animalia Deum esse sentirent' ('even the mute animals perceived God').[144] The lions are praised for their perception of God's divinity within the saintly men, demonstrated by their restored behaviour.

141 *VSG*, pp. 116–17.

142 *VSG*, pp. 116–17. Italics used to show source text.

143 *DMLBS*.

144 Jerome, *Vita Pauli*, p. 178. For a discussion of the relevance of the doctrine of theosis in early Anglo-Saxon England, see Leneghan, '*The Wanderer, Hesychasm* and *Theosis*'.

The raven-and-document episode further depicts this restoration by following the *VCA*'s focus on obedience within a delineated hierarchy, in both Guthlac's own faithful obedience and the obedience of the unnamed monk to Guthlac's *praecipio*.[145] Once again Felix first fixes the narrative to the physical English landscape with a description of the location to which the bird flies: 'Inter haec alitem longe in *austrum* volantem cernebant, cursumque suum inter *stagnosa paludis ligustra* deflectens' ('Meanwhile they saw the bird flying far to the *south*; turning in amidst the *growth in the marshy pools*').[146] Felix reaffirms Guthlac's spiritual progression in his response to this theft, where his stability of mind functions as the catalyst, 'firmam fidem firmo pectore' ('a firm faith and a firm heart'), in connection with his explicit *praeceptum* to the brother:

> fratri praefato praecepit, ut naviculam in contiguo portu positam conscendisset, et ut inter densas harundinum conpagines, quo via sibi monstraret, incederet.[147]
>
> ([he] ordered the aforesaid brother to get into a boat which was lying by the neighbouring landing-place and to make his way amid the dense clumps of reeds by the path which would reveal itself to him.)

This portion of the episode contrasts with Tatwine and Guthlac's original journey to Crowland, a journey through trackless bogs accomplished only by travelling under the protection of Christ.[148] Unlike his interactions with the landscape preceding his spiritual majority, here Guthlac and his fellow monk are not only unaffected by its difficulties, but are instead aided by Creation, with the fens implicitly revealing ('monstraret') the proper path for the monk. This support of Creation is connected to the saint himself in the next sentence where the obedience of the monk to Guthlac's *praeceptum*

[145] It must be said, however, that Felix is not as specific in his use of *praecipio*, compared with Anonymous. While Felix does use it in an equivalent manner and context here, as well as two other instances (pp. 118, 140), the direct connection to the restoration of Creation is not always present. For Felix *praecipio*, in both its noun and verb forms, is the lexical choice for almost any situation in which someone higher in the divine hierarchy gives an order to someone below. For other uses see *VSG*, pp. 60, 82, 90, 96, 106, 118, 136, 140, 144, 146, 158, 160.

[146] *VSG*, pp. 116–19.

[147] *VSG*, pp. 118–19.

[148] *VSG*, p. 88.

allows this to occur: 'Ille autem, *praeceptis* sancti viri *obtemperans*, quo se trames ducebat, perrexit' ('So he *obeyed* the *commands* of the saint and went whither the path led').[149] The trackless bogs of the fens are here drawn back into a prelapsarian relationship with the saint through Guthlac's firm faith, his complete lack of turbulence (*turbatum*), and through the subordinate brother's obedience to the saint's command (*praeceptum*).

The stable sanctity of Guthlac and the monk's obedience are united in like manner in the final portion of this episode to restore Creation. The parchment leaves are described as being found in the middle of a pool on top of a reed, placed there as though by human hands, yet are unharmed by elemental Creation: 'Mirabile dictu! tangi, non tactae, contiguis videbantur ab undis' ('marvellous to relate, [the parchment leaves] were apparently being touched by waves around them and yet were intact').[150] The miraculous nature of these events ('Mirabile') in relation to Guthlac's firm faith is reemphasised through the brother's reaction to the miracle: 'At ille frater [...] *cum magna admiratione* grates Deo persolvens, *venerantiam validae fidei* de eo quod contigit venerabili viro Dei Guthlaco conferens' ('And the brother [...] gave thanks to God in *much amazement*, at the same time showing great respect for the *steadfast faith* of the venerable man of God Guthlac with regard to what had happened').[151] The implication is that by Guthlac's firm faith the raven dropped the leaves in precisely the right way so that they landed upon the reed as if by human hands. Here we have the raven, while initially indifferent to the saint, acting in a manner helpful to him, just as the trackless bogs have given way to the obedient path, in which each aspect of the physical world is working to aid the saint and his fellow monk: the reed is holding the leaves in place, the wind is not blowing them off, the waves, though apparently touching the leaves, do not wet them in any way. Each of these actions and circumstances echo not only the behaviour of the lions in the *Vita Pauli*, but also the two sea animals and the sea in all three of Cuthbert's *vitae*; the elements and the raven here act without direction

149 *VSG*, pp. 118–19.
150 *VSG*, pp. 118–19.
151 *VSG*, pp. 118–19.

from God or the saint to aid the holy man. That they are behaving thus reveals the true miracle of this episode: everything acting in prelapsarian harmony, another holy moment. Guthlac has progressed in his sanctity, as evidenced by his firm faith, which starts this process; the monk, though troubled at first, obediently follows the saint's *praeceptum*, and Creation, in animal and elemental form, is thus restored into prelapsarian relationship.

In Ch. XXXVIII Felix presents an even clearer picture of how Guthlac's sanctity draws Creation back into prelapsarian relationship by utilising Bede's exegesis on Creation from the *VCP*. This episode has its roots in the hagiographical convention of the thieving raven more generally, with a combination of two episodes in Bede's *VCP* specifically: Ch. XIX, where a flock of birds steals Cuthbert's barley, and Ch. XX, where two ravens tear thatch from his guesthouse. The episode in the *VSG* centres on the actions of the ravens towards Guthlac and the saint's internal stability in response to these actions,[152] which leads into the Bedan quotation as explanatory exegesis concerning Guthlac's authority over the ravens. At first, the description of the ravens in this chapter appears incongruous with the overall depiction of Creation's indifference towards humanity, as they are presented as morally corrupt:

> quorum infesta nequitia fuit, ita ut, quicquid frangere, mergere, diripere, rapere, contaminare potuissent, sine ullius rei reverentia damnantes perderent.[153]
>
> (whose mischievous nature was such that whatever they could break, drop into the water, tear into pieces, steal or defile they would destroy, damaging everything without any respect.)

While *infesta* can be best translated as 'hostility' here, the adjective *nequitia* merits further consideration, as it is almost always used to refer to a moral sense of wickedness.[154] Scripturally it is consistently used this way, for example in Acts 3:26: 'Vobis primum Deus suscitans filium suum, misit eum benedicentem vobis: ut convertat se unusquisque a *nequitia* sua' ('For you first God, raising up his Son,

152 See above, p. 25 n.30.
153 *VSG*, pp. 118–21.
154 *DMLBS*, 1: 'wickedness, malice; 2: wicked deed, crime, sin'.

has sent him to bless you; so that every one may convert himself from his *wickedness*').[155] This shift from indifferent to potentially wicked Creation centres on the polyvalent nature of the raven as an actual and interpreted creature. It must be remembered that Bede's exegesis in *In Genesim,* distinct from Augustine, does describe Creation as falling together with humanity, and Creation therefore contains the possibility of wickedness itself.[156] Within this exegesis of the Fall ravens could, therefore, be interpreted as good, indifferent, or wicked, depending on how they are being employed; the important point here is their polyvalent ambiguity. Felix also seems to be echoing the expansive exegesis concerning the raven sent out by Noah from the ark that never returned in Gen. 8:6–7. As Gatch has discussed, there were varying interpretations of this episode, most taking as their premise that the ark represents the church, and the people and animals within as Christians:[157] Ambrose ignored the literal sense and instead saw the raven as 'the sinfulness of penitent man […] which is sent away from the vicinity of the ark';[158] while Isidore presents two possibilities: that the raven either drowns, and thus 'betokens rebaptism', a heretical practice, or the raven feeds 'on the bodies of those who were killed in the flood […] [and] represents the Christian seduced back to worldly things'.[159] Bede, on the other hand, presents a tropological reading of the raven as what Gatch terms a 'nominal Christian', one who is baptised but takes no advantage from it.[160] Felix gestures towards these various traditions by his negative description of the ravens' moral state here, but uses them to focus once again on experiential Creation, as physical ravens, to depict the glory of his saint in this world; Guthlac is

155 Used a total of thirty-two times in Scripture, always with the sense of internal wickedness. Bede, however, uses it only once in the *HE*.

156 See above, Introduction, pp. 13–15.

157 M. McC. Gatch, 'Noah's Ravens in Genesis A and the Illustrated Old English Hexateuch', *Gesta* 14 (1975), 3–15, at 5–6.

158 Gatch, 'Noah's Ravens', p. 5; referencing Ambrose, *De Noe* in *Ambrosii Opera pars prima*, p. 458.

159 Gatch, 'Noah's Ravens', p. 5; referencing Isidore, *Isidorus episcopus Hispalensis: Expositio in Vetus Testamentum. Genesis*, ed. M. M. Gorman and M. Dulaey (Freiburg, 2009), pp. 30–37. It should be noted that Aldhelm incorporates this interpretation into his *Enigmata* on the Raven, in *Variae Collectiones Aenigmatum Merovingicae Aetatis*, 2 vols, CCSL 133–133A (Turnhout, 1986), Ch. LXIII, p. 463.

160 Gatch, 'Noah's Ravens', p. 5.

here able to restore even one of the most notoriously mischievous portions of Creation into prelapsarian relationship. The raven in Ch. XXXVIII is therefore neither allegorical in a general sense nor specifically tropological, but is instead meant to be read literally, allowed by divine providence to act thus towards Guthlac to provide an avenue by which the saint's great patience and charity can be displayed not only to humanity, but to all of Creation:

> Supramemoratus autem Dei famulus, varias eorum iniurias perferens, longanimiter pio pectore sufferebat, ut non solum in hominibus exemplum patientiae ipsius ostenderetur, sed etiam in volucribus et in feris manifesta esset.[161]
>
> (This same servant of God bore their manifold injuries patiently and piously, so that the example of his patience was shown not only among men but was clear even among birds and wild beasts.)

Felix's emphasis here is clear in his reiteration of four terms for Guthlac's saintly patience ('perferens', 'longanimiter', 'sufferebat', and 'patientiae'); not only does Guthlac endure their injuries, he suffers patiently with a pious heart. The lexical choice of the adjective *longanimis* drives the point home, a term used biblically of God himself in the Psalms: 'Miserator et misericors Dominus, *longanimis*, et multum misericors' ('The Lord is compassionate and merciful: *long suffering* and plenteous in mercy') (Ps. 102:8). Likewise, *sufferre* connects Guthlac with Paul's famous passage on charity in 1 Cor. 13:7: 'omnia *suffert*, omnia credit, omnia sperat, omnia sustinet' ('*bears* all things, believes all things, hopes all things, endures all things'). Felix emphasises the link between suffering and charity, but unlike Paul, causally connects them with Creation's restoration:

> Erga enim omnia eximiae caritatis ipsius gratia abundabat, in tantum ut incultae solitudinis volucres et vagabundi coenosae paludis pisces ad vocem ipsius veluti ad pastorem ocius natantes volantesque subvenirent.[162]

161 *VSG*, pp. 120–21.

162 *VSG*, pp. 120–21. While it is uncertain to what degree Felix understood and utilised Bede's *VCM*, the depiction of the saint as *pastor* ('shepherd') of the ravens does echo that given by Bede of Cuthbert in the *VCM*. *Fontes* identifies only one use of the *VCM* by Felix: 'Domino in domino dominorum'. *Fontes*, Source Details: L.E.2.1.001.01, citing *VCM*, p. 56.

(For the grace of his excellent charity abounded to all creatures, so that even the birds of the untamed wilderness and the wandering fishes of the muddy marshes would come flying or swimming swiftly to his call as if to a shepherd.)

The structure is clear, the conjunction *enim* leading to the adverb *ut*, which leads to the birds, wild animals, and even the fishes being drawn back into prelapsarian relationship with the saint. The wickedness ('nequitia') of the ravens which began this episode is not a confusing contrast with this, but is instead allowed by both God and Guthlac to provide an avenue for this display, by which the rest of Creation can be transformed; the implication being that, had Guthlac so chosen, he could have rebuked the ravens in order to transform them as well.

This use of the raven's ambiguous status is reiterated in Felix's third raven miracle (Ch. XL), where the ravens steal two gloves from Æthelbald while he is visiting Guthlac. The roots of this particular miracle, as noted by Thacker, are from an episode in Jonas' *Vita Columbani*,[163] which not only involves the theft of the same item – gloves – but also possesses the same focus on prophecy as the heart of the miracle.[164] The *Vita Columbani* describes the act of thievery in similarly moral terms, 'corvus alis *rapax* advolavit, unumque ex eis rostro ferens, abstulit' ('a *thieving* raven flew up and carried off one of the gloves in its beak'),[165] utilising the adjectival form *rapax* to highlight the same internal moral corruptness as Felix's description of the raven's mischievous beak ('inprobo forcipe').[166] The *Vita Columbani* interprets the raven's theft in light of the exegetical tradition concerning the raven and the ark, arguing that such behaviour is to be expected from ravens.[167] Felix never makes this connection, but the undercurrent of moral depravity within the ravens which causes them to act in this manner is nonetheless apparent: 'veluti criminis sui conscius esset, ales vero manicam in culmine casae relinquens, velut *fuga* facta, occiduas in auras

163 *Vita Columbani*, I.15, pp. 178–79.

164 Thacker, 'Social and Continental Background', p. 310.

165 *Vita Columbani*, p. 178.

166 *VSG*, pp. 124–25.

167 *Vita Columbani*, p. 178.

volabat' ('as if conscious of its ill-doing, it left the glove on the top of the cottage and like a *fugitive* fled westwards').[168] Not only does the raven possess a mischievous beak and flee from the cottage with the connotations of banishment with the noun *fuga*, it is also aware of its wrongdoing. While Felix clearly has the episode from the *Vita Columbani* in mind, he is once again focusing on the physical aspect of the raven, and its use in presenting Guthlac's ability to restore Creation, even the troublesome raven.

Felix's inclusion of Bede's exegesis from the *VCP*, in its entirety, provides authorising support for Guthlac's prelapsarian relationship. He even includes Bede's prefatory description which functions as a backward-reaching delineation to include all the miraculous interactions with Creation, and utilises it to similar effect for the *VSG*:

Bede:	Felix:
Non sola autem aeris sed et maris animalia, immo et ipsum mare sicut et aer et ignis iuxta quod in superioribus exposuimus, uiro uenerabili praebuere obsequium.[169]	Non solum vero terrae aerisque animalia illius iussionibus obtemperabant, immo etiam aqua aerque ipsi veri Dei vero famulo oboediebant.[170]
(Moreover not only the creatures of the air but also of the sea, yes, and even the sea itself, as well as the air and fire as we have shown above, did honour to the venerable man.)	(Not only indeed did the creatures of the earth and sky obey his commands, but also even the very water and the air obeyed the true servant of the true God.)

As can be seen, Felix's changes are superficial at best, and the allusion to Bede lends authority to his depiction of Guthlac and Creation. Felix adjoins to this material the exegesis from the *VCP* almost without alteration:

> Nam *qui auctori omnium creaturarum fideliter et integro* spiritu *famulatur, non est mirandum, si eius imperiis ac votis omnis creatura deserviat.*

[168] *VSG*, pp. 124–25.
[169] *VCP*, pp. 224–25.
[170] *VSG*, pp. 120–21.

> *At plerumque idcirco subiectae nobis creaturae dominium perdimus, quia Domino* universorum *creatori servire negligimus.*[171]

> (For if a man faithfully and wholeheartedly serves the Maker of all created things, it is no wonder though all Creation should minister to his commands and wishes. But for the most part we lose dominion over the Creation which was made subject to us, because we ourselves neglect to serve the Lord and Creator of all things.)

Here Felix aligns Guthlac with the Cuthbert of the *VCP*; such a reading does not, however, necessarily imply an expectation for Felix's audience to make the intertextual connection. While it is certainly possible that a portion of his audience would recognise the lines from the *VCP*, their similarity to both Augustine and Bede's exegetical works on Genesis suggests that Felix's main intention here is legitimisation, and for most of the audience the lines would have appeared as an application of received exegesis concerning Creation. Regardless, the exegesis is functionally employed by Felix here to delineate the manner in which Guthlac is able to perform his miraculous interactions with animate and elemental Creation. Guthlac has already achieved spiritual majority, as was seen in the raven-and-document miracle, through his firm faith, as well as through the connected focus in the raven-and-thatch miracle on his immovable patience based on charity. The application of Bede's exegesis serves to support this portrayal of Guthlac as having progressed in sanctity to the point of restoring a portion of Creation to what it should be. His sanctity has restored the dominion that Adam forfeited, even with the problematic and thieving ravens.

The full development of Guthlac's ability to restore Creation occurs in the miracle of the obedient swallow in Ch. XXXIX. In keeping with Felix's focus throughout the *VSG*, the narrative vision centres on exquisite detail and description of enargaeic life. It should be noted, particularly given Felix's specificity regarding the demonic seeming in the bestial horde episode, that Felix frames this miracle initially as a 'spiritual' one: 'Libet etiam beatissimi Dei famuli Guthlaci quoddam spiritale miraculum explicare' ('It is also pleasant to describe a spiritual miracle of Guthlac the most blessed

[171] *VSG*, pp. 120–21. Italics are used to show Bede's text from the *VCP*.

servant of God').[172] His use of Latin terms like *spiritus* and *spiritalis* is wide ranging, covering everything from the positive spirit of prophecy, to the evil spirits that comprise the demonic horde.[173] Felix also employs it in his adaptation of Bede's exegesis, where he substitutes Bede's 'corde' ('heart') with 'spiritu' ('spirit').[174] As will be discussed below, his use of the term here is not meant as a description of the quality of the miracle, that the birds are mere metaphor or signifier but, in keeping with the *VSG* as a whole, instead highlights the quality of Guthlac's sanctity. The miracle itself has its roots in the penitent ravens of both the *VCA* and *VCP*, but Felix has made one crucial change: the birds here are swallows, not ravens, and their actions involve pure obedience towards the saint. As Guthlac and Wilfrid are conversing, two swallows fly in:

> velut cum indicio magnae laetitiae avino forcipe flexuosi gutturis carmen canentes, veluti ad adsuetas sedes devenissent, sese non haesitantes humeris viri Dei Guthlaci inposuerunt, ac deinde, cantulis vocibus garrulantes brachiis genibus pectorique illius insedebant.[175]
>
> (showing every sign of great joy, they [birds] opened their beaks and sang a song from their supple throats, as though they had arrived at their accustomed abode; without any hesitation they settled on the shoulders of the man of God Guthlac, and then chirping their little songs they settled on his arms, his knees, and his breast.)

Despite the mundane nature of this scene, another one of Robin Waugh's 'non-miracles', given that in the previous chapter we are told the saint fed the birds and beasts about him,[176] the response to this event by Wilfrid, and Felix's explanation for his response, nonetheless emphasise the miraculous nature of this event: 'Wilfrith vero stupefactus […] quid incultae solitudinis volucres humani successus insueti, illam propiandi fiduciam habuerunt' ('Wilfrid was indeed amazed […] how the birds from the wild solitudes, unused to the approach of human beings, had the confidence to

172 *VSG*, pp. 120–21.
173 See *VSG*, pp. 110–11, and 114–15.
174 *VSG*, pp. 118–19; Cf. *VCP*, pp. 224–25.
175 *VSG*, pp. 122–23.
176 *VSG*, pp. 118–19.

come near him').[177] Felix includes the Bedan exegesis within Guthlac's response, adapting it into an enargaeic vision of his saint's ability to restore Creation:

> Nonne legisti, quia, qui Deo puro spiritu copulatur, omnia sibi in Deo coniunguntur? et qui ab hominibus cognosci denegat, agnosci a feris et frequentari ab angelis quaerit? nam qui frequentatur ab hominibus, frequentari ab angelis nequit.[178]
>
> (Have you not read how if a man is joined to God in purity of spirit, all things are united to him in God? and he who refuses to be acknowledged by men seeks the recognition of wild beasts and the visitations of angels; for he who is often visited by men cannot be often visited by angels.)

It is here that Felix makes clear what his framing entails: that it is Guthlac's purity of spirit ('puro spiritu'), his extraordinary sanctity, that allows the miracle to take place. Felix's lexis here is also unique in hagiography, the use of the verb *coniungo* implying a level of relational intimacy far beyond that described by Bede or the author of the *VCA*.[179] This intimate union is furthered by Felix's novel joining of the two objects of seeking in this latter portion of the eremitic career: the wild animals ('feris') and the angels ('angelis'). While the saint seeks the company of angels for obvious reasons, the seeking of wild animals requires explanation. The verb *agnoscere*[180] denotes not only recognition and intimate knowledge of something, but also, as a result of this knowledge, to 'bring into close association, [or] unite'.[181] It is here that the true miracle is displayed (the real reason Wilfrid is so amazed); it is the miracle of the world as it should be. This notion finds further explication in a hitherto unnoticed parallel for the *VSG* from Augustine's *Ioannis Epistolam ad Parthos Tractatus Decem*, Homily 8.7:

177 *VSG*, pp. 122–23.

178 *VSG*, pp. 122–23.

179 *DMLBS*, 1: 'to yoke, join, hold together; b. to put (close) together'; '4 a. (of persons) to attach, unite, associate; b. to unite sexually in marriage; c. to unite spiritually'.

180 *DMLBS*, 1: 'a. to recognize (person or thing); b. to recognize or know (fact)'.

181 *Oxford Latin Dictionary*, p. 409.

> Intendite quid dicam: Deus, homo, pecora: verbi gratia, supra te Deus; infra te pecora. *Agnosce eum qui supra te est, ut agnoscant te quae infra te sunt*. Ideoque *cum Daniel agnouisset supra se Deum, agnouerunt illum supra se leones*.[182]
>
> (Listen to what I say: God, man, beasts: for example, above you, God; beneath you, the beasts. *Acknowledge Him that is above you, so that those beneath you may acknowledge you*. Therefore, *when Daniel acknowledged God above him, the lions acknowledged Daniel above them*.)

The recognition and understanding Felix includes in the *VSG*, therefore, are meant to depict the transformed relationship between Guthlac and Creation in direct response to his progression in sanctity; an image of Creation as it should be. The saint, by his humility and proper relationship with God, and subsequent holiness, has restored Creation, and in this state the animals recognise his dominion as flowing through him from God. Creation, which was once purely indifferent, now rejoices to interact with Guthlac; and Creation, which was once, at best, a crucible, is now a glimpse of the joyous Eden, an example of the proper way to live, and one of the goals of an eremitical life. This interaction is presented as the positive obverse to the ravens from the various *vitae* of Cuthbert, where they are allowed to remain on Farne only after they have returned penitent to the saint;[183] the permission to live alongside the saint is a gift for the ravens' humble recognition of their wrongs and their rightful behaviour in obtaining absolution. Felix, however, transforms this into a single vision of a portion of Creation behaving in a prelapsarian way, with these representatives of animate Creation drawing close to the saint out of joy and then humbly waiting for his guidance so that they may establish a nest.[184] This vision extends beyond the present moment, echoing the *VCA*'s focus on the continuing efficacy of the miraculous, as the swallows from this point onwards never presume to choose a nesting site

[182] Augustine, *Homélies sur la première Épître de Saint Jean – In Iohannis epistulam ad Parthos tractatus decem*, ed. J.W. Mountain, J. Lemouzy, and D. Dideberg (Paris, 2008), p. 332.

[183] See *VCA*, pp. 100, 102; *VCM*, ll. 431–50; *VCP*, pp. 222, 224.

[184] *VSG*, p. 122.

without Guthlac's permission.[185] Taken as a whole the scene depicts a complete prelapsarian moment, where Guthlac's sanctity allows dominion to be restored in the physically delineated fens. The swallows not only seek the saint for relationship, as evidenced by their initial joyous songs when alighting upon Guthlac, but also seek in all ways to defer to his commands regarding even one of the most basic and essential portions of their avian lives: nesting. Felix has thereby transformed the hagiographical trope of obedient birds to give a clear and sustained picture of how Guthlac is able, via his established sanctity, to restore a representative portion of Creation.

Guthlac's Death and Ascension

Felix gives a further display of his saint's ability to restore Creation in the death-and-ascension scene in Ch. L. While he bases the structure of the scene on the equivalent moment in the *VCP*,[186] he alters this derivative moment of Guthlac's career to focus once again on the physical landscape of Crowland and the fens. In this depiction, Felix uniquely elides the traditional illumination associated with the death of saints and their souls' ascension to heaven with the miraculous elements surrounding the translation of saints and the authentication of their relics. This combination, with its focus on Creation, can be seen in three miraculous portions of Guthlac's death: illumination surrounding Guthlac; the sweet fragrance emanating from him; and the sound of the angelic chorus above him.

The hagiographical trope of illumination in connection with the death of a saint has lengthy precedent,[187] and is ultimately derived from the wider symbolism of light in Christian tradition. The trope

185 *VSG*, p. 122.

186 See *VCP*, Ch. 37–39; this is noted by Colgrave, see *VSG*, p. 192.

187 For example, Jerome, *Vita Pauli*, p. 174; Cuthbert's vision of Aidan's soul in the *VCA*, pp. 68, 70; the *VCM*, ll. 120–41; the *VCP*, pp. 164, 166; and Adomnán's *Vita Columbae*, p. 224. It is also connected to the miraculous authentication of saints' relics, as mentioned by Colgrave, *VSG*, p. 193. See, for example, the story of Oswald's translation in Bede's *HE*, III.11, in which a column of light rests on them throughout the night; and the *Vita Sancti Onuphrii*, PL 73, col. 218D, which describes the saint being surrounded by light, and ascending in a flashing light.

is, at its base, a visual declaration of sanctity, and is most often used in hagiography to depict the ascension of the saint's soul to heaven with the help of angels, who are themselves often the source of that illumination. Felix's application of the trope is distinct, especially in his presentation of two different manifestations of light, perimortem and postmortem, instead of a single postmortem miracle. This separation, as well as several elements of the individual miracles, has a parallel in Adomnán's *Vita Columbae*.[188] In the perimortem miracle Guthlac is resting against the wall of his cell, while his companion Beccel engages in the vigils of the Night Office. At some unidentified point during those prayers, Beccel witnesses the entire cell filled with light. This mirrors the *Vita Columbae*, where Columba's attendant sees the church in which Columba is praying at midnight also filled with light:

VSG:

Nocte vero sequenti, cum praefatus frater nocturnis vigiliis incumberet, igneo candore a mediae noctis spatio usque in auroram totam domum circumsplendescere videbat.[189]

(Now on that night, when the same brother was engaged upon his nightly vigils, he saw the whole house lit up with a fiery brightness from midnight until dawn.)

Vita Columbae:

Tum proinde media nocte pulsata personante cloca festinus surgens ad ecclesiam pergit [...] Diormitius minister tardius prosequtus eodem momento eminus totam intrinsecus eclesiam angelica luce erga sanctum repleri uidet.[190]

(Then, when the beaten bell had resounded at midnight, he rose in haste and went to the church [...] In that moment Diormit, the attendant, following later, saw from a distance the whole church filled inside with angelic light about the saint.)

In both examples the saints are engaged in the Night Office, implicitly in the case of Guthlac, and there is emphasis on the total

188 I do not seek to argue here that the *Vita Columbae* was a source for Felix, as there is at present no evidence that he knew the *vita*; the comparison simply serves to highlight Felix's focus.

189 *VSG*, pp. 158–59.

190 *Vita Columbae*, pp. 224–25.

suffusion of light throughout the buildings: 'totam domum' for Guthlac, and 'totam [...] ecclesiam' for Columba. The visual manifestation of the function of the Night Office is powerful in both *vitae*, literally banishing night, but Felix's more ambiguous presentation of the light, simply appearing versus being specifically described as angelic, allows for a deeper connection between the saint and the light; it is his sanctity that implicitly provides the source of the illumination.

The postmortem illumination miracle in the *VSG* likewise parallels a similar miracle in the *Vita Columbae*. After Guthlac has received communion, via an allusion to Cuthbert in the *VCP*,[191] and proceeds to heaven, Beccel again sees the house filled with light, but also beholds a pillar of fire stretching from earth to heaven. This parallels one of the two postmortem vision miracles of the *Vita Columbae*, in which a monk fishing with some other men in the river Fendea sees not only all of heaven lit up, but also a pillar of fire when Columba ascends to heaven:

VSG:	*Vita Columbae*:
Inter haec praefatus frater subito caelestis luminis splendore domum repleri turremque velut igneam a terra in caelum erectam prospicit, in cuius splendoris conparatione, cum tunc sol in medio caelo steterat, velut lucerna in die pallidescere videbatur.[192]	Cuius miraculi subitatione permoti oculos ad orientem eleuatos conuertimus, et ecce quasi quaedam pergrandis ignea apparuit columna, quae in illa nocte media susum ascendens ita nobis uidebatur mundum inlustrare totum sicuti aesteus et meridianus sol[193]
(Meanwhile this same brother suddenly beheld the house filled with the splendour of heavenly light and a tower as of fire stretching from earth to heaven, in comparison with whose splendour the sun, though it was in mid heaven, seemed to grow pale like a lamp in daylight.)	(Startled by the suddenness of this miracle, we raised our eyes and turned them to the region of the rising sun; and behold, there appeared what seemed like a very great pillar of fire which, rising upwards in that midnight, seemed to us to illumine the whole world like the summer sun at midday.)

191 See *VSG*, p. 158; and *VCP*, pp. 282–85.
192 *VSG*, pp. 158–59.
193 *Vita Columbae*, p. 228.

While there are no lexical parallels, the descriptive similarities in the curious use of a pillar of fire in relation to the saints' ascensions is telling, as is the connection between the sun and the brightness of the pillar. Where Adomnán is using the sun as a comparative simile, Felix's description highlights the glory of sanctity with a pillar blazing to the extent that the sun is relegated to a single lamp. The image of the pillar of fire deserves further consideration, as its employment during the death and ascension of their respective saints is unique to the *Vita Columbae* and the *VSG*.[194]

Peter Lucas has argued, in relation to the image's use in *Guthlac B*, that the pillar can be symbolically connected to the paschal candle and the associated Easter liturgy. This argument is founded on the eight-day illness of Guthlac and his death on the fourth day of the Easter festival, which seems to echo the celebration of Christ's resurrection on the eighth day of the Jewish week.[195] Further, after the dispelling of darkness in the Easter service, the *Exultet* joins the biblical image of the pillar of fire from Exodus with the candle, which together eliminate the darkness of sin, as well as point to Christ as the source of the lights.[196] Further still, Lucas argues that this collocation of imagery, including the illumination, the sweet scent, and the angelic music, 'mirrors the Easter Vigil, with the light of the Paschal Candle, the music of the choir, the perfume of incense, and what they signify'.[197] While there are echoes of this in the *VSG*, I would suggest that Felix's imagery can be connected more readily with the image of Jacob's ladder in hagiographical tradition.[198] Butler notes how '[e]xegetes from the time of Origen [...]

194 Bede describes a similar scene in which the nuns of Barking are shown where they should bury the bodies of those taken by the plague with a luminous sheet. It therefore is not specifically about a saint's death and ascension. See *HE*, III.7, p. 356.

195 P. J. Lucas, 'Easter, the Death of St Guthlac and the Liturgy for Holy Saturday in Felix's *Vita* and the Old English *Guthlac B*', *MÆ* 61 (1992), 1–16, at 2.

196 Lucas, 'The Death of Guthlac', at 9: 'In the *Exultet*, through the use of the Latin word *columna* with a dual meaning (fire-pillar/Paschal Candle), the Paschal Candle is fused with the Pillar of Fire that protected the Israelites on their way to the Red Sea'.

197 Lucas 'The Death of Guthlac', 11, on the Easter Liturgy.

198 See B. Butler, 'The Whitby Life of Gregory the Great: Exegesis and Hagiography', Unpub. Ph.D. Thesis (University College Cork, 2005), pp. 35–36, for a discussion of this in relation to the Whitby Life.

viewed this story of Jacob's ladder as a symbol of the contemplative life', including Gregory the Great, St Benedict in his *Regula*, and St Augustine.[199] The *VCA* employs this same image when describing Cuthbert's vision of Aidan's soul being led to heaven by angels in a 'globe of fire',[200] as does Bede in his *VCP*.[201] The elision between Jacob's ladder and the pillar of fire also occurs in the earliest life of Gregory the Great by an anonymous monk at Whitby,[202] though it is employed for a different purpose than in both Adomnán and Felix: Gregory is depicted as fleeing to the forest to avoid his election to the papacy, but God reveals his location by a pillar of fire, which is interpreted by a nearby anchorite as Jacob's ladder.[203] This miracle is interpreted as evidence of Gregory's humility and necessary sanctity for the papacy. The connection, therefore, between the spiritual ascent of the ascetic via Jacob's ladder and the pillar of fire as evidence of the final ascent to heaven has hagiographical precedent.

Felix incorporates this conventional image but alters its use to highlight once again Guthlac's sanctity via his restoration of Creation. In the *Vita Columbae* the pillar of fire appears in the sky like a second sun, anchored to no fixed point in the landscape, its only spatial reference is the cardinal direction East; its intent is to focus on heaven, on the saint's ascension, with little concern for physical Creation. Felix, on the other hand, more clearly echoes the depiction in the Whitby *Vita*, and unites heaven with both the earth in general and Guthlac's cell in particular. The pillar is part of the illumination of Guthlac's cell, issuing from Guthlac himself rather than from angelic presence, thereby transforming the physical, shadowed night of his eremitic home on Crowland into a heavenly space centred on saint and cell, an image reminiscent of the elided heaven/earth cell of Cuthbert from the *VCM*. This anchoring

199 Butler, 'The Whitby Life', pp. 33–39. This includes Gregory's *Moralia in Iob*, and *Dialogus*; the *Regula Benedicti*; and Augustine's *Homilies*.

200 *VCA*, p. 69.

201 See *VCP*, p. 164, 166. In Bede's poetic rendering in the *VCM* he retains the illumination but leaves out the specific image of the ladder, *VCM*, ll. 120–41.

202 See Butler, 'The Whitby Life', pp. 38–39.

203 *The Earliest Life of Gregory the Great*, ed. B. Colgrave (Lawrence, KS, 1968; repr. Cambridge, 1985), p. 86.

connects the pillar to Felix's overall physical depiction of Crowland, now restored and suffused with heavenly light.

This process of adapting hagiographical convention to focus on physical Creation is equally apparent in the second miraculous portion of Guthlac's death scene: the profusion of sweet smells. In this sequence of miraculous moments, Felix adapts the hagiographical commonplace of sweet fragrances that accompany the discovery of the uncorrupt corpse as part of translation miracles.[204] Felix employs this trope three times in the short narrative of Guthlac's death, and in each case it serves to display his sanctity. Just as with the illumination miracles, the olfactory transformations are divided into perimortem and postmortem. The perimortem moment occurs in Guthlac's final hours when he is resting his head against a wall of his cell, immediately preceding the first illumination miracle:

> cum parumper anhelaret, velut melliflui floris odoratus de ore ipsius processisse sentiebatur, ita ut totam domum, qua sederet, nectareus odor inflaret.[205]
>
> (After he had breathed with difficulty for a short time, there seemed to proceed from his mouth the odour of sweet-smelling flowers, so that the scent of nectar filled the whole building in which he sat.)

Felix's unique addition to Guthlac's saintly life not only highlights the particular sanctity of the hermit in life, producing a heavenly aroma even before death, but also restores his immediate surroundings, the cell in which he and Beccel are praying. The second instance, and the first of the postmortem olfactory miracles, occurs in the same moment as the second illumination of Guthlac's cell and the appearance of the pillar of fire. It is the final item in the depiction of the saint's ascension, and instead extends outwards and sanctifies the olfactory space of the entire island: 'insulam etiam illam diversorum aromatum odoriferis spiraminibus inflari cerneres' ('one would have thought the island to be filled with the

[204] While most examples post-date Felix, the connection between the saint's incorrupt corpse and a sweet fragrance has its foundation early in the hagiographical tradition: see, for example, Jerome's *Vita Hilarion*, p. 296.

[205] *VSG*, pp. 156–59.

sweet scents of many kinds of spices').[206] Here Felix has included a novel description of the heavenly scent, comparing it to spices instead of flowers or perfume; yet with Guthlac's passing his sanctity exerts greater influence, restoring the entire island of Crowland as opposed to his single cell. The final odour-miracle echoes the previous, and depicts the scent lingering into the next day when Pega returns to bury him.[207] In this third variation on the same image Felix unites the previous two, displaying the eremitic cell and the entire island united in their restored state.

The final miracles surrounding Guthlac's death are the angelic songs that herald his ascension to heaven. These are used by Felix to mirror Guthlac's attainment of spiritual majority in the cleansing of the air in his return from hell in Ch. XXXIII, and thus once again focus on Guthlac's ability to restore the elemental world: 'Cantantibus quoque angelis spatium totius aeris detonari audiebatur' ('The whole air was heard to thunder with angelic songs').[208] Instead of the demonic din in the Aldhelmian *clangisonis*, the whole air is said to thunder with angelic song, the force of the verb *detono* providing a clear obverse here, as most hagiographical depictions involve a simple description of angelic singing.[209] For example, this distinction is evident by comparison with the *Vita Columbae*, in which Columba's luminous ascent to heaven with the help of innumerable angels includes a description of angelic song:

> Altisona quoque carminalia et ualde suauia audiui angelicorum coetuum cantica, eodem momento egressionis inter angelicos sanctae ipsius animae ascendentes choros.[210]
>
> (Also I heard, sounding on high, the songs, tuneful and very sweet, of the angelic hosts, at the very moment when his holy soul went forth among the ascending companies of angels.)

Here the music is depicted as sonorous and sweet, in keeping with traditional models, and is meant to highlight the honour granted to

206 *VSG*, pp. 158–59.

207 *VSG*, p. 160.

208 *VSG*, pp. 158–59.

209 See, for example, *Vita Sancti Onuphrii*, PL 73, cols. 218D–219A.

210 *Vita Columbae*, pp. 226–27.

the saint. In Felix's *VSG*, the image is slightly modified to focus once again on the physical landscape by way of mirroring Ch. XXXIII, as well as to provide a saintly status for Guthlac: he is not merely escorted to heaven bathed in heavenly song, but is lifted amidst an entire sky ('totius aeris') thundering with angelic song. The change is slight but, in the context of Felix's *VSG* as a whole, significant, representing another moment of connection between sanctity and Creation.

Felix's construction of Guthlac's death-and-ascension scene therefore follows his methodology for the entire *VSG*, incorporating a substantial amount of traditional material into his novel depiction of a saint tied to the landscape in which he lived. Creation has been restored in elemental and animate form through Guthlac's sanctity, and in the apogee of his saintly career the delineated physical landscape is transformed as well, the fens and Crowland becoming part of the prelapsarian space. When Pega buries Guthlac on Crowland, his body is interred within a landscape that is in that moment wholly restored and filled with the odour of heaven.

Conclusion

Felix's *VSG* represents several important changes in Anglo-Latin hagiography. While heavily indebted to his immediate sources (the *Vita Antonii*, *Vita Martini*, *VCP*, and *VCM* in particular) and more generally to hagiographical convention, Felix nonetheless creates a *vita* distinctively focused on the enargaeic landscape, on the saint's imaginative journey through a physical space bereft of significant landmarks or named sites. Felix builds on his sources by tying Guthlac's Antonian progression towards sanctity to this central physical landscape. Felix's debt to the exegesis of Augustine and Bede is also clearly evident throughout, as Guthlac's path to spiritual majority includes the thorns of the fens pushing him towards sanctity just as the thorns are meant to in Gen. 3. The *VSG* takes the restoration of Creation even further than Bede in either the *VCM* or *VCP*, by utilising Guthlac's death and ascension as the true apogee of his saintly career, whereby the entire island of Crowland becomes suffused with heavenly light.

5

Landscape Lexis and Creation Restored: The *Old English Prose Life of Guthlac* and *Guthlac A*

The date and context of the two Old English iterations of Guthlac's life discussed in this chapter, the *Old English Prose Life of Guthlac* (*OEPG*) and *Guthlac A*, have been the subject of much debate. The *OEPG* has proved difficult to date, and its author equally difficult to identify. This is in part due to the survival of the text in only two witnesses: the eleventh-century Cotton Vespasian D.xxi manuscript and, in abbreviated form, as the text now referred to as Homily XXIII in the late tenth-century Vercelli Book.[1] On the basis of Anglian traits in the Vercelli version and other internal evidence, Jane Roberts and others have argued for a lost Anglian original composed between the late ninth and early tenth centuries.[2] This chronological range is further supported by the text's descriptive landscape lexis which, as I explore below, is shared with boundary clauses of the mid tenth century. The *OEPG* is, on the whole, very

1 The dating of these manuscripts is based on critical consensus: Lapidge and Gneuss, *Anglo-Saxon Manuscripts: a Handlist*, date the Cotton MS to s. xi3/4 or s.xi2 (no. 657) and the Vercelli Book to s.x2 (no. 941); N. R. Ker, *Catalogue of Manuscripts Containing Anglo-Saxon* (Oxford, 1957; repr. with suppl. 1990), dates the Cotton MS to s.xi2 (p. 422, no. 344) and the Vercelli Book to s.x2 (pp. 460–64, no. 394); D. G. Scragg, 'The Corpus of Vernacular Homilies and Prose Saints' Lives before Ælfric', *ASE* 8 (1979) 223–77, at 258 and 225 respectively, dates the Cotton MS to s.xi2 and the Vercelli Book to s.x2/3, and in his edition, *The Vercelli Homilies and Related Texts* (Oxford, 1992), p. xxiii, Scragg dates the Vercelli Book to s.x2.

2 J. Roberts, '*The Old English Prose Translation of Felix's* Vita Sancti Guthlaci', in *Studies in Earlier Old English Prose: Sixteen Original Contributions*, ed. P. E. Szarmach (Albany, NY, 1986), pp. 363–79, at 375–76; E. G. Whatley, 'Lost in Translation: Omission of Episodes in some Old English Prose Saints' Legends', *ASE* 26 (1997), 187–208, at 193; Waugh, 'The Blindness Curse', 399, n. 2; Downey, 'Intertextuality in Guthlac', p. 107.

faithful to its main source text, Felix's *VSG*.[3] Gordon Whatley suggests the Old English author sought not only to provide a relatively complete, accurate, and often word-for-word translation, but also 'through close imitation, to give the vernacular reader the illusion of reading the Latin text itself'.[4] The author does, on occasion, streamline Felix's verbose Latin, however, as well as omit, amalgamate, or truncate various sections of the *VSG* which, as Whatley notes, 'might have seemed redundant, repetitious or frivolous'.[5] I build here on the work of Whatley and others in order to explore in detail how the OE author adapts the *VSG* to create a vernacular narrative centred on the physical landscape of Anglo-Saxon England with wider appeal and deeper focus than the Latin text.

The delicate strands of influence that shaped the Exeter Book poem *Guthlac A* have proved equally difficult to untangle. It is certainly indebted to Felix (however indirectly), as well as, potentially, other unattested Guthlac material resulting from the accretion of local legend.[6] Roberts posits in her edition that various features point to 'an earlier rather than later time of composition within the Anglo-Saxon period [...] for *Guthlac A*'.[7] Lexomic analysis, however, suggests that the poem belongs firmly in the English Benedictine Reform, which would place it in the mid- to late-tenth century.[8]

3 All quotations taken from *Das angelsächsische Prosa-Leben des hl. Guthlac*, ed. P. Gonser (Heidelberg, 1909), with reference to J. Crawford (now Roberts), 'Guthlac: an Edition of the Old English Prose Life together with the Poems in the Exeter Book', Unpub. D.Phil. Thesis (University of Oxford, 1967). The latter author will be referred to as Roberts in all future quotations. The Vespasian *Life* is the central text of this discussion, with reference to *Vercelli Homily XXIII*, also printed in Gonser's edition. The text will be referred to as the *OEPG* throughout.

4 Whatley, 'Lost in Translation', 193. For example, Whately notes that the translator includes Felix's prefatory material, which is often omitted in other vernacular translations.

5 Whately, 'Lost in Translation', 194.

6 See J. Roberts, '*Guthlac A*: Sources and Source Hunting', in *Medieval English Studies Presented to George Kane*, ed. E. D. Kennedy, R. A. Waldron, and J. S. Wittig (Cambridge, 1988), pp. 1–18, at 1–3.

7 *The Guthlac Poems of the Exeter Book*, ed. J. Roberts (Oxford, 1979), p. 70. All quotations from *Guthlac A* will be taken from this edition, and will be referred to as *Guthlac A*.

8 See S. Downey et al., '"Books Tell Us": Lexomic and Traditional Evidence for the Sources of *Guthlac A*', *Modern Philology* 110 (2012), 153–81.

This late dating rests in great part on the assumption that the poem must rely on later redactions of the *OEPG* and Vercelli Homily XXIII. Patrick Conner and Chris Jones also argue, primarily for reasons of theme and content, and from the apparent reliance of *Guthlac A* on the Benedictine Rule, that the poem's composition should be placed within the reform period of the late tenth century.[9] The heavy emphasis on cenobitic affairs within an ostensibly eremitic text need not, however, imply a reform context for composition, as this study has shown regarding Bede's two *vitae* of Cuthbert. The physical location for the poem's composition is equally difficult to determine. While Roberts posits Crowland as a possible site of origin for the poem,[10] Robert D. Fulk more generally assigns it to Mercia and Northumbria upon the basis of phonological, morphological, and syntactical features.[11] The content of *Guthlac A* is consonant with this assessment, as the presentation of the landscape shifts from the East Anglian fens of the *VSG* to a landscape of wooded green hills, more reminiscent of Mercia. Alvin Lee, citing this transformation, argues for a location closer to central or western Mercia, associating the text in broader terms with the Midlands.[12] With all this in mind, it seems safe to fix the composition to a non-Crowland milieu, within central to western Mercia. The date of the original composition, taking into account the prevalence of Anglian and Mercian forms, and the focus on a landscape much closer to that of the Midlands, could profitably be placed around the mid tenth century.

Guthlac A focuses solely on the saint's time as a hermit, his descent into hell at the hands of his demonic oppressors, and his journey to heaven. As numerous scholars have noted, the narrative centres on a verbal battle concerning the rightful ownership of the

9 P. W. Conner, 'Source Studies, the Old English *Guthlac A* and the English Benedictine Reform', *Revue bénédictine* 103 (1993), 380–413, at 383, 410; C. A. Jones, 'Envisioning the *Cenobium* in the Old English *Guthlac A*', *Mediaeval Studies* 57 (1995), 259–91 at 289. See also B. D. Weber, 'A Harmony of Contrasts: The Guthlac Poems of the Exeter Book', *JEGP 114* (2015), 201–18, who focuses on the reception context of the reform period rather than the date of composition.

10 Roberts, *Guthlac Poems*, p. 71.

11 R. D. Fulk, *A History of Old English Meter* (Philadelphia, 1992), pp. 399–400.

12 Lee, *Guest-Hall*, p. 116.

barrow on which Guthlac chooses to found his eremitic cell.[13] This centrality of the land, in combination with the land's non-fenland topography, has provoked numerous interpretations. *Guthlac A* has been seen as an externalisation of Guthlac's internal spiritual and mental state,[14] a metaphor for the saint's climb up the mountain to salvation,[15] a reflection of heaven,[16] a presentation of Anglo-Saxon nation building,[17] and an exemplification of the monastic ideal of *stabilitas*.[18] Stephanie Clark has argued that the spiritual contest between the devils and the saint for the wooded hill is presented as a 'land dispute between spiritually unworthy tenants who have held the land through temporary loan and a warrior of God who is granted permanent tenure as his reward for faithful service'.[19] I would argue, however, that the depiction of the relationship between the saint and the natural world in *Guthlac A* is better understood in terms of the restoration of Creation via Guthlac's sanctity, a theme which is central to both the *VSG* and *OEPG*. Like these texts, *Guthlac A* presents its saint typologically as the New Adam,[20] repairing the effects of the Fall between humanity and Creation. Uniquely within the Guthlac tradition, however, the poet connects Guthlac's restoration of Creation to the eschatological doctrine of replacement. This doctrine in its variations can be defined thus: the fall of Satan and his followers created an absence in heaven that needed to be filled, whether to sustain numerical perfection or to balance the numbers of angels and humans in heaven.[21] Part of the

13 For example, see Hill, 'The Middle Way', at 183; K. O'Brien O'Keeffe, 'Guthlac's Crossings', *Quaestio: Selected Proceedings of the Cambridge Colloquium in Anglo-Saxon, Norse and Celtic* 2 (2001), 1–26; and F. R. Lipp, '*Guthlac A*: an Interpretation', *Mediaeval Studies* 33 (1971), 46–62, at 60.

14 L. K. Shook, 'The Burial Mound in *Guthlac A*', *Modern Philology* 58 (1960), 1–10.

15 P. F. Reichardt, 'Guthlac A and the landscape of spiritual perfection', *Neophilologus* 58 (1974), 331–38.

16 Lee, *Guest Hall*, p. 116.

17 A. K. Siewers, 'Landscapes of Conversion: Guthlac's Mound and Grendel's Mere as Expressions of Anglo-Saxon Nation-Building', *Viator* 34 (2003), 1–39.

18 Jones, 'Envisioning the *Cenobium*'.

19 S. Clark, 'A More Permanent Homeland: Land Tenure in *Guthlac A*', *ASE* 40 (2011), 75–102, at 76.

20 See above, pp. 14, 135; also, see below, pp. 278–81.

21 For knowledge and influence of replacement doctrine in Anglo-Saxon England see D. Haines, 'Vacancies in Heaven: The Doctrine of Replacement and Genesis A', *Notes and Queries* 44 (1997), 150–54; M. Fox, 'Ælfric on the Creation

purpose for creating humanity, therefore, was to fill these vacancies in heaven. In the Fall Adam and Eve removed humanity from that teleological trajectory, thus necessitating Jesus' redemptive actions. Dorothy Haines traces this doctrine back to Augustine's *Enchiridion ad Laurentium*, in which the repopulation of heaven is conceived as 'a matter of numerical subtraction and addition. The citizenship of heaven has been depleted through the angelic rebellion and must be made whole (*integratis*) again.'[22] Haines notes that Augustine's interpretation of the repopulation of heaven was adapted by Gregory the Great, who 'states that as many humans will reach heaven as there are angels remaining there after the fall',[23] and subsequently informed Anglo-Saxon exegesis.[24] Clark argues that the struggle in *Guthlac A* over the contested *beorg* can be viewed as a replication of 'the cosmic exchange begun at the devils' first rebellion: as the devils lost heaven, so they are further losing the *beorg*',[25] and Guthlac's settling of the *beorg* is thus a regaining of the seats the demons forfeited in heaven. *Guthlac A* directly connects this theology with the more fundamental restoration of Creation observed in the Guthlac tradition. In this poem Guthlac emerges as the New Adam in two interrelated ways, repairing the Fall between humanity and Creation, and restoring the eschatological trajectory for humanity so that it might fill the glorious dwellings made vacant in heaven.

and Fall of the Angels', *ASE* 31 (2002), 175–200, at 194–96. J.-M. Pont, 'Homo angelorum decimus ordo', *Cahiers de civilisation médiévale* 31 (1988), 43–48; M. Bridges, 'The Heroic and Elegiac Contexts of Two Old English Laments of the Fallen Angels: Towards a Theory of Medieval Daemonization', *Swiss Papers in English Language and Literature* 4 (1988), 117–32, at 122; T. D. Hill, 'The Fall of Angels and Man in the Old English Genesis B', in *Anglo-Saxon Poetry: Essays in Appreciation for John C. McGalliard*, ed. L. E. Nicholson and D. W. Frese (Notre Dame, IN, 1975), pp. 279–90, at 287.

22 Haines, 'Vacancies in Heaven', 151.

23 Haines, 'Vacancies in Heaven', 152.

24 Haines, 'Vacancies in Heaven', 152–53.

25 Clark, 'Land Tenure in *Guthlac A*', 87.

OEPG: The Lexis of Landscapes

As we saw in the previous chapter, the lexis of the *VSG* is characterised by neologisms and obscure terminology which are employed to create an enargaeic physical landscape. In order to render Felix's complex Latin landscape into the vernacular, the OE author creates a number of novel compounds, including a number of *hapax legomena* in the initial description of the fens. A particularly striking feature of the *OEPG*'s style is the close resemblance between its lexis and that used in vernacular boundary clauses of the mid tenth century. As Scott Smith, Della Hooke, and Nicholas Howe have shown, the purpose of these clauses was to provide a locally intelligible delineation of the landscape, a mental map by which the people who lived within and around the boundaries might recognise novel divisions of land.[26] The use of the vernacular for these boundary clauses is necessary given their function: the clauses only work if local people have access to their meaning, something unlikely if written in Latin like the rest of the charter.[27] The author of the *OEPG* likewise utilises a precise and widely understandable lexis in the vernacular to create a locally intelligible topographical description. Hooke highlights the wide variety of specific vocabulary employed by the Anglo-Saxons and its remarkable precision:

> While a *burna* tended to be a substantial stream or small river, a *winterburna* was one marked with seasonal flow; really small watercourses might be described by the term *sic* [...] and a drainage channel in a marsh might be termed a *sihtran*.[28]

Functionally these terms needed to be precise enough that the boundaries of the properties could be walked, the clause's perambulatory detail providing an accurate narrative representation of the landscape. When translating Felix's prolix description of the fens, the OE author chooses to employ these terms of topographical

26 See S. T. Smith, *Land and Book: Literature and Land Tenure in Anglo-Saxon England* (Toronto, 2012), pp. 156–57; Howe, *Writing the Map*, pp. 31, 37; and D. Hooke, *The Landscape of Anglo-Saxon England* (London, 1998), pp. 11, 86–87, 92.

27 Howe, *Writing the Map*, p. 37. There are several examples of Latin used in boundary clauses, for example S 262 (see below, note 36), but the overwhelming majority were written in Old English.

28 Hooke, *Landscape*, p. 92.

precision, such that the depiction is highly enargaeic and precisely imaginable:

Vita Guthlaci:

Est in meditullaneis Brittanniae patribus inmensae magnitudinis aterrima palus, quae, a Grontae fluminis ripis incipiens, haud procul a castello quem dicunt nomine Gronte, nunc stagnis, nunc flactris, interdum nigris fusi vaporis laticibus, necnon et crebris insularum nemorumque intervenientibus flexuosis rivigarum anfractibus, ab austro in aquilonem mare tenus longissimo tractu protenditur.[29]

(There is in the midland district of Britain a most dismal fen of immense size, which begins at the banks of the river Granta not far from the camp which is called Cambridge, and stretches from the south as far north as the sea. It is a very long tract, now consisting of marshes, now of bogs, sometimes of black waters overhung by fog, sometimes studded with wooded islands and traversed by the windings of tortuous streams.)

OEPG:

Ys on Bretonelande sum fenn unmætre mycelnysse, þæt onginneð fram Granta ea naht feor from þære cestre, ðære ylcan nama ys nemned Granteceaster. Þær synd unmæte moras, hwilon sweart wætersteal, and hwilon fule eariþas yrnende, and swylce eac manige ealand, and hreod, and beorhgas and treowgewrido, and hit mid menigfealdan bignyssum widgille and lang þurhwunað on norðsæ.[30]

(There is in Britain a fen of boundless size, which begins from the River Granta not far from the city of the same name, named Grantchester. There are immense moors, sometimes dark stagnant-pools, sometimes foul rivulets running; and also many islands, and reeds, and barrows, and tree-thickets. It continues with vast and long windings to the North sea.)

The Old English author's translation method here is word-for-word, retaining almost every detail of Felix's highly descriptive passage. For example, he replicates the forward momentum of Felix's text by rendering the Latin *nunc* [...] *nunc* structure with

[29] *VSG*, pp. 86–87.
[30] *OEPG*, p. 113.

OE *hwilon* [...] *hwilon*. The immensity of the fens is evoked through a similarly exaggerated construction, 'fenn *unmætre mycelnysse*', which retains the double amplification of a noun and an adjective for size modifying *fen*. Notable OE expansions include the addition of the intensifying adjective '*unmæte* moras' for the Latin *stagnum*. The sense of Felix's descriptive neologism *flactris* ('marsh' or 'reed-bed') is rendered through the *hapax legomenon* 'sweart wæter-steal'. *Sweart* is part of the basic Old English colour lexis, meaning 'swarthy, black, dark',[31] with the secondary meaning of 'absence of light or brightness',[32] and a tertiary meaning, the absence of good, therefore 'dark' and 'dismal' in a moral and spiritual sense.[33] While the term is often used to gloss Latin *ater*,[34] in this instance it bears the full range of connotations outlined above to convey the range of meanings implied by Felix's description of the fens. The compound *wæter-steal* appears, on the surface, to be a straightforward literal rendering of Felix's 'flactris', but it draws its descriptive potency here from the shared landscape lexis with boundary clauses. The second element, *steall*, usually meaning 'a standing position' or 'position or place', provides the necessary sense of stasis.[35] A close analogue is provided by the compound *treow-steall*, solely attested in three charters from the early to mid tenth century: S 361, *c.* 900 for 904; S 402, for 929; and S 496, *c.* 944 for 942.[36] While compounds involving *steall* are not exclusive to the charter record, in most cases the term is applied to denote a physical place, such as in *Vainglory* l. 39: 'þæt he þæt *wigstæl* wergan sceolde' ('that he should defend that

31 *BT*. C. P. Biggam, 'Sociolinguistic Aspects of Old English Colour Lexemes', *ASE* 24 (1995), 51–65, at 53, defines a 'basic colour term' as a word that would have 'been regularly used by all adult native speakers'.

32 See R. E. Woolf, 'The Devil in Old English Poetry', *RES* 4 (1953), 1–12, at 9, who discusses the use of *sweart* as a modifier for the flames in hell which bring no illumination.

33 *BT*.

34 For example, Hymn 23, l. 1, in I. B. Milfull, *The Hymns of the Anglo-Saxon Church: A Study and Edition of the Durham Hymnal* (Cambridge, 1996), p. 159.

35 *BT*; Roberts, in 'Guthlac: an Edition', notes a similar construction in the seventh homily of the Vercelli Book, *meresteallum*, p. 397.

36 Charter references are labelled S by their number in *The Electronic Sawyer*, P. Sawyer, *Anglo-Saxon Charters: an Annotated List and Bibliography*, Royal Historical Society Guides and Handbooks 8 (London, 1968).

bastion').[37] The unique combination of *sweart*, and its connotations, with *wæter-steal* in the *OEPG* creates a vernacular image not only close to that of the *VSG*, but also one of precise topographical detail. When read or heard the enargaeic image conjured would have been widely imaginable and evocative: the dark, stagnant, and familiar peaty pools of the fens.

The OE author is consistent in his method of translation, seeking to depict the landscape of the fens and Crowland as plausibly as possible through the implementation of vocabulary shared with landscape features in boundary clauses. For example, he renders Felix's 'flexuosis rivigarum anfractibus' ('the windings of tortuous streams') with the phrase 'fule *ea-riþas*' ('foul *streamlets*'). While the compound is itself another *hapax*, the elements used to construct it, as well as the intensifying adjective applied to it, are common to landscape vocabulary. *Ful* is applied here with its entire range of connotations, including 'foul' and 'physically loathsome', but with special emphasis on its olfactory function, 'putrid' and 'foul-smelling',[38] and thus translates the pungent imagery employed by Felix with 'flactris' and 'stagnis'. Old English *ful* also appears frequently as a descriptor of landscape features in boundary clauses such as S 414, a charter from King Æthelstan in 931: 'of þam gete [ond] lang herpoþes on þæt *fule sloh*, of þam [*slo*] west be wyrttruman to Cocggan hylle' ('from the gate along the high-road to the *foul slough*; from the *slough* west along the woodbank to Tog Hill').[39] The collocation of *fule* with *sloh* ('mire', 'slough,' or 'a pathless, miry place')[40] is representative of the kind of physical depiction found in *OEPG*, suggesting the author drew from a pool of landscape vocabulary shared with the authors of boundary clauses, to convey the physicality of the East Anglian landscape in terms which were evocative, detailed, and accurate.

This use of landscape lexis shared by boundary clauses can likewise be seen in another *hapax*: *ea-riþas*. The first element, *ea*, is

37 Text cited from *The Exeter Anthology of Old English Poetry: an Edition of Exeter Dean and Chapter MS 3501*, ed. B. J. Muir, second edition (Exeter, 2000).

38 *DOE*, 1; 1a; 2a.

39 I have expanded the 7 to ond. Also, see S 584 (King Eadwig to St Mary's abbey, Abingdon, 956) as well as S 617 (King Eadwig to Brihtnoth at Tadmarton, Oxfordshire, 956).

40 *BT*.

common enough,[41] but the second, *riþig* or *riþ(ð)*, is an extremely rare word outside of its use in charters, appearing primarily as a gloss or within translations of Latin texts; for example, *riþ* appears as a gloss for Latin *rivus* twice in Aldhelm's prose *De Laude Virginitatis*; as a sense translation of Num. 20:11; and as an expansion of the image 'fons bonorum' in the Old English translation of Boethius' *De consolatione philosophiae*: 'manige brocas and *riðan*' ('many brooks and *streamlets*').[42] Ælfric, likely following Aldhlem, employs the word in his Homily XXX *Assumptio Sanctae Mariae Virginis*, where he writes 'streamlicum riðum' ('rivers of waters') to translate the phrase 'rivos aquarum' ('rivers of waters') from the *Responsoria sive Antiphonae de Assumptione sanctae Mariae*, a text linked to Gregory the Great.[43] The use of the noun *riþ* as a translation for Latin *rivus* appears in another text connected to Gregory the Great, his *Cura Pastoralis*, where the Old English gives 'riðum' as a translation for the Latin's 'riuus'.[44] It appears once more in the Metrical Epilogue, likely following the phrasing from the prose *Regula*: 'riðum torrinan' (l. 18).[45] With these few exceptions, again almost entirely related to translating or glossing Latin terminology, the noun *riþ* is rare outside of the use of charters. Including the glosses, there are no examples outside of the *OEPG* and the charter record where this

41 *DOE*, *Ea*: 1,'river, large body of running water'.

42 Cf. *OE Corpus*. See *The Old English Glosses of MS. Brussels, Royal Library, 1650 (Aldhelm's De laudibus virginitatis)*, ed. L. Goossens (Brussels, 1974), ll. 1442, 3467; *The Old English Heptateuch and Ælfric's Libellus de Veteri Testamento et Novo*, ed. R. Marsden (Oxford, 2008), p. 149; *The Old English Boethius: an Edition of the Old English Versions of Boethius's* De consolation philosophiae, ed. M. Godden and S. Irvine, with contributions from Mark Griffith and Rohini Jayatilaka (Oxford, 2009), vol. 1, Ch. 34 (B-Text), p. 318. Compare with Boethius, *De consolatione philosophiae; Opuscula theologica*, ed. C. Moreschini (Munich, 2000), 3. p.10, section 3. See also Milfull, *Hymns of the Anglo-Saxon Church*, p. 145, l. 6, where Hymn 16 has *riþ* as a gloss to *rivulus* ('a small book, rivulet'), a word related to *rivus*.

43 Ælfric, *Assumptio Sancte Marie Virginis*, in *Ælfric's Catholic Homilies: the First Series*, ed. P. Clemoes (Oxford, 1997), p. 433; Gregory the Great, *Responsoria sive Antiphonae de Assumptione sanctae Mariae*, *PL* 38, Col. 798A–798B. The attribution to Gregory the Great is found in *PL*, but the source is listed as Anonymous in the *Fontes* database.

44 *King Alfred's West-Saxon Version of Gregory's Pastoral Care*, ed. H. Sweet, EETS, original series 45, 2 vols (1871, repr. New York, 1978), p. 277.

45 'Metrical Epilogue to the *Pastoral Care*', in *The Anglo-Saxon Minor Poems*, ed. E. V. K. Dobbie, ASPR 6 (London, 1942).

particular form of the word, *riþig*, appears. By contrast there are more than fifty examples of *riþig* and *riþ* in boundary clauses as a landscape word meaning 'tributary streamlet'.[46] For example, S 673, a grant from King Edgar to Abingdon Abbey in 958 (for 959), reads: 'of þa furh þæt innan þæt *riþig*, of þam *riþige* þæt innan Land broc' ('From the furrow to that *streamlet*, and from that *streamlet* to the land brook').[47] Thus the *hapax legomenon ea-riþas*, and its adjective *ful*, anchor the *OEPG* in a more immediately recognisable landscape than Felix's *VSG*, drawing on the kind of specific terminology with which the Anglo-Saxons in the tenth century delineated their own physical landscape.

The OE author continues to employ language which echoes stock terminology for describing landscapes in charters throughout the translation. Another example is the *hapax* compound 'treow-gewrido'. As with the majority of the novel compounds in the *OEPG*, the first element is a common word, *treow* ('tree'),[48] while the second element, *gewrid*, is only found in the *OEPG*, where it appears four times, and as a gloss to *frutectum* and *glumulum* in the *Latin-Old English Glossaries*.[49] *Frutectum* ('a thorn-bush, briar; or land covered with thorn-bushes, thickets'),[50] a term which can indicate a dense gathering of plants or trees, most closely resembles the usage of *gewrid* in the *OEPG*. Old English *gewrid* is derived from another rare noun, *wrid* ('plant' or 'bush'), which is found only in the *Old English Herbarium* (twice), the *Lacnunga* (once), and in three boundary clauses (all from the mid tenth century).[51] In the *Herbarium* and *Lacnunga*, *wrid* appears to refer to a single plant, modified by a name or other term. For example:

46 Cf. *OE Corpus*; Hooke, *Landscape*, p. 87.

47 Cf. S 737, a grant from King Edgar in 966, 'of þan up heafdan on mær dene middewearde to þan *riþige*, of þan *riþie* be þæra æcera heafdan to þan ealdan dic' ('From the heads up on the boundary of the valley, middle-wards to the *streamlet*, from the *streamlet* by the acres' head to the old ditch').

48 *BT*.

49 Cf. *OE Corpus*. For glosses see W. G. Stryker, 'The Latin-Old English Glossary in MS. Cotton Cleopatra A.III', Unpub. Ph.D. Diss. (Stanford University, 1951); corrections by M. Voss, 'Strykers Edition des alphabetischen Cleopatraglossars: Corrigenda und Addenda', *Arbeiten aus Anglistik und Amerikanistik* 13 (1988), 123–38.

50 *DMLBS*.

51 Cf. *OE Corpus* and *Sawyer*: S 412, S 444, and S 495. The term appears five times in these charters.

'Gif hwa forbærned sy genim þas ylcan wyrte *serpillum* ond æscþrote anne *wrid*' ('If anyone is burned then take the same plant *thyme* and a whole ashthroat *plant*').[52] In S 495, a grant made by King Edmund in 944, however, the usage is much closer in meaning to that found in the *OEPG*, where it twice depicts a dense collection of hazel: 'þurh þone lea to þam miclan *hæsl wride*. of þam *hæsl wride* adun on þa blacan rixa' ('Through the meadow to the great *hazel-thicket*; from the *hazel-thicket* down towards the black rush'). The second element of the compound *hæsl-wride* is used as a collective gathering noun implying dense intertwining shoots as a means of granting expanse, especially when modified by *micel*. The same compound + adjective phrase appears in a slightly earlier grant (with a variant spelling of -ð for -d), S 412 (AD 931): 'of witan hamme on ðone *miclan hæslwrið* wið neoðan þæt grafet . of þam *hæslwriðe* on ge rihte wið þæs lytlan stan beorges up on hæsl hille' ('From Wittenham up to the *great hazel-ride* under the ditch; from that *hazel-ride* on the right towards the little stone mound up on hazel hill'). The compound use of *wrid* in these examples from the charter corpus to denote a thick collection of trees is paralleled by the *OEPG*'s novel *treow-gewrid*, further strengthening the connection between the lexis of the *OEPG* and the boundary clauses of the early to mid tenth century.

The *OEPG* follows a similar pattern in translating Felix's account of Guthlac's journey with his guide Tatwine through the fens to Crowland itself:

VSG:	*OEPG*:
Quo audito, vir beatae recordationis Guthlac illum locum monstrari sibi a narrante efflagitabat. Ipse enim imperiis viri annuens, arrepta piscatoria scafula, per invia lustra inter atrae paludis margines Christo viatore ad praedictum locum usque pervenit; Crugland dicitur.[53]	Mid þam þe se halga wer Guðlac þa word gehyrde, he bæd sona, þæt he him þa stowe getæhte, and he þa sona swa dyde. Eode þa on scip, and þa ferdon begen þurh þa rugan fennas, oþ þæt hi comon to þære stowe, þe man hateð Cruwland.[54]

52 *The Old English Herbarium and Medicina de quadrupedibus*, ed. H. J. de Vriend (London, 1984), CI. 3, p. 148; this is similar to *Lacnunga*, in *Anglo-Saxon Magic and Medicine: Illustrated Specially from the Semi-pagan text 'Lacnunga'*, ed. J. H. G. Grattan and C. Singer (London, 1952), Ch. LXXXIII. p. 156.

53 *VSG*, pp. 88–89.

54 *OEPG*, p. 114.

(Guthlac, the man of blessed memory, on hearing this, earnestly besought his informant to show him the place. Tatwine accordingly assented to the commands of the man and, taking a fisherman's skiff, made his way, travelling with Christ, through the trackless bogs within the confines of the dismal marsh till he came to the said spot; it is called Crowland.)

(When the holy man Guthlac heard these words, he bade him immediately show him the place, and he immediately did so; he climbed on a ship, and they both fared through the uncultivated fens until they came to that place, which men call Crowland.)

Here the Old English again follows the Latin closely, though it abbreviates Felix's lengthy depiction of the fens to the simple statement: 'þurh þa rugan fennas'. The use of the term *rugan/ruh/ruwan* ('rough, hairy, shaggy'; 'untrimmed, uncultivated'),[55] while covering much of Felix's depiction in its semantic range, is most likely being employed here with its more specific landscape description of 'wildness'.[56] Aside from here in the *OEPG*, this latter usage appears in the extant corpus to charters and *Bald's Leechbook*. In the *Leechbook* it is used twice to signify an uncultivated form of a specific plant, '*ruwan* wegbrædan' ('*uncultivated* greater plantain'),[57] while in the charter evidence it is always used to describe a wild and untended portion of the landscape. For example, in S 649, from King Eadwig to Wulfstan in 957, *ruh* is used specifically to modify the noun *fen*: 'ond lang broces on *gyruwan fen* þa eft on bugla fen' ('Along the brook to the *uncultivated fen*, then to the inhabited fen'). Of the more than seventy-five uses of *rugan/ruh/ruwan* in the charter evidence, it is used nearly thirty times to refer to a mound, designated by the often interchangeable terms *beorg* and *hlæw*. This is telling given the centrality of the *tumulus* in the *VSG*, which is usually translated by the OE author here as *hlæw*. A good example can be seen in S 482, from King Edmund to Sæthryth in 942: 'on *ruwan beorg*, of *ruwan*

55 *BT*.

56 *BT*.

57 Cf. *OE Corpus*. *Bald's Leechbook*, in *Leechdoms, Wortcunning and Starcraft of Early England*, ed. T. O. Cockayne, Rolls Series 35 (London, 1864–6), vol. 2, I.41, p. 106. The second use of the same phrase is in II.65, p. 292.

beorge on teappan treow' ('to the *uncultivated barrow,* and from the *uncultivated barrow* to the tree of Teappa'). When considered in the light of the charter corpus, the translator's use of the adjective *rugan/ruh/ruwan* contributes to a physical depiction of extraordinary depth and accuracy. This lexical choice not only includes Felix's earlier description of Crowland as uncultivated (*inculta*) and physically rough, but also evokes one of the most common landscape features associated with *rugan/ruh/ruwan,* OE *beorg/hlæw,* the very location which Guthlac is arriving to occupy and, through his saintly presence, restore to its prelapsarian state.

The OE author's depiction of Crowland is likewise constructed by employing a lexis shared with contemporary boundary clauses to create a more physically delineated and realisable location than the *VSG*. This can be demonstrated by the OE author's use of OE *hlæw* for Latin *tumulus* in reference to the site of Guthlac's hermitage:

> Wæs þær on þam ealande sum *hlaw* mycel ofer eorðan geworht, þone ylcan men iu geara for feos wilnunga gedulfon and bræcon. Þa wæs þær on oþre sidan þæs *hlawes* gedolfen swylce mycel wæter-seað wære; on þam seaðe ufan se eadiga wer Guðlac him hus getimbrode.[58]
>
> (There was on that island a great *barrow* built over the earth, which men long ago had delved and broken desiring treasure. There was also, on one side of the *barrow,* a great well. Over this well the blessed Guthlac constructed a house for himself.)

This choice of *hlæw* is a departure from the more common *beorg,* which is the preferred term in both of the poetic renditions of Guthlac's life. While OE *hlæw* does cover similar lexical ground as *beorg* ('a constructed mound' or 'funeral mound'),[59] the extant evidence places it firmly in the category of terms connected to the description of the physical landscape in charters.[60] The term *beorg* appears more than 400 times in the extant corpus in everything from poetry to homiletic prose, to charters and grants,[61] while *hlæw* appears only

58 *OEPG,* pp. 117–18.

59 *BT.*

60 Hooke, *Landscape,* p. 99, argues that the term was primarily applied in charters to refer to burial mounds built or reused by the Anglo-Saxons.

61 Cf. *OE Corpus* and *BT.*

about 156 times, of which 142 attestations are in charters; the remaining fourteen are found in a few specific texts, mostly poetry.[62] This evidence suggests that, outside of a few instances in poetry, *hlæw* was a term tied directly to the same common landscape lexis from which boundary clauses were drawn. The OE author's decision to use this term provides another example of his efforts to ground the physical landscape of the *OEPG* by utilising language otherwise employed to describe the concrete, delineated natural landscape. The charter evidence suggests that OE *hlæw* was a very stable descriptive term, appearing in both early and late sources ranging from S 141, a grant made by Offa of Mercia between 757 and 774, to S 1351, a grant from Oswald, bishop of Worcester, in 985. Another example is provided by S 544 from 949:

> Æræst on ða halgan ac, swa ollonc þæs gemær heges to *þæm slo* æt *þæm more* ufewerdan, of *ðæm more* on dunes pyt [...] on þone oðerne weg to *þæm Rugan hlawe*, þonan on gerihte on ðæt lytle *riðig*.
>
> (First to the holy oak, along the boundary hedge to *the slough* at the upper part of *the marsh*, from *the marsh* to the hill's pit [...] onto the other way to *the rough barrow*, then straight on into the little *streamlet*).

In this single boundary clause we find four landscape terms used in the *OEPG*: *sloh*, *riþige*, *rugan*, and *hlæw*. This dense collocation of landscape lexis highlights the method by which the OE author has rendered the enargaeic descriptions in Felix's *VSG* for a vernacular audience. The specific use of *hlæw*, so central to the Guthlac legend, further fixes Guthlac's progression in sanctity to the physical landscape.

Spiritual Progression in the OE Landscape

As we have seen the *OEPG* simplifies and abbreviates the text of the *VSG* whenever possible, in particular Felix's prolix and often poetic descriptions. But the *OEPG* nonetheless preserves the main themes

[62] Cf. *OE Corpus*. For example, see *Beowulf*, ll. 1119, 2293, 2409, 2773, 2802, 3169; and *Maxims II*, in *The Anglo-Saxon Minor Poems*, ed. Dobbie, l. 26b. The term *hlæw* also appears in *The Phoenix*, the *Meters of Boethius*, the *Metrical Charms*, and the *Paris Psalter*. It also appears in the *Latin-Old English Glossaries* for the word *aggerens* 'earthwork'.

and structure of its primary Latin source, including Felix's emphasis on Guthlac's Antonian progression in sanctity in connection with Creation. While this progression is still primarily depicted through Guthlac's response to his devilish tormentors, the *OEPG* places more emphasis on the hardship of the hermit's loneliness, as demonstrated in the initial description of Crowland:

VSG:	*OEPG:*
quidam de illic adstantibus nomine Tatwine se scisse aliam insulam in abditis remotioris heremi partibus confitebatur, quam multi inhabitare temtantes propter incognita heremi monstra et diversarum formarum terrores reprobaverunt.[63]	þa wæs Tatwine gehaten sum man, sæde þa, þæt he wiste sum ealand synderlice digle, þæt oft menige men eardian ongunnon, ac for menigfealdum brogum and egsum *and for annysse þæs widgillan westenes* þæt hit nænig man adreogan ne mihte, ac hit ælc forþan befluge.[64]
(a certain man among those standing by, whose name was Tatwine, declared that he knew a certain island in the more remote and hidden parts of that desert; many had attempted to dwell there, but had rejected it on account of the unknown portents of the desert and its terrors of various shapes.)	(There was a certain man named Tatwine, who said that he knew a remote and secret island, that many men had often tried to settle, but for the manifold terrors and fears, *and the loneliness of that wide wilderness*, none could endure it, but they all fled.)

The inclusion of the novel hardship of loneliness ('for annysse') here is a striking divergence from the Latin source,[65] transforming the central struggle of Guthlac's eremitic life, the physical and spiritual battle over the *tumulus*, into one characterised by his personal difficulty of pursuing God alone, in isolation from the cenobitic

63 *VSG*, pp. 88–89.

64 *OEPG*, p. 114.

65 Roberts, 'Guthlac: an Edition', p. 400: 'Loneliness (*annysse* 181) is not mentioned by Felix as one of the terrors of the island and where his settlers *reprobauerunt* [331] the island, the *Life* tells us that no man could endure its loneliness, but *hit ælc forþan befluge* 183.'

milieu. As discussed in chapter 2 of this book, this unease with the strictly eremitic life is present even in the works of Bede. This concern, however, becomes particularly heightened from the mid-tenth century onwards as a result of the English Benedictine Reform. As Mary Clayton, Tom Licence, and others have shown, the reformers had a strained relationship with hermits, exacerbated, no doubt, by their apparent popularity.[66] Ælfric, in particular, has been noted for his deliberate 'avoidance of eremitic saints' in general,[67] and his exclusion of Guthlac and other hermits from his *Lives of Saints*.[68] The OE author appears to be addressing a similar sense of unease about the eremitic life by emphasising that solitude itself can be perilous. By highlighting the difficulty inherent in eremitic isolation, the OE author provides a reminder of the benefits of membership in a church community, while still retaining the central message of Felix's text.

The *OEPG*'s focus on the dangers of isolation alters Guthlac's initial temptation, where the devil appears as a roaring lion with venomous arrows,[69] by interpreting it as centred on loneliness:

VSG:	*OEPG:*
tum miles Christi totis sensibus turbatus de eo, quod incoeperat, *desperare coepit* et huc illucque turbulentum animum convertens, quo solo sederet, nesciebat. Nam cum sua ante commissa crimina inmensi ponderis fuisse meditabatur.[70]	ða wæs his mod þæs eadigan weres swiðe gedrefed on him *be þam onginne, þe he ongan, þæt westen swa ana eardigan.* Mid þam he þa hine hider and þyder gelomlice on his mode cyrde, and gemunde þa ærran synna and leahtras, þe he gefremede and geworht hæfde.[71]

66 Clayton, 'Hermits in Anglo-Saxon England', p. 158; T. Licence, *Hermits and Recluses in English Society 950–1200* (Oxford, 2011), pp. 47–49.

67 Clayton, 'Hermits in Anglo-Saxon England', p. 162.

68 See Licence, *Hermits and Recluses*, p. 47; also, see C. A. Jones, 'Ælfric and the Limits of "Benedictine Reform"', in *A Companion to Ælfric*, ed. H. Magennis and M. Swan (Boston, 2009), pp. 67–108, at 98–103. As Licence mentions, it has been argued that this type of adaptation for St Guthlac occurs in *Guthlac A* as well. See Jones, 'Envisioning the *Cenobium*'. For Ælfric's treatment of Cuthbert in his *Catholic Homilies*, see below, pp. 289–90.

69 *OEPG*, p. 119.

70 *VSG*, pp. 96–97.

71 *OEPG*, p. 120.

(Then every feeling of the soldier of Christ was disturbed by it, and he *began to despair* about what he had undertaken, and turning things over in his troubled mind he knew not in what place to rest. For when he remembered that the sins he had committed in the past were of immense weight.)	(Then the mind of the blessed man was greatly disturbed about the *task he had began, to live alone in the wilderness.* Then he turned himself hither and thither continually in his mind, and thought of his old sins and crimes which he had committed and wrought.)

The saint's ambiguous despair of the *VSG* is more precisely defined in the *OEPG* as total isolation and separation from community of any kind. This echoes Tatwine's account of how Crowland had remained uninhabited due in great part to its loneliness.[72] The *OEPG* interprets Guthlac's ensuing turbulence of mind, the image which in the *VSG* displays Guthlac's novice status, as directly caused by this loneliness. The OE author causally connects this loneliness with Guthlac's instability by rendering the multivalent adjectives of Felix, *turbatus* and *turbulentus*, through a simple visual image: 'Mid þam he þa hine *hider and þyder* gelomlice on his mode cyrde' ('Then he turned himself *hither and thither* continually in his mind').[73] This physical image renders the complex internal turbulence of the *VSG* in a much clearer and more easily imaginable way, aiding the *OEPG*'s vivid description of the idea of the Latin text for a vernacular audience. The instability of mind displayed in this image also echoes wider patristic tradition, as discussed in chapter 1 concerning the *VCA*,[74] where the wandering of both mind and body are perceived as potentially sinful. Benedict in his *Regula* cautions against becoming like the unstable Gyrovagues, 'semper vagi et numquam stabiles' ('always wandering and never stable'),[75]

72 *OEPG*, p. 114.

73 *OEPG*, p. 120. The focus on loneliness and its associated mental disturbance extends to one further character in the *OEPG*, King Æthelbald, who is described with a similar image of turning his mind hither and thither: 'Ða he þa on þam huse inne wæs, *þa wæs he on þam unrotan mode hider and þyder þencende*' ('Then when he was in that house, *then with his sorrowful mind hither and thither thinking*'), p. 171. Compare *VSG*, p. 164, which has a similar image.

74 See above, p. 25 n. 30.

75 Benedict, *Regula*, Ch. I.10–11, pp. 438–40.

and Evagrius Ponticus delineated how demonic thoughts, *logismoi* (λογισμοι), function to distract the ascetic from the stability necessary for divine contemplation.[76] In the *OEPG*, these complex notions have been rendered with the evocative but simple image of a mind turning back and forth to aid in its wider appeal for a vernacular audience.

The *OEPG* likewise adapts Guthlac's external reaction, his retreat and prayer for help, to focus more specifically on loneliness:

VSG:	*OEPG:*
Igitur vir praefatus, veluti miles inter densas acies dimicans, cum caeleste adiutorium *angelicae lucis* adventasse persensisset, extimplo discussis *nefandarum cogitationum nebulis, inluminatio* turbulenti pectoris gremio, velut triumphali voce psallebat aiens.[77]	And he þa sona se eadiga wer Guðlac swiþe bliþe wæs þæs heofonlican cuman; and him sona *his heorte and his geþanc eall wæs onlihtod;* and he þa hrædlice þa yfelan and *þa twyfealdan geþohtas forlet*.[78]
(So when this same man, like a soldier fighting in the serried ranks, had realized that the heavenly aid and angelic light had reached him, immediately the *clouds of impious thought* were dissipated, his troubled heart was *enlightened* and he sang triumphantly.)	(And immediately the blessed man Guthlac was very happy for the heavenly arrival; and immediately *his heart and mind were all enlightened*; and he *swiftly abandoned* the evil *double-thoughts*.)

Here the juxtaposition between the illumination of St Bartholomew and the clouds of 'impious thoughts' in Felix is adapted to create a juxtaposition between the loneliness of the physically-described Crowland and the banishing of that loneliness, and its troubling

76 Evagrius, *On Thoughts*, in *Evagrius Ponticus*, ed. and trans. A. M. Casiday (London, 2006), p. 91; for Greek text see *Évagre le Pontique: Sur Les Pensées*, ed. P. Géhin, C. Guillamont and A. Guillamont (Paris, 1998). For a wider discussion of these issues see Leneghan, '*The Wanderer, Hesychasm* and *Theosis*', at 125–27.

77 *VSG*, pp. 96–97.

78 *OEPG*, pp. 121–22. This section is only in the Vespasian *Life*, not in Vercelli Homily XXIII, though the sense of the passage remains the same.

of Guthlac's internal state (*heorte* and *geþanc*), by the illumination of Bartholomew. The OE author presents the saint's conquest of the mental turmoil with the word *twyfealdan* ('two-fold', 'doubtful', and 'irresolute'),[79] as opposed to the Latin's 'nefandarum cogitationum nebulis' ('clouds of impious thought'). In this way, the *OEPG* adapts Felix's account of Guthlac's internal instability in terms tied to the enargaeic landscape, in order to focus more on the hermit's state of loneliness and isolation. The effect is, on the one hand, to place more emphasis on the saint's inner spiritual progression, but also, on the other hand, to highlight the inherent dangers of a life lived in isolation from the ecclesial community. In both cases the narrative is fixed to a recognisable Anglo-Saxon landscape, which renders Guthlac's spiritual progression more vividly for a vernacular audience.

The OE author adapts Felix's presentation of Guthlac's progression towards spiritual majority in similar ways to the passage treating his initial depiction of the fens, combining the lexis shared by boundary clauses outlined above with Felix's images and their connected patristic exegesis of the Fall. This is evidenced in *OEPG* Ch. 5, which depicts the most heightened moment of Guthlac's spiritual progression, when he is physically immersed in the landscape, by utilising this landscape lexis together with terminology connected to Bede and Augustine's exegeses:

> Næs þa nænig yldend to þam þæt, syþþan hi on þæt hus comon, hi þa sona þone halgan wer eallum limum gebunden, and hi hine tugon and læddon ut of þære cytan, and hine þa læddon on þone *sweartan fenn*, and hine þa on þa *horwihtan wæter* bewurpon and besencton. Æfter þon hi hine læddon on þam *reðum stowum* þæs *westenes* betwux þa *þiccan gewrido* þara *bremela*, þæt him wæs eall se lichama gewundod.[80]
>
> (There was not then any hesitation, so that, after they had come into that house, they immediately bound the holy man in all his limbs, and dragged and led him out of the cottage, and then they led him into the *dark fen*, and threw him into the muddy water and submerged him. After that they led him into the *wild places* of the

79 *BT*.

80 *OEPG*, p. 129.

wilderness amongst the *dense thickets of the brambles*, so that his body was wounded throughout.)

Felix describes the fens as dark ('atrae'), and the waters as muddy ('coenosis'), terms which have here been rendered in OE with the familiar adjective *sweart*, and the *hapax legomenon* 'horwihtan'. The inclusion of the adjective *sweart*, collocated with *fenn*, as well as landscape terms discussed above such as *gewrid*, echoes the OE author's initial description of the fens and Crowland; an internal reflexivity which itself follows Felix's structure. This vocabulary can also be connected with the exegeses of the various Cuthbertine texts discussed above, as well as that of the *VSG*.[81] Where Felix employs the vocabulary of the Vulgate *Genesis* to describe the thorns Guthlac is dragged through, 'densissima *veprium vimina*' ('dense *thickets of brambles*'), the OE author chooses a term connected to both the shared lexis of boundary clauses, and the same moment in the Old English translation of Genesis, 'þa þiccan *gewrido* þara *bremela*'. The noun *bremel* is used to refer to a 'thorny plant' or 'bramble' within medical texts like the *Herbarium*,[82] and to translate Latin *vepris* in Ælfric's *Glossary*.[83] Old English *bremel* is also used as a landscape term in charters such as S 416, from 931: 'on *bremeles sceagan* easte weardne' ('to the *bramble thicket* eastward'). The noun *bremel* is also, importantly, applied to translate the central verse in the description of the Fall in Gen. 3:18 taken up by Augustine and Bede, as well as Felix: 'spinas et tribulos' becomes '[ð]ornas and *bremelas* heo a(sp) rit ðe' ('[i]t will bring forth thorns and *brambles* for you').[84] Although the Old English translation of Genesis postdates the *OEPG*,[85] the connection to the use of *bremel* biblically here is striking, especially given the context of this hagiographical moment and its relation to Augustinian and Bedan exegesis. Together with the internally reflexive use of the landscape adjectives *sweart* and *reþige*, and the noun *gewrido*, the choice of the noun *bremel* adds to the causal

81 See above, Introduction, pp. 9–15.

82 For example, *Old English Herbarium*, p. 128.

83 Ælfric, *Ælfrics Grammatik und Glossar: Text und Varianten*, ed. J. Zupitza; intro. H. Gneuss (Hildesheim, 2003), p. 312, ll. 11–12: '*uepres* bremlas'.

84 *OE Heptateuch*, Gen. 3:18, p. 13.

85 See R. Marsden, 'Ælfric as Translator: The Old English Prose *Genesis*', *Anglia* 109 (1991), 319–58, at 319–20.

connection between the highly enargaeic physical world and Guthlac's progression in sanctity. *Bremel* likewise connects this moment to the exegetical models of Augustine and Bede, similarly to Felix, where the thorns are being used by God to push Guthlac towards spiritual majority, allowing for the restoration of Creation. These adaptations once again open up the complexities of the *VSG* to a vernacular audience, by locating Guthlac's spiritual progression and the connected Augustinian/Bedan exegesis firmly in Anglo-Saxon soil.

The *OEPG* likewise carefully follows Felix's transition in the *VSG* from a physical landscape to one delineated by the metaphoric, utilising the airy regions the demons lift Guthlac into as a transitional space:

VSG:	*OEPG:*
Cum ergo ad ardua aeris culmina adventasset, horrendum dictu! ecce septentrionalis caeli plaga *fuscis atrarum nubium* caliginibus *nigrescere* videbatur.[86]	Þa he þa wæs on þære heannysse þære lyfte, þa geseah he ealne norðdæl heofones, swylce he wære þam *sweartestan wolcnum* ymbseald swiðlicra *þeostra*.[87]
(Now when he had reached the lofty summit of the sky, then, horrible to relate, lo! the region of the northern heavens seemed *to grow dark* with gloomy mists and *black clouds*.)	(Then when he was in the highness of the air he saw all the north-part of heaven surrounded as if by the *blackest clouds* of profound *darkness*.)

The OE author follows Felix by mirroring the reflexive use of vocabulary employed in both texts for the fens in the first part of the passages: just as 'fuscis nubium' echoes 'fuscis nubibus',[88] 'sweartestan wolcnum' alliteratively echoes 'sweart wætersteal',[89] further anchoring the depictions to the physical landscape. The *OEPG* then follows Felix's transition from the literal to the metaphoric by rendering the description of the demonic horde in the

86 *VSG*, pp. 104–05.
87 *OEPG*, p. 130.
88 *VSG*, pp. 104, 102.
89 *OEPG*, pp. 130, 113.

northern heavens with the noun *þeostru*.[90] While *þeostru* and its adjectival form *þeostre* are commonly used to mean simply 'without light' in a physical sense,[91] in most instances it denotes moral or spiritual depravity, an absence of internal light manifested by the absence of physical light. Hence, for example, *þeostre* is used to describe hell in *Christ and Satan*, '[þ]is is *ðeostræ* ham, ðearle gebunden / fæstum fyrclommum' ('[t]his is a *dark* home, firmly bound with fire-bonds') (ll. 38–39a),[92] and in *Genesis A* as the obverse of light, 'ða com oðer dæg, / leoht æfter *þeostrum*' ('then came the next day, / light after *darkness*') (ll. 143b–144a). Ælfric is very fond of the noun, and uses it in a similar sense: 'He is soð leoht. þe todræfde þa *þeostra* [...] ic eom leoht ealles middangeardes: Se ðe me fyligð ne cymð he na on *þystrum*. ac he hæfð lifes leoht' ('He is the true light, who banishes the *darkness* [...] I am the light of all the world: he who follows me will not come to *darkness*, but have the light of life').[93] The implementation of this term brings with it a colouring of moral and spiritual darkness, as well as physical. The OE author uses *þeostru* in the sense of moral and spiritual darkness again in a passage in the prologue of the *OEPG* which Waugh refers to as the 'blindness curse':[94]

> Forþon la, þu leornere, gif þu mid þan þeawe tælendra me hleahtrige, warna þe sylfne, þær þu þe hleahtres wene, þæt þu þær semninga ne wurðe mid dymnysse *þystro* ablend. Þæt bið blindra þeaw, þonne hi on leohte beoð, þæt hig sylfe nyton, buton hi on *þeostrum* dwelion.[95]
>
> (Therefore, you learner, if you scorn me after the way of slanderers, beware for yourself, lest while you scorn, you become suddenly

90 C. Giliberto, 'The *Descensus Ad Inferos* in the Old English Prose Life of St Guthlac and Vercelli Homily XXIII', in *Hagiography in Anglo-Saxon England: Adopting and Adapting Saints' Lives into Old English Prose (c. 950–1150)*, pp. 229–53, at 245–46, highlights how the locating of the devils in the northern heavens is connected to the tradition of that being the part of heaven Satan and his followers 'attempted to inhabit' before God banished them. Giliberto further highlights how the seat of Satan is placed in the northern heavens in *Genesis A*.

91 *BT*.

92 *Christ and Satan: an Old English Poem*, ed. M. D. Clubb (New Haven, 1925).

93 For example, Ælfric uses the related terms more than thirty times in his *Catholic Homilies* alone (Cf. *OE Corpus*); Cf. Ælfric, *In Purificatione Sanctae Mariae*, in *Catholic Homilies, the First Series*, p. 253, ll. 140–41.

94 Waugh, 'The Blindness Curse', 399–426.

95 *OEPG*, pp. 101–02. Felix uses the noun *tenebrae* here; see *VSG*, p. 63.

> blinded by the shadows of *darkness*. It is the way of blind men, when they are in the light, that they don't know they wander in the *dark*.)

The term *þeostru* is used here as an external description of an internal transformation. When *þeostru* is brought in to modify the dark clouds of night, the image created becomes a spiritually charged darkness that prefigures the imminent descent into hell. The distinction here between a descriptive term like *þeostru* and words such as *ful* and *rugan* used earlier to construct the physical landscape is evident in the consistent use of *þeostre* to mean internal depravity as opposed to external appearance; a distinction further supported by *þeostre*'s conspicuous absence from the corpus of boundary clauses which are filled with terms like *ful* and *rugan*. Both *þeostru* and *þeostre* are used a further three times in the *OEPG*'s rendering of this transformative episode in Guthlac's spiritual progression: first, by Guthlac himself to describe the demons ('*þeostra* bearn', 'children of *darkness*'); second, in the narrator's descriptive juxtaposition of Bartholomew's luminous entry and the state of hell itself ('þa dimnysse *þeostru* þære sweartan helle', 'the dim *darkness* of the black hell'); and finally, again by the narrator to describe hell into which the demons flee from Bartholomew ('ac hi sylfe on *þeostre* gehyddon', 'but they hid themselves in the *darkness*').[96] The OE author's departure from the landscape lexis shared with boundary clauses to metaphoric terms like *þeostru* represents an expansion of the transition in the *VSG* from the physical landscape of the fens to a metaphoric hell, which creates a more delineated separation of the two spaces.

The OE author follows Felix in depicting the landscape of hell in direct contrast to, but also intimately connected with, the delineated physical landscape of the fens. Where the *OEPG* differs from its main source, the *VSG*, is in its innovative use of boundary clause vocabulary to ground the description of the physical landscape. This allows an even greater contrast with the primarily metaphorical language used to describe hell. The depiction of hell in the *OEPG* also further connects with the specific moment of the Fall and its effects. The OE author amplifies the chaos of hell from the

[96] *OEPG*, pp. 133–34.

VSG, with its formlessness and flames, by subsuming the classical references to the names and features of Hades (Erebus, Styx, and Acheron) into a single descriptive line: 'on þas witu þisse *deopnysse*' ('into the torments of this *abyss*').[97] This succinct translation can be connected with the first description of hell in the *OEPG*, and its depiction of hell as smoke, fire, and darkness, bereft of spatial referent: 'þa *fulnysse* þæs *smyces* and þa *byrnenda lega* and þone *ege* þære *sweartan deopnysse*' ('the *foulness* of the *smoke* and the *burning flames* and the *terror* of the *black abyss*').[98] The use of the intensifying word *deopnes* in both passages utilises the terms' capaciousness, meaning both 'measureable depth' with its relative limitation, as well as the endless and formless depth of 'abyss'.[99] Thus it can be employed by Ælfric to refer to the great depths of the Gospel,[100] as well as a translation for Latin *abyssos* as seen in the Lambeth Psalter.[101] Likewise the term is applied in other texts to describe the abyssal nature of hell, for example in the Old English *Gospel of Nicodemus*: 'On þære hellican *deopnysse*' (in the hellish *abyss*').[102] The OE author is therefore presenting, in vivid vernacular form, the chaos at the centre of Felix's depiction of hell in his *OEPG*. It is a horror ('ege') of lightless ('sweartan'), unspecified fire and foul smoke, which at its heart is without form, boundary, or fixity. The lexical contrast with the carefully delineated natural landscape, recognisable to an Anglo-Saxon audience via the landscape vocabulary shared with boundary clauses, with this riotous amalgamation of fire and the abyss, creates a much more intense, vivid, and frightening depiction.

97 *OEPG*, p. 132. Cf. *VSG*, p. 106: 'en tibi patulis hiatibus igniflua Herebi hostia patescunt; nunc Stigiae fibrae te vorare malunt, tibi quoque aestivi Acherontis voragines horrendis faucibus hiscunt' ('Behold! the fiery entrances of Erebus gape for you with yawning mouths. Behold! the bowels of Styx long to devour you and the hot gulfs of Acheron gape with dreadful jaws.').

98 *OEPG*, p. 131.

99 *DOE*.

100 *Ælfric's Catholic Homilies*, p. 266, ll. 3–4.

101 *Der Lambeth-Psalter: eine altenglische Interlinearversion des Psalters in der Hs. 427 der erzbischöflichen Lambeth Palace Library*, ed. U. L. Lindelöf (Helsinki, 1909), p. 68, Ps. 41:8, 'deopnys deopnisse [...] *Abyssus abyssum*'.

102 *Gospel of Nichodemus*, in *Two Old English Apocrypha and their Manuscript Source: 'The Gospel of Nichodemus' and 'The Avenging of the Saviour'*, ed. J. E. Cross (Cambridge, 1996), Ch. XVIII, p. 201.

The author of the *OEPG* unites this depiction of hell with the Fall through his lexical choices in similar manner to his use of *bremel* discussed above. This connection is exemplified in one of the many juxtapositions between the formless fire and shadow of hell, and the brightness and implied fixity of heaven: 'þa *dimnysse þeostru* þære *sweartan* helle' ('the *dim darkness* of that *black* hell').[103] Here the noun *þeostru* is used again, but now in a dense collection of adjectives reminiscent of Felix. Darkness is modified by the adjective *dim*, and the shadowed landscape of the fens is modified by the adjective *sweart*. This collocation of three adjectives for darkness, *dim*, *þeostru*, and *sweart*, is representative of the OE author's attempt to present hell as the darkest, most morally corrupt and terrifying place conceivable. *Dim* ('without light', 'wretched and grevious', 'wicked'),[104] is often used with *sweart* in descriptions of that which is opposite to heaven.[105] Tellingly, the only other time in the extant corpus where the three terms are collocated is in *Genesis B*, where they are used to describe the Tree of Death: 'þonne wæs se oðer eallenga *sweart*, / *dim* and *þystre*. þæt wæs deaðes beam' ('the other tree was utterly *black*, *dim*, and *dark*: that was the tree of death') (ll. 477–79a). This collocation is associated with the death brought about by rebellions against God, either on the part of the demons in hell, or by humanity's action in prelapsarian Eden. Thus, when applied together, the contrast between that which is heavenly and that which is infernal is enhanced, and the connection to the overarching exegesis of the Fall and the saint's role in the restoration of Creation is maintained.

In the *OEPG*, Guthlac's path to victory over his devilish tormentors and his attainment of spiritual majority are tied to the physical landscape of the fens and Crowland to a greater extent than in Felix's *VSG*. The final confrontation, wherein the saint directly engages the demons, draws its narrative strength in the *OEPG* from the contrast between, on the one hand, the redeemable terrestrial landscape, and on the other, the boundless fire and ash of hell. With

103 *OEPG*, p. 133.

104 *BT*; *DOE*.

105 For example, *Christ and Satan*, ed. Clubb, ll. 110–11; and *The Old English Poem 'Judgment Day II': A Critical Edition with Editions of De die iudicii and the Hatton 113 Homily De domes dæges*, ed. G. D. Caie (Cambrige, 2000), ll. 104–06.

Guthlac's victory over the demons in this episode he has attained spiritual majority, and is able to progress towards restoring Creation; but in contrast to the *VSG*, it is a Creation delineated through a landscape lexis shared with boundary clauses. These various adaptations serve to highlight and delineate what the *VSG*'s complex Latin often alludes to, and as such create a text of wider appeal for its vernacular audience.

Restoration of Creation

The OE author's method of simplifying and abbreviating his source is further demonstrated in his treatment of Guthlac's interactions with Creation after he has attained spiritual majority. In each of the restoration miracles that follow the complexity of Felix's depictions has been untangled, the descriptions abbreviated, and the exegesis by which this restoration can occur given more emphasis. The most striking change is the truncation and amalgamation of Ch. XXXVII of the *VSG* (on the theft of parchment by ravens) and Ch. XXXVIII (Guthlac's endurance of the deprivations of two other ravens), into a single episode entitled 'Hu þæt gewrit begiten wæs' ('How the writing was found').[106] The initial portion of the episode, corresponding to the events of *VSG* Ch. XXXVII, is rendered faithfully in its basic content and progression.[107] The OE author has, however, removed any mention of the role of fate in this miracle, which is fundamental to Felix's depiction. Similarly no trace of Felix's discussion of the role of obedience in the restoration of Creation remains, save for the brief description of how the monk implicitly follows Guthlac's suggestion.[108] This omission is followed by another one, in which the OE author removes Felix's use of Mtt. 10:29 to explain Guthlac's faith in God's omnipotence, and his connected faith in the recovery of the document. The *OEPG* instead presents Guthlac assessing his own miracle in humble terms, but only after the miracle has occurred, and in no way directly connected to Felix's biblical

[106] *OEPG*, p. 140.
[107] *OEPG*, p. 141.
[108] *OEPG*, p. 141.

reference. These omissions of passages of clear importance to the *VSG*, didactic or otherwise, merit further consideration.

The OE author's amalgamation of *VSG* Chs. XXXVII–XXXVIII provides the rationale for the aforementioned omissions. Immediately following Guthlac's final assessment of the parchment miracle, the *OEPG* presents a substantially truncated adaptation of *VSG* Ch. XXXVIII, concerning the saint's endurance of two ravens. An entire chapter in the *VSG* is transformed into the final portion of the parchment miracle in the *OEPG*:

> Wæron on þam ylcan yglande twegen hrefnas gewunode, to þæs gifre, þæt swa hwæt swa hi mihton gegripan, þæt hi þæt woldon onweg alædan; and he þeahhwæþere heora gifernysse ealle *abær and geþolode*, þæt he eft sealde mannum bysene his geþyldes; and nalæs þæt an þæt him þa fugelas *underþeodde* wæron, ac eac swa þa fixas and wilde deor þæs westenes ealle hi him hyrdon, and he hym dæghwamlice andlyfene sealde of his agenre handa, swa heora gecynde wæs.[109]

> (There were living on that island two ravens, so greedy that they would carry away whatever they could seize; nevertheless *he bore and endured* all their greediness, so that he again gave to men the example of his patience. And not at all that alone, that the birds were *subject* to him, but likewise the fish and wild beasts of the wilderness all obeyed him, and he daily gave them nourishment from his own hand, as was natural for them.)

Felix's extended description of the ravens' thievery is omitted, as is the vitally important exegesis in this section. The material that is retained and adapted appears at first even more contradictory than in Felix; the first half of the passage above depicts how Guthlac suffered under two ravens, and the second seems to propound the opposite: that all of animate Creation is now subject ('underþeodde')[110] to the saint. Adapting Felix's convoluted presentation of Guthlac's saintly endurance, the OE author juxtaposes these two

[109] *OEPG*, p. 142.

[110] While the term *underþeodan* is common and used in a wide variety of ways, it is used by Ælfric to describe the hierarchical relationship of God-Humanity-Creation. See, for example, Ælfric's homily for the *Common of a Confessor*, in *Angelsächsische Homilien und Heiligenleben*, ed. O. B. Assmann (Kassel, 1889), p. 64: 'Nu habbað deor ege him betwynan, and þa wildan deor be ures drihtnes gesetnysse syndon mannum *underþeodde* and hi mannum þeowiað, ge wilde,

sections in order to present a more striking contrast between the saint's predicament and its solution, brought about through his sanctity. The reference to Guthlac's patience ('geþyldes') emphasises the exemplary nature of his actions. Similarly, the use of the verb *aberan* to translate Felix's 'perferens [...] longanimiter [...] sufferebat' evokes the lexis of homiletic literature, where it is often used in a negative sense.[111] Likewise the verb *geþolian* usually implies the willing choice to suffer, as can be seen in *Christ III*: 'Ic ðæt for worulde *geþolade*' ('I *suffered* for that world') (l. 557b [1423b]). Guthlac's willingness to suffer, to bear the ravens' destructive deeds, sets an example of saintly endurance for humanity.[112]

The omissions highlighted above, therefore, are not only a product of the OE author's consistent tightening of Felix's depiction, but also reflect his efforts to provide a more unified central focus on the restoration of Creation. Where Felix has a complex narrative focusing on the interaction between fate/providence, the obedience of the brother, and the Creation miracle, the *OEPG* presents a more streamlined version of the miracle of the parchment. The amalgamated chapter in its entirety now centres on Guthlac as restorer of prelapsarian order.

This same condensing approach is likewise evident in the *OEPG*'s treatment of the raven-and-glove miracle of *VSG* Ch. XI. Where the *VSG* engages with the tradition of moral depravity attached to the ravens on Noah's ark, the *OEPG* adapts this episode to function as an exemplum of Guthlac's restored authority over animate Creation:[113]

---

ge tame, and we sceolon beon gode *underþeodde*, þe ealle þing gesceop, se ðe ana rixað on ecnysse'.

111 For example, *Ælfric's Catholic Homilies*, p. 4, l. 32: 'Þa gif he come on ðære godcundnysse buton menniscnysse. þonne ne mihte ure tyddernys *aberan* his mihte'.

112 For a similar construction concerning imitative sanctity, see Wærferth's OE translation of Gregory's Dialogues, *Bischof Wærferths von Worcester Übersetzung der Dialoge Gregors des Grossen: über das leben und die wunderthaten italienischer Väter, und über die unsterblichkeit der seelen*, ed. H. Hecht, H. Johnson, and J. Zupitza, 2 vols (Leipzig, 1900–07), p. 99, ll. 11–14.

113 Cf. *VSG*, pp. 118–21.

He þa sona se halga wer Guðlac þone hrefn mid his worde *þreade* for his reþnysse, and he þa *his worde hyrsumode*; swa fleah se fugel west ofer þæt westen.[114]

(Immediately the holy man Guthlac *reproached* the raven with his word for its cruelty, and it *obeyed his word*; so the bird flew west over the wilderness.)

The raven's immediate obedience exemplifies how all the wild beasts and birds are now subordinate ('underþeodde') to the saint. Where the didactic explanation in Ch. IX of the *OEPG*, that all of animate Creation is now subordinate to Guthlac, functioned to give the necessary tension for Guthlac's saintly endurance to be displayed, here the explication is being displayed in the narrative itself: the saint orders, and animate Creation obeys ('hyrsumode'). While this depiction lacks the subtlety of Felix's depiction, including references to fate/providence, the *OEPG* neatly ties together this last example of restored animate Creation with the initial raven miracle to create an envelope pattern that unites this entire section.[115]

In keeping with Felix's narrative, the most vivid depiction of Guthlac's restoration of Creation in the *OEPG* is found in Ch. X's adaptation of the wild swallow miracle.[116] The OE author transforms this episode to create a clearer focus on the miracle of the prelapsarian relationship between the saint and the birds. For example, where Felix's chapter is entitled '[q]ualiter hirundines eius imperiis obtemperabant' ('how the swallows obeyed his orders'), focusing on the exemplary obedience in the birds' actions, the *OEPG* has the more simple and descriptive title, 'Hu þa swalawan on him sæton and sungon' ('How the swallows sat on him and sang'),[117] concentrating instead on the Edenic moment itself. The *OEPG*'s omission of the entire first sentence, which includes Felix's frame wherein

114 *OEPG*, p. 144.

115 This last raven miracle is the final in the series of Creation miracles in the *OEPG*, and thus envelopes them with the first miracle about thieving ravens.

116 Cf. *VSG*, pp. 120–23.

117 *VSG*, pp. 120–21; *OEPG*, p. 142. The titles are found in the Vespasian MS, and are therefore considered authorial. The material in Vercelli Homily XXIII begins after a title and a paragraph in Vespasian, and includes no titles during the section they share, and thus its lack of titles is not evidence against the authorial nature of them in Vespasian; see *OEPG*, pp. 117–34.

he describes the ensuing events as a spiritual miracle ('spiritale miraculum'),[118] sharpens this focus.[119] The *OEPG* instead launches directly into the narrative proper, describing Wilfrid's arrival.[120] Noticing this shift, Roberts has argued that it recentres the episode on 'the reason why wild creatures were so friendly with Guthlac'.[121] I would argue further that this omission also foregrounds Guthlac's miraculous restoration of Creation instead of Felix's spiritual interpretation in the *VSG*:

> þa comon þær sæmninga in twa swalewan fleogan, and hi efne blissiende heora sang upahofon; and þa æfter þon hi setton unforhtlice on þa sculdra þæs halgan weres Guðlaces, and hi þær heora sang upahofon, and hi eft setton on his breost and on his earmas and on his cneowu. Ða hi þa Wilfrið lange þa fugelas wundriende beheold, þa frægn hine Wilfriþ, forhwon þa wildan fugelas þæs widgillan westenes swa eadmodlice him on sæton. He þa se halga wer Guðlac him andswarode and him to cwæð: Ne leornodest þu, broðor Wilfrið, on halgum gewritum, *þæt se þe on godes willan his lif leofode, þæt hine wilde deor and wilde fugelas þe near wæron; and se man, þe hine wolde fram woruldmannum his lif libban, þæt hine englas þe near comon*.[122]

> (Suddenly two swallows came flying in there, and they lifted up their song as if rejoicing; and after that they then sat fearlessly on the shoulders of the holy man Guthlac, and lifted up their song there, and then sat on his chest, and on his arms, and his knees. When Wilfrith had long wondering seen the birds, then Wilfrith asked why the wild birds of the wide wilderness sat so humbly on him. Then the holy man Guthlac answered him and said: "Have you not learned, brother Wilfrid, in holy Scripture, *that he who has lived his life in God's will, the wild beasts and wild birds were near to him; and the man who lives his life apart from worldly men, to him the angels came near?*").

Equivalent to the adaptation of the raven-and-document miracle in Ch. IX, the *OEPG* removes any mention of fate or chance, which again serves to centre the miracle on physical Creation. This adaptation is also evident in the manner in which the *OEPG* engages

118 *VSG*, pp. 120–21.
119 Roberts, 'Guthlac: an Edition', p. 465, notes to ll. 482–98.
120 *OEPG*, p. 142.
121 Roberts, 'Guthlac: an Edition', p. 465, notes to ll. 482–98.
122 *OEPG*, p. 143.

with Felix's complex and potentially problematic depiction of the relationship between Creation and Guthlac. Felix's account of the union of all things with the man whose spirit is joined to God (with the terms *cognosci, agnosci,* and *coniunguntur* discussed in chapter 4, above) becomes in the *OEPG* a much simpler, logically sequential statement that the man who has led his life after God's will is able to restore Creation so that even the wildest of creatures came near to him ('þe near wæron'). This restored relationship with animate Creation is directly connected to the communion of angels through echoing with the previous phrase, 'þe near comon'; in each it is Guthlac's sanctity, evidenced in his dedication to go, that allows the two relationships to occur. While in the *VSG* the saint is said to seek the recognition of the wild animals with the verb *agnoscere,* the OE author presents the restored relationship between Guthlac and the animals as one of the two causal effects of the holy life. This causality is evidenced in the grammar of the passage, where the subjunctive verb 'to be' states that if a holy life is led, then wild animate Creation will be restored and come near to the saintly person leading that life. The *OEPG*'s account in Ch. X of the miracle of the swallows' prelapsarian relationship with Guthlac presents a highly condensed and vivid depiction of the restoration of Creation as described in the raven episode of Ch. IX. The saint's spiritual majority is evidenced here by his ability not simply to order the wild beasts and expect obedience, but by his capability of reversing the effects of the Fall to such an extent that he is able to enter into a prelapsarian relationship with the swallows.

Guthlac A: Hidden to Verdant Creation

As many scholars have noted, the depiction of Guthlac's eremitic landscape in *Guthlac A* is distinct from that of any of the other iterations, involving wooded hills and green fields rather than watery fens.[123] Roberts, for example, sees closer parallels to the

[123] For example, see C. A. M. Clarke, *Writing Power in Anglo-Saxon England: Texts, Hierarchies, Economies* (Cambridge, 2012), pp. 46–47; and J. Roberts, 'Hagiography and Literature: The Case of Guthlac of Crowland', in *Mercia: an*

conventional depictions of hagiographical wilderness, like those of the Evagrian *Vita Antonii*.[124] This change of topography has led to a number of different interpretations for the landscape in the poem, with many arguing that the poet's depiction of the land is deliberately ambiguous, or that the portion of Creation displayed is purely metaphorical or symbolic.[125] I would suggest, however, that the depiction of the natural landscape in *Guthlac A*, despite its apparent novelty, is better understood in terms of the wider Guthlac tradition and its exegetical background, in particular the Augustinian-Bedan interpretation of the restoration of Creation. The manner in which this poem explores the landscape and this theme is, nonetheless, distinct, and can be summarised as follows: the poet presents the saint entering not into a shadowed and thorny wasteland, but instead into a potential landscape, one that is inherently desirable but existing in a kind of stasis. This space possesses no negative characteristics, and instead is defined by its hiddenness and fundamental lack of a rightful guardian.[126] This depiction can be connected with both God's initial creative actions and Adam's prelapsarian position; the conquest of the *beorg*, therefore, remains centred on the restoration of Creation by Guthlac.

The poet's depiction of eremitic space in the introduction frames the ensuing presentation of the forested *beorg*. In this section, he delineates the various kinds of holy lives that may be led, foregrounding the imitative intention of his poem within hagiographical tradition.[127] He exemplifies this with a contrastive pairing of worldly men who seek earthly riches and scorn the saints, and saints and other holy men whose vision is firmly fixed on heaven, in which he describes eremitic space in imagery analogous to

Anglo-Saxon Kingdom in Europe, ed. M. P. Brown and C. A. Farr (London, 2001), pp. 69–86, at 83–84.

124 Roberts, 'Sources and Source Hunting', 10–11.

125 See above, pp. 231–33.

126 For other explorations of the positive depiction of the landscape here see: Appleton, 'Anglo-Saxon Landscapes', p. 56; Magennis, *Images of Community*, pp. 181–83; connected to this Magennis highlights how forests are places of refuge rather than danger, in comparison with fens, moors, and mountains which are usually perilous, p. 130.

127 *Guthlac A*, ll. 30–37a.

antecedent and contemporary depictions of the creation narrative in OE verse.[128] The parallel is most evident in *Genesis A*:

Guthlac A:

Sume þa wuniað on westennum
secað ond gesittað sylfra willum
hamas on *heolstrum,* hy ðæs heofoncundan
boldes bidað.

(ll. 81–4a)

(Some dwell in the wilderness. They willingly seek and settle homes in the *shadows*. They await heavenly dwellings.)

Genesis A:

Ne wæs her þa giet nymþe *heolstersceado*
wiht geworden ac þes wida grund
stod deop and dim, drihtne fremde,
idel and unnyt.

(ll. 103–106a)

(Nor was there yet, except *shadowed-darkness*, any thing made, but this wide ground stood deep and dark, far from the Lord, empty and useless.)

The emphasis on the hiddenness of these shadowed spaces has its origin in Gen. 1:2: 'Terra autem erat inanis et vacua, et *tenebrae* erant super faciem abyssi: et spiritus Dei ferebatur super aquas' ('And the earth was void and empty, and *darkness* came upon the face of the abyss; and the spirit of God moved over the waters'). In the prose *Old English Genesis*, however, the same image is rendered with the more general Old English term for darkness, *þeostru*, commonly used to translate Latin *tenebra*: 'Seo eorðe soþlice wæs ydel and æmtig and *þeostru* wæron ofer þære niwelnisse bradnisse' ('The earth was void and empty, and *darkness* was over the surface of the deep').[129] While the terms *þeostru, sceado,* and *heolstor* all denote the idea of shadows that inherently conceal,[130] the use of *heolstor* in *Genesis A* and *Guthlac A* over the other two terms suggests more specifically the idea of concealment rather than obscuring darkness: an intentional hiddenness for the purpose of revelation at a specific time and for a specific reason. The *Guthlac A* poet repeats God's concealment of the wooded *beorg* within the green *wong* a number of times, including in the initial depiction of Guthlac's specific eremitic landscape, where the poet presents Guthlac arriving as a builder (ll. 136b–140a).

128 *Guthlac A*, ll. 60–62; 65b–66.

129 *OE Heptateuch*, Gen. 1:2, p. 8.

130 *BT* gives the following for *hleostor*: 'That which covers or conceals, darkness, a veil, covering, place of concealment'. Cf. *þeostre*: 'Dark, without light; metaphorically, absence of spiritual or mental light, or of cheerfulness'.

The imagery in *Guthlac A* and *Genesis A* presents pre-created space in terms of potential, ripe for light to come upon it and bring it to life. The pre-created space of *Genesis A* has, however, often been interpreted in negative terms, implying that its potential status is rather one of antagonism towards God:

on þone eagum wlat
stiðfrihþ cining and þa stowe beheold,
dreama lease, geseah *deorc gesweorc*
semian sinnihte, *sweart under roderum,*
wonn and weste oðþæt þeos wourldgesceaft
þurh word gewearð wuldorcyninges.
(ll. 106b–111).

(The firm-minded king beheld with his eyes the place *without joys;* he saw *dark clouds* hanging in endless night, *black under the firmament, dim and empty,* until this world-creation was done through the word of the Glory-King.)

The passage presents an extraordinary collocation of images for darkness ('deorc gesweorc', 'semian sinnihte', 'sweort [...] roderum') as well as describing pre-Creation in agrarian terms as uncultivated ('won and weste'). Neville argues that this presents a shadowed hiddenness (*heolstru*) full of hostility towards God, 'not a void or neutral chaos', but instead an inimical darkness echoing poems like *The Wanderer*: 'God looks upon an apparently pre-existing land that has from the beginning antagonistic power, a land that is *drihtne fremd* "alien to the Lord"'.[131] I would suggest, however, that while the dense collection of synonyms for darkness used to describe pre-Creation is certainly notable, the negative sense comes not from the darkness' antagonism towards God, but instead from its separation, its distance from God and His ordering presence. Thus the adjective *fremde,* which Neville translates as the antagonistic 'alien', is more likely being used with its common spatial meaning in *Genesis A,* 'strange or situated abroad',[132] to designate pre-Creation's separateness from God. Similarly, the clouded darkness is emphasised in *Genesis A* for the purpose of juxtaposing it

[131] Neville, *Representations,* p. 59.
[132] *DOE.*

with the light that comes with God's presence. God's first act after He creates heaven and earth is to send His spirit over the newly created earth and banish the blackness (*sinniht*) with his light: 'him wæs halig leoht / ofer westenne' ('there was a holy light over the emptiness) (ll. 124a–125). Directly connected to this separateness is the depiction of pre-Creation as uncultivated and lacking in life ('wonn and weste'). Pre-Creation is, however, described in lines 116b–117a as brimming with potential, 'folde wæs þa gyta / græs ungrene' ('the earth was then un-green with grass') (ll. 116b–117a), where the teleology of 'gyt' and 'græs-ungrene' declares pre-Creation's embryonic status. This space will be brought into cultivation by God who separates the waters, creates dry land, and causes life to blossom. This argument depends on the provision that the lacuna in the manuscript of *Genesis A*, given the narrative of the Genesis account itself, would likely have included an account of the earth in full bloom.[133] This is further supported by the description of Eden as lush after the lacuna, including during God's instructions to Adam, 'eorðan ælgrene' ('the all-green earth') (l. 197), as well as in the description of God's overview of Eden: 'Fægere leohte / ðæt liðe land lago yrnende, / wylle-burne' ('A running pool, a wellspring, fairly irrigated that pleasant land') (ll. 210b–212a), and 'hwæðre wæstmum stod / folde gefrætwod' ('nonetheless the earth stood adorned with fruits') (ll. 214b–215a). The depiction of pre-Creation is therefore centred on God's separateness, and the ensuing effects of that separateness: darkness and a location bereft of life and the ordering that brings about cultivation.

The *Guthlac A* poet picks up on these twin notions of deliberate hiddenness in shadow and stasis in the Genesis narrative, eliding the pre-Creation and Eden: the saint's *beorg* is part of the

133 Doane, *Genesis A*, argues that the first lacuna, occurring after 168, represents the loss of three leaves and would have included the following material: 'Gen. 1:10 (conclusion) Naming of the waters; Gen. 1:11–25 The third through sixth days up to the creation of man; Gen. 2:7 Creation of Adam', p. 9. Also, note the second lacuna, after line 205 of only one leaf, which he argues includes material covering Genesis 1:29–30, which covers the explicit granting of dominion to humanity over animate and portions of inanimate Creation. I am uncertain why Doane does not include the material in Gen. 1:28, as the verse is part of what he says is excluded, where God gives 'authority to Adam over all living things', p. 9.

postlapsarian world. The hiddenness and potentiality of pre-Creation as depicted in the Genesis narrative is transferred in *Guthlac A* onto the eremitic space. For example, directly after the demons rage at Guthlac's appearance, the *beorg* is described as secret ('dygle'), hidden by the Lord, and uncultivated ('idel'):

> Stod seo dygle stow dryhtne in gemyndum
> *idel and æmen,* eþelriehte feor,
> bad bisæce betran hyrdes.
>
> (ll. 215–17)

> (The secret place stood in the Lord's protection, *empty and uninhabited,* far from hereditary right, waiting for the arrival of a better guardian.)

This presentation of the landscape connects the physical space of the hermitage to that of the pre-created world of *Genesis A*. The phrase 'idel and æmen' is a transformation of the well-attested phrase *idel and unnyt,* found throughout the Old English prose and verse corpus:[134] for example, it is used in *Beowulf* to describe the state of Heorot under Grendel's deprivations;[135] in a rogation homily to describe God's relationship with the earth;[136] and in the *Old English Bede,* where it appears as a phrase uttered by a bishop concerning horse racing.[137] In each case, the phrase carries a similar meaning: *idel,* covering a range of meanings from 'empty' to 'destitute, devoid, void (of something, when used with the genitive)', with a tertiary, and less well attested, definition of 'idle, useless, and without purpose'.[138] The adjective *unnyt* tends to convey a much smaller semantic range: 'useless, vain, idle, and unprofitable'.[139] The resulting phrase therefore covers everything from the

134 The *OE Corpus* lists thirteen uses of the specific phrase *idel and unnyt,* with a further thirty-six that include slight variations on the phrase (for example, 'idel oððe unnyt' and 'idel ofer unnyt').

135 *Beowulf,* ll. 411b–413a. *Klaeber's Beowulf and The Fight at Finnsburg,* ed. R. D. Fulk, R. E. Bjork, and J. D. Niles, 4th edn (Toronto, 2008). All references to *Beowulf* are taken from this edition.

136 *The Vercelli Homilies,* p. 206, ll. 175–76.

137 *The Old English Version of Bede's Ecclesiastical History of the English People,* ed. T. Miller (London, 1890–98), p. 400.

138 *BT.*

139 *BT.*

potential space of the pre-created world in *Genesis A* to the more negative uses found in other contexts, where it depicts something that is, at present, useless and unprofitable.

The alteration by the *Guthlac A* poet of the stock phrase 'idel and unnyt' to the novel phrase 'idel and *æmen*' ('idle and *uninhabited*') places extra emphasis on the absence of people. The adjective *æmen* is defined as 'uninhabited, desolate, and devoid of something', with the editors of the *DOE* defining the use in *Guthlac A* as 'devoid of + genitive'.[140] *Æmen* is rare in the extant corpus, occurring in only two other texts besides *Guthlac A*: twice in the Old English translation of *Orosius*, and twice in the Old English translation of Augustine's *Soliloquies*. In the *Orosius*, the word is employed to mean simply unpeopled,[141] and in the case of the OE *Soliloquies*, the word is used to denote the idea of a positive secret location ('dygela stowe') being devoid of something; in this case worldly distractions.[142] The evidence suggests that the adjective *æmen* was used to indicate a space devoid of something, most commonly of people, and when used in the Genesis narrative itself, devoid of ordered matter. Whereas the more common phrase *idel and unnyt* was used to present the idea of idleness and uselessness, the novel phrase *idel and æmen* in *Guthlac A* suggests a landscape waiting not only to be shaped by its Creator, but also occupied by its steward, Guthlac (l. 217), echoing the depiction of Adam's relationship with prelapsarian Creation in both Scripture and OE biblical verse. This is evidenced by the *Guthlac A* poet's use of the term *hyrdes* to describe Guthlac as the rightful steward in line 217b, a term also employed to describe Adam in *Genesis A* when God first places him in Paradise:

> ne þuhte þa gerysne rodora wearde
> þæt adam leng ana wære
> neorxnawonges niwre gesceafte
> *hyrde* and healdend.
>
> (ll. 169–172a)

140 *DOE*.

141 *The Old English Orosius*, ed. J. Bately (London, 1980), pp. 44, 73.

142 *King Alfred's Version of St. Augustine's Soliloquies*, ed. T. A. Carnicelli (Cambridge, MA, 1969), p. 49.

(The ruler of heaven did not think it fitting that Adam should be alone any longer, *the guard* and holder of paradise, the new Creation.)

The prose *Old English Genesis* likewise emphasises Adam's role as guardian and steward with the verbs *wyrcan* and *begiman* in its translation of the scene from Gen. 2:15: 'God genam þa þone man and gelogode hine on neorxenawange, þæt he þær *wircean* sceolde and þær *begiman*' ('God took the man then and placed him in paradise, so that he might *work* and *guard* it').[143] In light of these parallels, we can conclude that the *Guthlac A* poet has carefully constructed his eremitic space as a place deliberately hidden in shadow by God, resting in a kind of stasis, as yet untilled but full of potential. God's revelation of the location of the hermitage is simultaneous with Guthlac's arrival, and evokes His placement of Adam in Eden; both Adam and Guthlac are the rightful and expected guardians of the spaces they will inhabit.

Guthlac A's depiction of the landscape as a space awaiting its proper guardian is furthered by the introduction of ordered cultivation, again echoing the vernacular interpretation presented in *Genesis A*. Magennis notes that *Genesis A* places a great deal more emphasis on 'the life of agriculture' than Scripture or standard patristic exegesis of this section of the creation narrative, and that *Genesis A* presents 'an approving picture of Noah tilling the good land' without any 'hint of a typological interpretation'.[144] *Guthlac A* shares with *Genesis A* this focus on cultivation. For example, in the first description of the hermitage's location, the poet highlights its uncultivated and marginal nature:

Him wæs engel neah,
fæle freoðuweard, þam þe feara sum
mearclond gesæt. Þær he mongum wearð
bysen on Brytene, siþþan *biorg* gestah.
(ll. 172b–175)

(An angel was near him, a faithful peace-guardian for him, as one of the few who *settled the marchland*. There he became an example for Britain when he ascended *the hill*.)

143 *OE Heptateuch*, Gen. 2:15, p. 11.

144 Magennis, *Images of Community*, p. 153. See above, pp. 33–34, 45.

Here the location is not simply a *beorg*, but is located explicitly in the *mearclond*, a rare word denoting border-lands or march-lands outside the area of civilised cultivation which occurs mostly in verse.[145] The depiction of Guthlac's as yet unoccupied hermitage therefore focuses not only on its position on the margins of cultivated places, but also hints at its potential for cultivation with the emphasis on Guthlac's settling ('gesæt') of the landscape, which brings the previously hidden area within the realm of general knowledge: his climbing of the hill is said to be an example ('bysen') for many in Britain. When interpreted in light of the inherited exegetical tradition of the Fall, we can begin to appreciate how Guthlac, in his contest with the demons, seeks to transform the previously hidden and uncultivated space into a cultivated and known location, stewarding it, as Adam formerly was meant to do for prelapsarian Creation.[146]

Loneliness, Eorðlic Eþel, and Stewardship

In one of the most striking divergences from the Guthlac tradition, *Guthlac A* not only dislocates the narrative from its roots in Crowland, but also removes nearly all the description of the eremitic location's asperities. These changes transform the nature of the relationship between the saint's progression and the portion of Creation in which it is taking place.[147] While the Latin and OE prose *vitae* clearly follow Augustinian and Bedan exegesis, emphasising the role of postlapsarian Creation in pushing Guthlac onwards in

[145] *Mearclond* appears twice in *Andreas* (ll. 19–21a), once in *Exodus* (ll. 63–67), once in *Riddle 3* (ll. 21b–24a), and in the boundary clause of S 849. *Andreas: an Edition*, ed. R. North and M. Bintley (Liverpool, 2016); *Exodus*, ed. P. J. Lucas (London, 1977). With the exception of *Andreas*, in which the term is used to refer to Mermedonia, the word appears to be neutral in connotation. Cf. *BT, DOE*. Appleton, 'Anglo-Saxon Landscapes', pp. 27, 64, and 69, discusses *mearclond* as a liminal term, which inherently connects two distinct regions. My argument focuses here on the region itself rather than its uses as a connective space.

[146] See above, pp. 14, 135.

[147] See *Guthlac A*, ll. 530–46, where the poet delineates the way in which Guthlac's soul is refined through a process of demonic temptation and testing, especially ll. 535b–536: 'þær his sawl wearð / clæne 7 gecostad' ('There his soul became pure and refined').

sanctity, the landscape in *Guthlac A* retains only a single feature that functions this way, its loneliness brought about by its lack of people. This emphasis on loneliness resonates with the *OEPG*, as well as numerous Old English poetic texts, suggesting a particularly vernacular concern, which might tentatively be connected to the so-called elegiac verse tradition.[148] Discenza notes how the term *weste*, and the related *westen*, are employed in *Beowulf*, *Andreas*, and Guthlac for 'spaces outside civilzation'.[149] She also notes how the landscapes depicted with the word *weste* in these three poems focuses on their lack of people, including ll. 1158–59 of Andreas, which depicts a 'dwelling *made* waste, empty of inhabitants'.[150] In *Guthlac A*, the contestation of the *beorg* and the saint's attendant spiritual progression is particularly focused on loneliness, and the effects of that loneliness.

The forested *beorg* is initially bereft of people ('æmen'), waiting for its rightful steward, while the eremitic site is located within a wilderness (*westen*, *anad*). The demonic tormentors and the physical landscape interact to challenge the saint through the associated loneliness. *Westen* is a common term in prose and verse for 'desert' or 'wilderness',[151] but is only used once in *Guthlac A* as a general statement concerning the landscape context in which the location exists, rather than to describe the contested *beorg*. Given the text's focus on loneliness noted above, it is noteworthy that OE *westen* is also used to translate Latin 'solitudo' ('1. state of being alone; 2. state of being without a friend or protector, isolation, loneliness').[152] In *Guthlac A*, the term appears in a passage describing the demon's highlighting of Guthlac's separation from his kindred in his occupation of the

148 This can be seen in elegiac poetry, especially *The Wanderer*, *The Seafarer*, *Deor*, and the *Wife's Lament*. For a discussion of the theme of exile and its relationship to elegiac poetry, see A. L. Klinck, *The Old English Elegies: A Critical Edition and Genre Study* (Montreal, 1992), pp. 225–30, 233.

149 Discenza, *Inhabited Spaces*, p. 143.

150 Discenza, *Inhabited Spaces*, p. 143.

151 *BT*. Cf. *OE Corpus*: for example, *Exodus*, l. 123; *The Ruin*, l. 27; Ælfric, *Assumption of St John*, in P. Clemoes, *Catholic Homilies: First Series*, p. 216, l. 279; and *OE Heptateuch*, Ex. 3:18, p. 93.

152 *DMLBS*. *Westen* translates *solitudo* primarily in Scripture; for example, in *Exodus* ('to Sinai westene' for 'insolitudinem Sinai'), *BT*. The *DOE Corpus* lists twenty-nine instances in various MSS of the *Psalms* where *westen* glosses *solitudo*.

wilderness, and the lack of defence and sustenance that implies. Guthlac's victorious resistance of this temptation here echoes that of Cuthbert,[153] and similarly displays Guthlac as triumphing where Adam failed. This resistance is depicted by Guthlac's answering with the necessary saintly fixity of mind,[154] and in the speech that follows Guthlac describes the greater space as *westen*: 'Wid is þes *westen*, wræcsetla fela, / eardas onhæle earmra gæsta' ('Wide is this *wilderness*, with many exile abodes, secret places for wretched spirits') (ll. 296–97). The point of this description is to highlight how even in his isolation, in the uttermost solitude of the wilderness (*westen*), the demons will be unable to overcome him, even if they gather all the other demons residing in similar places throughout the wilderness.[155] In the series of refutations that follow, which constitute the main method of battle for the forested barrow, Guthlac declares that God will provide for his faithful servant both physical sustenance and protection; difficulties brought on by the isolation of the barrow. Guthlac's response here is meant to display that characteristic so central to sanctity: steadfastness.[156] As Thomas D. Hill argues, this monastic virtue of *stabilitas* is a central theme in *Guthlac A*.[157] Uniquely, however, in *Guthlac A*, the saint's *stabilitas* is centred on the loneliness of the eremitic site.

The noun *anad* is a rare word which, as Roberts notes, appears only here and in *Riddle 60*.[158] Roberts defines *anad* in her glossary as 'solitude', and she suggests that in both *Guthlac A* and *Riddle 60* the primary aspect of the locations depicted is their loneliness. Discenza likewise notes the use of the noun *anad* to describe the landscape in *Guthlac A* as functioning 'to emphasize its hero's remove from society'.[159] The contested *beorg* is described as located within the solitary

153 See above, pp. 106–07 for the *VCM*, and pp. 146, 155 for the *VCP*.

154 *Guthlac A*, ll. 292b–295.

155 *Guthlac A*, ll. 299–302a.

156 See Hill, 'The Middle Way', p. 182.

157 Hill, 'The Middle Way', p. 182. Hill further points to the Rule of Benedict where the virtue is highlighted; see Benedict, *Regula*, Ch. LVIII, pp. 626–32.

158 *Guthlac A*, p. 141, n. 333: 'The word, apparently from *an* + Germanic **odus*, occurs again in l. 356 and otherwise only in *Riddle* 60, l. 5. Cognate are OS. *enodi*, OHG. *einoti* and modern German *Einode*'. Roberts gives 'solitude' as the definition in her glossary: *Guthlac Poems*, p. 185.

159 Discenza, *Inhabited Spaces*, p. 153.

wilderness through this use of *anad*, together with the description of Guthlac's courageous settling of the hill: 'þonne he to eorðan on þam *anade* / hleor onhylde' ('When in the *solitude* he lowered his face to the ground') (ll. 332–35). The second use of the noun *anad* occurs just under thirty lines later, where loneliness is depicted as a direct attack by the demons. Guthlac is, however, the unperturbed holy man, and is resolute in his solitude:

> Symle hy Guðlac in Godes willan
> fromne fundon þonne flygereowe
> þurh nihta genipu neosan cwoman
> þa þe onhæle *eardas* weredon,
> hwæþre him þæs *wonges* *wyn* sweðrade.
> Woldun þæt him to mode *fore monlufan*
> *sorg* gesohte þæt he siþ tuge
> eft to *eþle*: ne wæs þæt ongin swylc
> ðonne hine engel on þam *anade*
> geornast grette.
>
> (ll. 348–357a)

> (Ever they [the demons] found Guthlac firm in God's will, when the wild-fliers came through the darkness of night, those ones who held secret *abodes*, seeking whether his *joy* in the *place* had diminished. They wished that *sorrow* would arise in his heart *on account of human love*, so that he would undertake a journey again to his *homeland*. There was no such action when the angel in that *solitude* eagerly visited with Guthlac.)

The demons hope the inherent loneliness of the solitude ('anade') will inspire a sorrow for human companionship, but instead find Guthlac fully committed to his eremitic life. The poet presents Guthlac as having delight in the place ('wonges wyn'), as well as juxtaposing his present settlement with his previous rightful homeland ('eþel'). This juxtaposition, between the human *eþel* Guthlac left behind and his present battle to establish a novel eremitic homeland, represents the *Guthlac A* poet's careful adaptation of his source material in order to emphasise the saint's spiritual progression.

As Howe and others have observed, the Anglo-Saxons had a relatively large lexis for defining home in its multitude of senses.[160] For example Howe, quoting Anita Riedinger, lists fourteen synonyms for home in Old English poetry alone: '*eðel, eard, geard, ærn, bold, reced, cnosl, cyðð, worðig, wic, eodor, hoc, hus,* and *ham*'.[161] The range of meanings covered by this list is extensive and one sense often bleeds into another. Poetic usage of one term instead of another may, at times, be based purely on the necessities of alliteration. While the *Guthlac A* poet employs a wide range of synonyms for home, including *eþel, eard, geard, bold, wic, hus,* and *ham,* he uses *eþel* in the distinct sense of a homeland that provides both shelter and identity.[162] Despite the frequency of *eþel* in the Old English corpus,[163] as well as its range of application, the term is not widely applied by the *Guthlac A* poet, appearing only five times as a standalone term, and twice in the compounds *eþel-lond* and *eþel-rieht,* in comparison with *eard,* used nine times, and *ham,* used ten times.[164] *Eþel* is far more restricted in its use than the *Guthlac A* poet's other terms for home, referencing only three distinct locations: Guthlac's original secular homeland, which he left to become a hermit; the contested barrow itself; and Guthlac's eternal homeland of heaven. This specificity can be distinguished from the poet's use of the other broadly synonymous terms for home: for example, *hus* is used once to denote Guthlac's proposed built space (ll. 250–51), twice to describe hell (ll. 562, 677), and finally to describe the physical bodies of saintly men (l. 802), in keeping with traditional enclosure

160 N. Howe, 'Looking for Home in Anglo-Saxon England', in *Home and Homelessness in the Medieval and Renaissance World*, ed. N. Howe (Notre Dame, IN, 2004), pp. 143–63, at 147.

161 Howe, 'Home in Anglo-Saxon England', p. 147, quoting A. R. Riedinger, '*Home* in Old English Poetry', *Neuphilologische Mitteilungen* 96 (1995), 51–59, at 51.

162 Several scholars have noted the importance of *eþel,* including Smith, *Land and Book,* who interprets this in the context of hereditary land tenure, p. 202; J. J. Cohen, *Medieval Identity Machines* (Minneapolis, 2003), who sees the use of *eþel-riehte* as Guthlac turning the empty wilderness into an 'inheritable domain', p. 142; and Clarke, 'Land Tenure', who sees the use of *eþel* connoting 'the emotional idea of home, family security, belonging, and an amorphous set of rights to protection and sustenance', 83.

163 *OE Corpus* includes 117 uses of the term, not including compounds. This search included both spellings of the word, *eþel* and *eðel.*

164 Cf. *Guthlac A* and *OE Corpus.*

metaphor;[165] while *eard*, a general term for a place in which someone dwells,[166] is used to denote numerous locations in *Guthlac A*, from the generalised home which the demons cannot have on earth (l. 220), to the contested space as claimed by Guthlac (l. 256), the places where the demons reside in the context of the wide wilderness (l. 297), Guthlac's eremitic location in full bloom in (l. 744), and heaven (l. 656). When contrasted with the wide range of uses for terms like *eard*, the more specific usage of the term *eþel* in *Guthlac A* is particularly striking, as it emphasises Guthlac's transition from an identity rooted in a secular home to an identity rooted in a religious home (whether eremitic, cenobitic, or heavenly). The term is used twice to refer to Guthlac's pre-monastic familial homeland: first, in l. 277a, where the devils point out that Guthlac will have no physical support from other people, because he is not in his homeland, 'Ne þec mon hider mose fedeð […] gif þu gewitest swa wilde deor / ana from eþele' ('No man will feed you nourishment here […] if you depart as a wild beast, alone from your homeland') (ll. 274, 276–277a); and second, in l. 355 of the section discussed above, where the devils hope the loneliness of the solitude will drive him to seek human companionship in his homeland. In both instances, *eþel* evokes the wider Guthlac legend, in which the saint is depicted as an aristocratic leader of a war band before taking up his monastic life. Guthlac's pre-monastic life would have included not only strong kinship ties in the contemporary aristocratic social structure of Anglo-Saxon England, but also Guthlac's claim to associated ancestral land within those ties.[167] His resignation of his previous

165 See, for example, P. Clemoes, *Interactions of Thought and Language in Old English Poetry* (Cambridge, 1995); E. Jager, 'Speech and the Chest in Old English Poetry: Orality or Pectorality?', *Speculum* 65 (1990), 845–59; B. Mize, 'The Representation of the Mind as an Enclosure in Old English Poetry', *ASE* 35 (2006), 57–90; and S. Low, 'Approaches to the Old English Vocabulary for "Mind"', *Studia Neophilologica* 73 (2001), 11–22. Of the various meanings for *hus* the wide range of the poet's use consistently focuses on its more general designation as a built structure in which dwelling occurs. See *BT*, esp. definitions I and VI.

166 *DOE* presents a similarly wide coverage for *eard*, but the principal meaning is still 'a dwelling-place'.

167 See Clarke, 'Land Tenure', 79–83, for a discussion of terminology in relation to ancestral land; Smith, *Land and Book*, pp. 8–15, for a discussion of the distinctions between *bocland, folcland,* and *lænland*, in relation to familial inheritance;

manner of life is conceptualised as a removal from the kind of *eþel* with which most Anglo-Saxons could readily identify. This resignation of worldly wealth and security echoes not only the poet's introduction where he describes the obverse of sanctity in those whose primary concern is for glory in society, and who focus on worldly wealth and goods instead of eternal life,[168] but also the initial description of Guthlac, who gives away all earthly possessions: 'þurh eaðmedu ealne gesealde / þone þe he on geoguðe bigan sceolde / worulde wynnum' ('through humility had given everything, that which he might have enjoyed in youth, the joys of the world') (ll. 103–105a). Guthlac's initial abandonment of this *eþel* is, therefore, the necessary initial step in his progression in sanctity.

The second location which the poet describes as an *eþel* is the contested space itself. As both Smith and Clark have noted in relation to their arguments concerning land tenure,[169] the poet presents the contested land as separate, not just hidden but distant from both known lands and hereditary ownership. This distinction can be seen in the poet's use of the term *eþel-rieht*, which Smith notes occurs only three times in the extant corpus;[170] he suggests that the term as applied in *Guthlac A*, in the phrase 'eþelriehte feor', denotes the contested land as 'not only distant from any existing *eðel* [...] but also far removed from the condition of being an ancestral territory attached to any kindred or community'.[171] This depiction of the *beorg* (existing outside any existent hereditary claim) can be connected directly to the presentation of the space as pre-Creation in the Genesis narrative, and allows Guthlac's claim once again to connect with that of Adam.[172] The poet presents Guthlac making his

and H. R. Loyn, 'Kinship in Anglo-Saxon England', *ASE* 3 (1974), 197–209, at 201, for a brief discussion of land ownership in relation to kinship laws.

168 *Guthlac A*, ll. 60, 62.

169 See Smith, *Land and Book*, pp. 198–99, and Clark, 'Land Tenure', 83.

170 Smith, *Land and Book*, p. 199. Clark, 'Land Tenure', 79–80, n. 20, notes this as well, that it only appears here and in *Beowulf* and *Exodus*, but points out that in each the term carries no specific legal meaning.

171 Smith, *Land and Book*, p. 199.

172 In *Genesis A* the term *eþel* is used in a greater variety of senses, but does include uses where it references heaven and postlapsarian earth. For example, in ll. 92–97a the poem describes how after the fall of the angels God ruminates on how to replace the seats in heaven vacated by them with the compound 'eðelstaðolas'; and in ll. 925–31, God's command to Adam after the Fall begins with a

strongest verbal claim over the eremitic location in the entire poem within this structure:

> Ic me anum her eaðe getimbre
> Hus ond hleonað
>
> . . . .
>
> Gewitað nu awyrgde, werigmode,
> from þissum earde þe ge her on stondað,
> fleoð on feorweg
>
> . . . .
>
> Her sceal min wesan
> *eorðlic eþel,* nales eower leng.
> (ll. 250–251a; 255–257a; 260b–261)

(Here I shall easily build for myself house and shelter alone [...] Depart, now, accursed and weary-minded ones, from this earth on which you here stand; flee far away [...] Here shall my *earthly homeland* be, yours no longer.)

The addition of the adjective *eorðlic* to modify the noun *eþel* is unique to the poem, and suggests a distinct sense from the other two locations referred to as *eþel*. Here Guthlac is declaring that this particular portion of postlapsarian Creation is his true earthbound homeland, the place most important for his life before heaven, because it is where his progression in sanctity can take place.[173] The attainment of the contested space is essential for his forward journey to the final use of the term *eþel*: his seat or home in heaven in the doctrine of replacement.

The description of heaven as an *eþel* occurs twice in *Guthlac A*: one, in the final lines of the poem describing the generalised holy soul's experience of journeying to heaven, 'fusne on forðweg to fæder *eðle*' ('eager to go forth to the Father's homeland') (l. 801); and two, within the compound *eþel-lond* in Guthlac's final speech against the demons before St Bartholomew rescues him from hell.

---

reference to the postlapsarian world as an *eþel*. See also *The Dream of the Rood*, ed. M. Swanton (Exeter, 1970; rev. 1987; rev. repr. 1996), ll. 150–56, which uses *eþel* similarly, as well as the OE *Dialogues*, p. 260, ll. 29–31, for a description of those who hear and understand the teachings being enflamed towards heaven (*eðel*).

173 See Clark, 'Land Tenure', 102, who argues that '[f]or Guthlac, the poet communicates the extent of his purification through the concept of a land cleared of unworthy tenants, restored to peace and granted a better tenant'.

As he stands before the doors of hell, Guthlac makes his second explicit claim about his relationship to a place as his true homeland:

Eom ic soðlice
leohte geleafan and mid lufan dryhtnes
fægre gefylled in minum feorhlocan,
breostum inbryrded to þam betran ham,
leomum inlyhted to þam leofestan
ecan earde þær is *eþellond*.
(ll. 651b–657)

(Truly, in my chest I am beautifully filled with the *light of faith* and with the love of the Lord, inspired in my breast towards a better home, *illuminated with light*, towards that most loved eternal land, where is the *homeland*.)

The density of words for internal illumination, despite the saint's current location in hell, echoes the stark contrasts between the chaotic darkness and the brightness of the heavenly as seen in the *VSG* and the *OEPG*, and their combined descriptive weight joins with Guthlac's claim that his true home, his *eþellond*, is heaven. While the teleology of heaven as the end of the Christian life is, of course, both a commonplace in hagiography and fundamental to Christian theology in general, the *Guthlac A* poet is here meaningfully displaying the progression of Guthlac's sanctity based on his affiliation with, and explicit connection to, his various homelands. In this version of the narrative, Guthlac leaves his familial *eþel* to become a hermit and refine his soul within the eremitic *eþel* provided for him by God. His sanctity within the eremitic homeland as depicted through his defence of the space allows him to migrate seamlessly into his true, heavenly, homeland. The poet points to this succession of homelands with further verbal repetition, whereby the contested terrestrial space, the 'leofestan / earde on eorðan' ('most beloved land on earth') (ll. 427b–428a) is followed by heaven, 'leofestan / ecan earde' ('most beloved eternal land') (ll. 655b–656a), just as 'eorðlic eþel' ('earthly homeland') (l. 261a) is followed by 'eþellond […] in fæder wuldre' ('homeland […] in the Father's glory') (ll. 656b, 657b). This repetition focuses Guthlac's progression on the specific locations he claims as his true home during his life, and likewise furthers the depiction of Guthlac in his role as the New Adam, connecting his tripartite *eþel* with his ability to restore Creation and attain a seat in heaven. The journey begins with Guthlac's renunciation of his secular *eþel* and journey towards his eremitic one, a choice that allows him to be the rightful guardian for which the hidden space waits.[174] Finally, the successful defence of the eremitic space is the necessary contingency upon which Guthlac's attainment of his heavenly seat is founded. His trajectory becomes assured through his faithful declaration in the darkness of hell, where he is firmly set upon his true *eþellond*.

Guthlac's Victory and Heavenly Seat

During the contest between the demons and Guthlac, Creation as a whole is depicted by the metonymic *wong/beorg* elision as initially hidden, unpeopled, and uncultivated, awaiting the arrival of its true tenant, the renewed Guthlac/Adam. When Guthlac defeats the demons, with the help of God via Bartholomew, the relationship between saint and Creation is radically redefined; the effects of the Fall are reversed and the land bursts into life. In this section, the poet unites a host of disparate traditions, including heroic verse, the classical topos of the *locus amoenus*, Antonian hagiographical precedent, and Christological typology, all subsumed in this moment of victory displaying Guthlac's restoration of Creation. The victorious scene begins in distinctly heroic idiom, '*sigehreðig* cwom', a compound only found elsewhere in *Beowulf* (ll. 94a; 490; 1597a; 2756a), which implies a heroic resonance for Guthlac's deed. The lexical choice is telling, as in *Beowulf* it is only ever used in non-ironic circumstances, where the person being described is truly victorious within a heroic setting: first describing how God set up the moon and sun triumphantly in the scop's creation narrative (l. 94); during Hrothgar's speech to the Geats before the feast, in which he tells them to think on their glorious victories (l. 490); then to describe Beowulf after he has beheaded Grendel, but before he has returned

[174] Clark, 'Land Tenure', notes this, how Guthlac's choice to leave his 'natural *eþel* and all the rights and safety, the *eþelrieht* (line 216) that go along with it, [begins] the process of replacement', 89.

from the mere (l. 1597); and finally to describe Beowulf after his slaying of the dragon (l. 2756). A similar compound, the adjective *sigehremig*, occurs in the heroic poem *Christ I* to describe Christ after his Ascension: 'Gesæt *sigehremig* on þa swiþran hand / ece ead-fruma agnum fæder' ('He sat *triumphant* in the stronger hand of his own father, the eternal author of happiness'). Yet the specific kind of heroism at work in *Guthlac A* is clearly that of the *Cristes cempa* ('the warrior of Christ'), the Old English rendering of the Latin *miles Christi*.[175] Guthlac's martial role is immediately connected in the following line, 733a, with his more fundamental role, that of the rightful tenant of the land, here expressed in his capacity as builder ('bytla'). The victorious scene begins with Guthlac as *Cristes cempa* (contextually related to his typological position of the New Adam), able to restore Creation based on his attainment of spiritual, and here also physical, majority:

Sigehreðig cwom
bytla to þam beorge; hine bletsadon
monge mægwlitas meaglum reordum,
treofugla tuddor tacnum cyðdon
eadges eftcyme; oft he him æte heold
þonne hy him hungrige ymb hond flugon,
gradum gifre geoce gefegon.
Swa þæt milde mod wið moncynnes
dreamum gedælde, dryhtne þeowde,
genom him to wildeorum wynne siþþan he þas woruld forhogde.
Smolt wæs se sigewong ond sele niwe,
fæger fugla reord, folde geblowen;
geacas gear budon; Guþlac moste,
eadig and onmod, eardas brucan.
(ll. 732b–741)

(Victorious came the builder to the hill; they blessed him, the many kinds with earnest voices, the children of tree-birds

[175] *Guthlac A*, l. 153a. The phrase appears in several Old English texts, most often referencing saints and sanctity: in *Andreas*, ll. 990–91, 'Hæfde þa se æðeling in geþrungen, Cristes cempa, carcerne neh'; in *The Old English Life of Machutus*, ed. D. Yerkes (Toronto, 1984), p. 55, 17.4, 'Soþlice he wæs æþela *Cristes cempa*'; and *Ælfric's Lives of the Saints*, ed. and trans. W. W. Skeat, EETS, 2 vols (London, 1966), in the *Passio s. Dionysii et socorum eius*, p. 176: 'swa swa *cristes cempa*'. See above, pp. 205–06, for the *miles Christi* tradition in the *VSG*.

> proclaimed with signs the blessed man's return; often he held food for them when they to him, hungry, flew near his hand, starving and eager, rejoiced for his help. Thus that mild heart separated from mankind's joys served the Lord, found his joy in wild beasts after he forsook the world. Peaceful was the victory-plain and the dwelling renewed, fair the voices of birds, the earth blooming; the cuckoos proclaimed the year; Guthlac could, blessed and bold, enjoy the dwelling-place.)

This descriptive passage is the most elaborate of all the restoration of Creation scenes in the Guthlac tradition, particularly in connection with Bede's exegetical image of the relationship between birds and saints as an example of prelapsarian Creation.[176] Contextually the presentation focuses on Guthlac's ability to bring the site into full efflorescence, acting as the repairing Adam and transforming the postlapsarian world. The passage itself is formed by the combination of formulaic elements used in Old English poetry to describe Paradise, focusing primarily upon greenness, which are helpfully summarised by Kathleen Barrar:

> A composite picture would show it as a wide open plain, decorated by trees permanently in leaf; irrigated by streams; filled with blossom and fruit; scented; filled with birdsong and characterized by a list of negative formulas.[177]

Barrar acknowledges the wider debt to the classical *locus amoenus* tradition, but posits that the focus on spaciousness is an addition from Germanic secular tradition.[178] Magennis also connects this focus on green plains to wider Germanic tradition,[179] while Ananya Kabir delineates the specific elements of the tradition, arguing that they make up a 'vernacular equivalent of the *locus amoenus*' with formulaic elements including nouns for 'green open spaces': *wang*,

176 See above, pp. 13, 49, 115, 159, 163.

177 K. Barrar, 'A Spacious, Green and Hospitable Land: Paradise in Old English Poetry', *Bulletin of the John Rylands University Library of Manchester* 86 (2004), 105–25, at 108.

178 Barrar, 'Hospitable Land', 110–11.

179 See Magennis, *Images of Community*, p. 148: 'The *Genesis A*-poet's favoured adjective *ælgrene*, "all-green", is paralleled in Old Norse tradition: in the immediately preceding chapter we have seen the cognate word *algrœn* applied to the great plain of Gnitaheiðr, and that same plain is also celebrated as *víðr* "broad"'.

folde, land, grund, eorðe; 'spatial markers which [...] accompany them': *rume, wide, side*; and 'adjectives of brightness': *beorht, torht, leoht, wlitig, fæger*.[180] The *Guthlac A* poet's use of the above tradition is readily apparent ('smolt [...] sigewong', 'folde geblowen', 'grena wong'), but he creates the scene here for a specific purpose, employing the lexis of the vernacular Eden to present the restoration of Creation to its prelapsarian state which underlies both the Cuthbert and Guthlac traditions. Hence he includes all the elements from the *VSG*, retaining even the connected depiction of the saint interacting with Creation as a steward or shepherd by feeding various birds from his own hand. The poet describes how a chorus of birds praise Guthlac's victorious return, 'meaglum reordum [...] cyðdon', which is a poetic rendering of the ideas expressed by Felix's 'agnosci'. In both texts there is a sense of recognition by animate Creation, with the birdsong of *Guthlac A* giving a more powerful example of Creation's response to Guthlac; they are described as consecrating ('bletsadon'), and making known or proclaiming with words ('cyðdon'), Guthlac's return. This prosopopoeic description of the birds is not simply a fitting welcome for the *miles Christi*, but is also a match with the joy depicted relationally between Guthlac and the swallows in both the *VSG* and the *OEPG*. In *Guthlac A*, however, the poet furthers this tradition by creating a causal link between Guthlac's victorious eremitic life of forsaking the world, and the locating of his joy in relationship with wild beasts (ll. 739–41); the 'agnosci' of Felix has become instead relational joy centred in beasts that remain wild ('wildeorum wynne'). This presentation borders on heterodoxy in its emphasis on the relationship between Cuthbert and Creation, but with the firm exegesis of Guthlac as the typological New Adam, its highlighting of the joy found in God's created world presents a prelapsarian glimpse of Eden.

The joy that can be found in Guthlac's earthly *eþel* is explored further by the second half of the restoration scene in *Guthlac A* where the land itself, initially depicted as unpeopled and uncultivated, fundamentally static and yet full of potential, is brought

180 A. J. Kabir, *Paradise, Death and Doomsday in Anglo-Saxon Literature* (Cambridge, 2001), pp. 144–45.

into bloom: 'Smolt wæs se sigewong and sele niwe, / fæger fugla reord, folde geblowen' ('Peaceful was the victory-plain and the dwelling renewed; fair the voices of birds, the earth blooming') (ll. 742–744b). The victory-plain is at peace, and the dwelling, Guthlac's *eþel*, is renewed; in it the fair birds sing, and the earth itself is in bloom. This renewal is heralded as a new year by the singing of cuckoos, 'geacas gear budon' ('The cuckoos proclaimed the year') (l. 744a).[181] Guthlac, in his capacity as saint and New Adam, is allowed to find joy in his relationship with Creation, just as humanity should have done in the prelapsarian world: 'Guþlac moste, / eadig and onmod, *eardes brucan*' ('Guthlac could, blessed and bold, *enjoy* the *dwelling-place*') (ll. 744b–745). The word 'brucan' occurs widely throughout the Old English corpus,[182] and covers a wide lexical range, from 'use' to 'possess' and even 'fulfil', but within the text itself the poet uses it for consistent reasons and with consistent meaning. Of the eleven uses in *Guthlac A*, the word is used in only two of its myriad senses: 'to use, or employ', and 'to enjoy'.[183] Of these two meanings, the poet's predominant one is 'to enjoy', with only two uses employing the former definition of 'use, or employ'. Both the uses of *brucan* to mean 'use' or 'employ' are found in the poet's most explicitly didactic portions of *Guthlac A*: first, in the introduction directly following his discussion of imitating holy men on earth; and second, in his post-narrative conclusion where he defends both the verity and usefulness of Guthlac as an imitative exemplar. In the introductory section, the poet follows his didactic call for imitation with a connected declaration that the wise man can use well the opportunity of these exemplars:

> Mæg nu snottor guma   sæle *brucan*
> godra tida   ond his gæste forð
> weges willian.
>
> (ll. 35–37b)

(The wise man can now *use* the prosperity of the good hours, and his spirit go forth, desiring the way.)

[181] Cf. *The Seafarer*, ll. 53–4, in which the cuckoo is referred to as the guardian of summer.

[182] *OE Corpus* has 189 uses.

[183] *DOE*.

In the concluding section the poet argues that these deeds could profit the soul if heeded: 'þonne hy in gesihþe soþes *brucað*' ('When they might *profit* by the truth in their sight') (ll. 756–59). The implication is that *Guthlac A* provides a saintly example for emulation, especially concerning the saint's internal stability; this imitative function of the text includes the nine other uses of *brucan*, which depict various but ultimately related moments of enjoyment. In particular, the poet presents two parallel pairs of moments, both of which are connected to the two points of connection between Adam and Guthlac: his restoration of Creation to its prelapsarian state of perfection and harmony, and his subsequent regaining of a celestial seat.

The first pair of parallel moments presents the joy that will be experienced by the faithful in heaven. The initial moment occurs within the first twenty-five lines of the poem, where the angel is describing the glories of heaven itself:

> ðæt sind þa getimbru þe no tydriað,
> ne þam fore yrmþum þe þær in wuniað
> lif aspringeð ac him bið lenge hu sel;
> geoguþe *brucað* and Godes miltsa.
>
> (ll. 18–24)

> (Those are structures that do not decay, nor for those who dwell within will life fail, but the longer they remain the better; they will *enjoy* youth and the mercies of God.)

The poet foreshadows here the upcoming narrative, with Guthlac as the example of this kind of holy life. The object of enjoyment in this moment is the eternal youth that will be the primary experience for those in heaven, the obverse of the entropy of the physical world. This use of the verb *brucan* is paralleled by Guthlac himself in his response to the demons' threats of physical violence,[184] which include having his body tossed into surging fire,[185] as well as other

[184] These threats are implicit given that this speech is in response to a section in a lost folio. Roberts, *Guthlac A*, suggests it 'must have contained some sixty verse lines, probably continuing the speech begun in l. 361', p. 141, note to 368.

[185] *Guthlac A*, ll. 374–375b.

ambiguous physical torments.[186] Guthlac's faithful reply focuses on the joy his soul will experience in his promised home in heaven:

nis me þæs deaþes sorg.
ðeah min ban and blod   butu geweorþen
eorþan to eacan,   min se eca dæl
in gefean fareð   þær he fægran
botles *bruceð*.

(ll. 379b–383b)

(Of death there is no sorrow in me. Though my bones and blood both aid the earth's increase, the eternal part of me will go forth into joy, where it may *enjoy* a dwelling more fair.)

The focus for both moments is the eternal enjoyment of heaven, each highlighting how the juxtaposed transitory physical matter may be subject to the torments of the postlapsarian present, but Guthlac, and those who emulate him, are able to gain not only salvation, but also the seats which Adam and Eve forfeited in heaven.

The second pair of moments focuses on Guthlac's ability to restore Creation. The first appears in the description of the devils' tenancy of the *beorg*, how even in its untilled and potential state they find great joy in it as a sublunary respite. Immediately after they complain that Guthlac has broken their forested hill ('beorgas *bræce*'), the poet presents the demons' enjoyment of the land twice with the term *brucan*:

þær hy bidinge
earme ondsacan   æror mostum
æfter tintergum   tidum *brucan*
ðonne hy of waþum   werge cwoman
restan ryneþragum.

(ll. 209b–213a)

(The wretched adversaries had before been able to *enjoy* rest for a time after their torments, when they had come to rest, weary from wandering.)

ne motun hi on eorþan   *eardes brucan*
ne hy lyft swefeð   in leoma ræstum
ac hy hleolease   hama þoliað.

(ll. 220–22)

([These devils] may not *enjoy* a home on earth, nor will the air send them to sleep, into the body's rest, but instead they will suffer shelter-less, without a home.)

[186] *Guthlac A*, l. 377.

Once again the poet has incorporated a scene from the wider Guthlac tradition, here in reference to the demons' sublunary respite in the *VSG*. The use of *brucan*, however, highlights the tension in this sublunary imprisonment of the demons; while the demons are allowed momentary respite on earth, their true home awaits them in hell, and the earth, however fleeting, is meant to be the physical homeland (*eþel*) for humanity. The poet sets up a parallel between this depiction of the devils enjoying Creation temporarily, and the presentation of Guthlac's rightful enjoyment of restored Creation: 'Guðlac moste, / eadig and onmod, *eardes brucan*' ('Guthlac could, brave and humble, *enjoy the land*') (ll. 744b–745). In both instances the enjoyment (*brucan*) is centred upon the physical enjoyment of the *beorg* within the surrounding *wong*. In the case of the demons, the location functioned as a place of refuge, a moment of rest from their exilic wanderings, which can only end in hell. In Guthlac's case, because he has defeated the demons and attained full spiritual majority through that act, the eard is restored to its prelapsarian verdant state, now that its true steward has arrived, and is capable of being enjoyed by him.

The final line of this section stands in enigmatic contrast to the equivalent portion of Felix's *VSG* and the *OEPG*; here the entire restoration episode is encapsulated in the emphatic summation of the realisation of a desire: 'Hwylc wæs fægerra / *willa* geworden in wera life' ('What more beautiful *desire* has been attained in the life of men') (ll. 748b–749). What exactly this desire is, however, is more problematic, given the multitude of interpretations highlighted earlier. Within the framework of Guthlac appearing as the New Adam, however, the realisation speaks directly to the saint's restoration of Creation; a glimpse of how that relationship might be. Kabir has noted how the description of the landscape's efflorescence is framed by an envelope pattern linked through cross alliteration in the phrases 'Smolt wæs se sigewong', and 'Stod se grena wong'.[187] The full second line, however, 'Stod se grena wong in Godes wære' (l. 746), coming directly before the rhetorical question, brings to completion the depiction based on the Genesis

[187] Kabir, *Paradise, Death and Doomsday*, p. 145.

narrative. The landscape here is not only claimed and settled by its rightful guardian, but is also fully within God's protective perceptual purview. It is no longer hidden in shadow, as pre-Creation was, but is in full flower and is fully known ('wære') by God and Guthlac. The answer to the rhetorical question, therefore, is that it is an image of how things should have been, as holy and beautiful as Cuthbert and the sea animals.

The second focus of the poem, Guthlac's connected ability to fulfil replacement doctrine, is brought to its conclusion in the depiction of Guthlac's ascent to heaven:

> Swa wæs Guðlaces gæst gelæded,
> engla fæðmum in uprodor
> fore onsyne eces deman:
> læddon leoflice. Him wæs lean geseald,
> *setl on swegle,* þær he symle mot
> awo to ealdre eardfæst wesan,
> bliðe bidan.
>
> (ll. 781–787a)

> (So Guthlac's spirit was led into the heavens in the arms of angels, leading him lovingly before the face of the Eternal Judge. The reward was granted to him, a *seat in heaven,* where he will always be settled in eternal life, and endure, joyful, forever.)

In due reward for his holy life, by whose virtue he was able to repair the Fall and exist in the postlapsarian world in prelapsarian relationship with animate and inanimate Creation, Guthlac is given one of the seats in heaven for which mankind was originally created. His successful establishment of a holy earthly *eþel* has allowed God to grant him his hoped-for heavenly one. The *Guthlac A* poet ends much as he began, presenting a didactic interpretation for the poetic *vita,* suggesting that the faithful who emulate, in whatever measure, the saintly life of Guthlac, might partake in the reward of the heavenly seats or heavenly abodes ('hames in heahþu') (l. 796a).

Conclusion

These two vernacular *vitae* of Guthlac therefore present distinct, though interrelated, depictions of the saint's life. In the *OEPG*, the enargaeic landscape of Felix is taken a step further, where the fens and Crowland are depicted through a landscape lexis shared with contemporary boundary clauses. This precise language allows for a much more accessible and appealing depiction of Guthlac's Antonian spiritual progression within that landscape. The restoration of such a thoroughly experiential Creation takes centre stage in the *OEPG*. In *Guthlac A*, however, the restoration of a portion of Creation by a saint is given its most detailed presentation. With Guthlac's successful conquering of Crowland the once potential landscape, echoing pre-Creation, is brought into full bloom, a scene echoing both God's arrival into pre-Creation and Adam's placement in Eden. Uniquely, however, this victory by Guthlac not only restores Creation, but also grants him the eschatological trajectory lost by Adam and Eve in the garden: he can rise to heaven and attain the seat or home made vacant by the fallen angels.

Conclusion: Afterlives of Cuthbert and Guthlac

At the heart of the various *vitae* of Cuthbert and Guthlac explored in this study lies the relationship of the saints with Creation which, I have argued, was fundamental to their role as saints. While sometimes regarded as only an allegorical signifier for heavenly ideas, Creation in these early *vitae* functioned in a much more complex and physical way. Sanctity was achieved and experienced within delineated portions of the Anglo-Saxon landscape. Though each text under consideration engages with Creation in different ways, their shared Augustinian/Bedan exegesis frames their saints' interactions with Creation, and allows for a comparative analysis that lends the distinctions weight and colour. Given the constraints of space, this study necessarily excludes the later traditions of both saints. The findings outlined here, however, invite further analysis on the role of Creation within the later literary iterations of the Cuthbertine and Guthlacian cults, pointing the way for further study. I will consider briefly some of them below.

Saint Cuthbert's popularity continued throughout the medieval period, and the central role of the relationship between Cuthbert and Creation remained. This is evidenced, for example, in the late Old English poem *Durham*,[1] which focuses on the city that by the time of the poem's composition was the resting place for both Cuthbert and his cult. While the poem has often been interpreted as an

1 Several scholars have argued that *Durham* was written in the early twelfth century; see, for example, J. Grossi, 'Preserving the Future in the Old English *Durham*', *JEGP* 111 (2012), 42–73; P. Evan, 'Word-Play as Evidence for the Date of *Durham*', *MÆ* 82 (2013), 314–17. However, H. S. Offler, 'The Date of *Durham* ("Carmen De Situ Dunelmi")', *JEGP* 61 (1962), 591–94, suggests an eleventh-century date. For a review of the evidence, see H. Appleton, 'The Old English *Durham* and the Cult of Cuthbert', *JEGP* 115 (2016), 346–69, at 346–47.

encomium urbis for the city, Helen Appleton has persuasively argued that though it utilises this form, *Durham* is better understood as focusing on Cuthbert himself.[2] *Durham* opens with a description of the city that emphasises the physical landscape in which it resides, highlighting two distinct features that echo the depictions of Creation in the *vitae* of Cuthbert: first, with regard to the landscape itself, Durham is surrounded by water, an echo of both Lindisfarne and Farne; second, the waters, wooded hills, and dales are overflowing not simply with animate life, but specifically with wild animals that are part of the positive description. The description of the encircling Wear, 'Weor ymbeornad, / ea yðum stronge [...] on floda gemonge' ('the encircling Wear, a river of strong waves [...] in the thronging waves') (ll. 3b–4a, 5b),[3] echoes Bede's use of the term *oceanus* in his descriptions of Farne in both the *VCM* and *VCP*, encircling the monastic island with its boundless strength. The wildness of this water is, however, part of the panegyric, and the proliferation of different kinds of fish points to a kind of prelapsarian abundance. The wooded hill and the dales surrounding it are likewise filled with creatures, but again, specifically wild creatures: 'wuniad in ðem wycum wilda deor monige, / in deope dalum deora ungerim' ('many wild beasts dwell in that place, / in the deep dales, beasts without number') (ll. 7–8). While the emphasis on positive wild Creation echoes *Guthlac A* to a greater extent than the *vitae* of Cuthbert, his interaction with the sea animals/otters and the eagle, both untamed creatures, can be connected with the depiction here. The city itself is brought into focus through association with Cuthbert who, as Appleton suggests, is the focus of the poem: 'Is in ðere byri eac bearnum gecyðed /ðe arfesta eadig Cudberch' ('There is also in that famous city, the merciful, blessed Cuthbert') (ll. 9–10). Cuthbert unites the city with the wild Creation encircling it, and it is his implicit connection to Creation that allows the *Durham* poet to present it in such positive terms. The entire depiction echoes a kind of prelapsarian harmony, one founded not on containment but on

2 Appleton, 'The Old English *Durham*', 351–52.

3 *The Anglo-Saxon Minor Poems*, ed. Dobbie. All references to *Durham* are from this edition.

the balanced relationship with the monastic city and wild Creation, centring on Cuthbert.

The function of Creation in the later Cuthbertine tradition was not, however, so one-sided. As mentioned in the introduction, Ælfric's single exploration of Cuthbert, Homily II.X,[4] removes the framing exegesis completely, as well as most of the landscape detail. While the latter move is in part due to Ælfric's use of the *VCM* as his primary source for the homily,[5] the former merits further consideration, especially considering the structural emphasis Homily II.X places on Cuthbert's animal miracles. As Mechthild Gretsch notes, Ælfric retains all but one of the animal miracles,[6] and the miracle involving two sea-animals drying his feet is not only 'one of the least abbreviated episodes in Ælfric's *uita*', but also 'one of its most ornate passages, embellished by an elaborate diction, rhythmical prose and heavy alliteration'.[7] With such an emphasis resting at the centre of Ælfric's text, it is surprising that he removes the explanatory exegesis, especially considering the text's intended lay consumption. The most likely reason for this approach to the Cuthbert tradition has to do with that intended lay audience.[8] While the restoration of Creation via sanctity could be a pastoral tool, the specialised knowledge and training involved in properly interpreting it posed a potential problem for the reformer Ælfric. His solution is, as expected, a simplification of the miracles to their most basic element: Cuthbert is a saint, and God supports his saints. This is evidenced by Ælfric's addition of statements, in nearly all of the animal miracles, that God is the primary agent of the miracles, rather than the more complicated ability of sanctity to bring about prelapsarian relationships. For example, in the horse-and-loaf miracle Cuthbert thanks God for the food: 'He ða geðancode

4 *Ælfric's Catholic Homilies: the Second Series*, ed. M. Godden, EETS (London, 1979).

5 Gretsch, *Ælfric and the Cult of Saints*, p. 103.

6 Gretsch, *Ælfric and the Cult of Saints*, p. 108.

7 Gretsch, *Ælfric and the Cult of Saints*, pp. 108–09.

8 For a discussion of the problem of Ælfric's audience for his homilies, and his general method of adaptation of material for homiletic purposes, see H. Magennis, 'Ælfric's Scholarship', in *A Companion to Ælfric*, pp. 5–34, esp. 30–31.

gode þære sande' ('He then thanked God for [what] was sent');[9] in the eagle-and-fish miracle Cuthbert states that it was God who fed them through the bird: 'Sy lof ðam ælmihtigan þe unc þurh ðisne fugel fedan wolde' ('Praise be to the Almighty who through this bird wished to feed the two of us');[10] in the well miracle the water appears because God desired it: 'ða ða hit wolde god' ('Even as God wished it');[11] and in the barley miracle Cuthbert declares it will only flourish if God grants it: 'gif hit swa geuðe. se ælmihtiga god. þæt he mid his foton. hine fedan moste' ('If Almighty God so grants it that he [Cuthbert] might feed himself with his feet').[12] This simplifying strategy is most evident in the sea-and-beam miracle, which in the *VCM* and *VCP* is where the Augustinian/Bedan exegesis is positioned. Unlike every other iteration where the sea acts as a nearly autonomous agent, honouring Cuthbert's sanctity by its prelapsarian service, Ælfric specifies that it is God who directs the beam on the tide:

> ac se ælmihtiga god his wæs gemyndig. and him ða sylle sylf asende mid þam sælicum flode. and þæt flod hi awearp ðær he sylf smeade þæt hus to arærenne. on ðam sealtum ofre.[13]
>
> (But the Almighty God was mindful of it and sent to him [Cuthbert] the gift himself with the tide of the sea. And that tide cast them [beams] there where he himself desired to construct that house on the shore of the sea.)

The emphasis caused by the use of the reflexive pronoun *self* drives the point home: it is God who is the agent, and the sea merely the means by which he supports his faithful saint. The end result of these changes is a text where Creation plays a much less central role, and where the demand for pastoral clarity overruled the complex exegesis of the earlier tradition.[14]

9 *Catholic Homilies II.X*, ed. Godden, p. 82, l. 57.
10 *Catholic Homilies II.X*, ed. Godden, p. 84, ll. 110–11.
11 *Catholic Homilies II.X*, ed. Godden, p. 86, l. 176.
12 *Catholic Homilies II.X*, ed. Godden, p. 86, ll. 178–79.
13 *Catholic Homilies II.X*, ed. Godden, p. 87, ll. 206–09.
14 Di Sciaccia argues that Ælfric's changes are in great part an attempt to present Cuthbert as the perfect Gregorian balance between the active and contemplative lives: see, '*Concupita, quaesita, ac petita solitudinis secreta*: The Desert Ideal'.

The role of Creation in the later tradition of Saint Guthlac, on the other hand, appears to dwindle. This is evidenced in the adaptation of Felix's *VSG* in both Orderic Vitalis' *Historia Ecclesiastica*,[15] as well as the Guthlac Roundels in British Library MS, Harley Roll Y 6.[16] In the former, Orderic not only substantially truncates material dealing with Creation, but also removes the explicit exegesis itself. Felix's complex depiction of prelapsarian moments, like that of the raven and the document of *VSG* Ch. XXXVII, have been stripped to their barest essentials, simply that by Guthlac's merits ('meritis uiri Dei') a miracle occurs:[17] the raven drops the document into a pool, but it is undamaged. While Orderic retains the description of the birds and fish coming to Guthlac like sheep to their shepherd ('ad uocem eius uelut ad pastorem'),[18] the central role of Creation in Guthlac's spiritual progression is diminished. In its place Orderic emphasises Guthlac's healing miracles, as well as his death and ascension scene. This shift of focus is even more pronounced in the Guthlac Roundels which, akin in form to Orderic's *Historia*, present a summarised account of the *VSG*. The Roundels depict the key events of Guthlac's life with almost no space given to Creation. In fact, the Roundels completely ignore the eremitic element of Guthlac's life, instead presenting him living within a monastic community of large buildings.[19] Creation has been excised and retains only its role as a medium by which the narrative moves: thus, in Guthlac's voyage to Crowland, folio 4r depicts the saint travelling over water in a boat with two men; and likewise in folio 15r the water is depicted again, this time with St Pega travelling by boat to Crowland. The diminished role of Creation is most evident in folio 7r, which depicts Guthlac's attainment of spiritual majority in the devils' final assault which drags him down to hell. While hell

15 Orderic's *Historia* was composed between 1123 and 1137; see *The Ecclesiastical History of Orderic Vitalis*, ed. M. Chibnall, vol. 1 (Oxford, 1969–80), p. 32.

16 The Harley Roll was likely produced in the early thirteenth century; see, J. R. Black, 'Tradition and Transformation in the Cult of St. Guthlac in Early Medieval England', *The Heroic Age* 10 (2007), 1–22, at 11; M. T. Clanchy, *From Memory to Written Record: England, 1066–1307*, 3rd edn (Oxford, 2013), p. 394, n. 1.

17 *EH of Vitalis*, p. 329.

18 *EH of Vitalis*, p. 329. Cf. *VSG*, pp. 120–21.

19 Black, 'Tradition and Transformation', at 12, argues that the Roundels emphasise Guthlac's role as founder of the monastery at Crowland.

is depicted with the elided hell mouth/gate from the *VSG*, Guthlac's isolated eremitic cell is depicted as a stone monastery with Beccel sitting by a table; the fens, Crowland, and Guthlac's hermitage are all subsumed. The end result is an image of Guthlac disconnected from his eremitic roots and the rich tradition of his relationship with Creation.

Restoring Creation: Anglo-Saxon Lives of Cuthbert and Guthlac has demonstrated that Creation as understood through Augustinian/Bedan exegesis was the frame by which these early Anglo-Saxon hagiographers constructed the relationship between Creation and the saints Cuthbert and Guthlac. I would suggest that this central role of the physical world and its connection to sanctity speaks to wider Anglo-Latin and Old English traditions more generally, and that such a focus is likely to have influenced depictions of Creation in other contexts. This study has approached these questions with a syncretic approach, considering Anglo-Latin and Old English hagiography together, and has sought within this also to address the scholarly lacuna in pre-Ælfrician hagiography. It is my hope that this study will prove a useful step in re-centring the role of the physical world in early medieval literature, and that as a result subsequent studies might engage with the literary treatment of the non-human world in a more nuanced and theologically-informed way.

Bibliography

Primary Sources

Adomnán, *Adomnán's* Life of Columba, ed. Alan Orr Anderson and Marjorie Ogilvie Anderson, 2nd edn (Oxford: Clarendon Press, 1991).

Ælfric, *Ælfric's Catholic Homilies: The First Series: Text*, ed. Peter Clemoes, EETS (Oxford: Oxford University Press, 1997).

_____, *Ælfric's Catholic Homilies: The Second Series*, ed. Malcolm Godden (London: Published for the Early English Text Society by Oxford University Press, 1979).

_____, *Homily for the Common of a Confessor*, Angelsächsische Homilien und Heiligenleben, *ed. Oskar Bruno Assmann (Kassel: G. H. Wigand, 1889).*

_____, *Ælfric's Lives of the Saints: Being a Set of Sermons on Saints' Days Formerly Observed by the English Church*, ed. and trans. Walter W. Skeat, 2 vols (London: Published for the Early English Text Society by Oxford University Press, 1966).

Alcuin, Epistolae Karolini Aevi (II), ed. Ernst Dümmler, MGH, Epistolae 4 (Berlin: Weidmann, 1895).

Aldhelm, *Aldhelmi Opera*, ed. Rudolfus Ehwald. MGH AA 15 (Berlin and Charlottenburg: Weidmannsche Verlagsbuchhandlung, 1919; repr. 1961).

_____, *Aldhelm: the Poetic Works*, ed. and trans. Michael Lapidge and James L. Rosier (Cambridge: D. S. Brewer, 2009).

_____, *Aldhelm: the Prose Works*, ed. and trans. Michael Lapidge and Michael Herren (Cambridge: D. S. Brewer, 1979; repr. Woodbridge, 2009).

_____, *Variae Collectiones Aenigmatum Merovingicae Aetatis*, ed. F. Glorie and Maria De Marco, 2 vols, CCSL 133–133A (Turnhout: Brepols, 1986).

Ambrose, *Sancti Ambrosii Opera pars prima, qua continentur libri Exameron ...*, ed. Carole Schenkl, CSEL 32.1 (Vienna: Austrian Academy of Sciences, 1896).

Arator, *Aratoris subdiaconi De actibus apostolorum*, ed. Arthur Patch McKinlay, CSEL 72 (Vienna: Hoelder-Pitchler-Tempsky, 1951).

Augustine, *Confessionum Libri XIII*, ed. Lucas Verheijen, CCSL 27 (Turnhout: Brepols, 1981).

_____, *Sancti Aurelii Augustini De Civitate Dei*, ed. Bernhard Dombard and Alphonsus Kalb, 2 vols, CCSL 47–48 (Turnhout: Brepols, 1955).

_____, *Sancti Aurelii Augustini De Doctrina Christiana. De vera religione*, ed. Klaus D. Daur and Josef Martin, CCSL 32 (Turnhout: Brepols, 1962; repr. 1996).

_____, *De Genesi contra Manichaeos*, ed. Dorothea Weber, CSEL 91 (Vienna: Austrian Academy of Sciences, 1998).

_____, *De moribus Ecclesiae Catholicae et De moribus Manichaeorum libri duo*, ed. Johannes Baptist Bauer, CSEL 90 (Vienna: Hoelder-Pichler-Tempsky, 1992).

_____, *Sancti Aurelii Augustini De Trinitate Libri XV*, ed. William John Mountain and François Glorie, 2 vols, CCSL 50–50a (Turnhout: Brepols, 1968).

_____, *Enarrationes in Psalmos*, ed. Eligius Dekkers and John Fraipont, 3 vols, CCSL 38–40 (Turnhout: Brepols, 1956; repr. 1990).

_____, *Homélies sur la première Épître de Saint Jean. In Iohannis epistulam ad Parthos tractatus decem*, ed. John William Mountain, Jeanne Lemouzy, and Dany Dideberg (Paris: Institute of Augustinian Studies, 2008).

_____, *Sancti Aurelii Augustini In Iohannis Evangelium Tractatus CXXIV*, ed. D. Radbod Willems, CCSL 36 (Turnhout: Brepols, 1954; repr. 1990).

_____, *On Genesis: Two Books: On Genesis against the Manichees and On the Literal Interpretation of Genesis: An Unfinished Book*, trans. Roland J. Teske (Washington, D.C.: Catholic University of America Press, 1990).

_____, *Sancti Aureli Augustini De Genesi ad litteram libri duodecim*, ed. Joseph Zycha, CSEL 28.1 (Vienna: F. Tempsky; Leipzig: G. Freytag, 1895).

_____, *Sancti Aurelli Augustini Epistulae LVI–C*, ed. Klaus D. Daur, CCSL 31A (Turnhout: Brepols, 2005).

Bately, Janet, ed., *The Old English Orosius* (London: Published for the Early English Text Society by Oxford University Press, 1980).

Bede, *Bede's Ecclesiastical History of the English People*, ed. Bertram Colgrave and Roger A. B. Mynors, Oxford Medieval Texts (Oxford: Clarendon Press, 1969).

_____, *Bedas metrische Vita Sancti Cuthberti*, ed. Werner Jaager, Palaestra, 198 (Leipzig: Mayer and Müller, 1935).

_____, *Bedae Venerabilis opera: pars 3, Opera homiletica; parts 4, Opera rhythmica*, ed. David Hurst and Johannes Fraipont, CCSL 122 (Turnhout: Brepols, 1955).

_____, *Bedae Presbyteri Expositio Apocalypseos*, ed. Roger Gryson, CCSL 121A (Turnhout: Brepols, 2001).

_____, *In Genesim, Bedae Venerabilis opera. Pars 2, Opera exegetica*, ed. Charles Williams Jones, CCSL 118A (Turnhout: Brepols, 1967).

_____, *On Genesis*, ed. Calvin B. Kendall (Liverpool: Liverpool University Press, 2008).

_____, *Vita Sancti Felixi, Venerabilis Bedae Opera IV*, ed. John Allen Giles (London: Whittaker and Co., 1893).

Benedict, *La règle de saint Benoît*, ed. Adalbert de Vogüé and Jean Neufville, Sources chrétiennes 181 (Paris: Les Éditions du Cerf, 1972).

Biblia Sacra iuxta Vulgatam versionem, ed. Robert Weber and Roger Gryson, 5th ed. (Stuttgart: Deutsche Bibelgesellschaft, 2007).

Bischof Wærferth von Worcester Übersetzung der Dialoge Gregors des Grossen: über das leben und die wunderthaten italienischer Väter, und über die unsterblichkeit der seelen, ed. Hans Hecht, Henry Johnson, and Julius Zupitza, 2 vols (Leipzig: G. H. Wigand, 1900–7).

Boethius, *Anicius Manlius Severinus, De consolatione philosophiae; Opuscula theologica*, ed. Claudio Moreschini (Munich: K. G. Saur, 2000).

Breviarium Romanum ex decreto Sacrosancti Concilii Tridentini Restitutum S. PII V. Pontificis Maximi jussu editum Clementis VIII and Urbani VIII. Auctoritate recognitum, cum officiis sanctorum novissime per Summos Pontifices usque ad hanc diem concessis; in Quatuor Anni Tempora divisum (Rome: Typographia Balleoniana, 1568; repr. facs. Libreria Editrice Vaticana, Città del Vaticano, 1999).

Caie, Graham D., ed., *The Old English Poem Judgment Day II: a Critical Edition with Editions of* De die iudicii *and the Hatton 113 Homily* Be domes daege (Cambridge: D. S. Brewer, 2000).

Carnicelli, Thomas A., ed., *King Alfred's Version of St. Augustine's Soliloquies* (Cambridge, Mass.: Harvard University Press, 1969).

Cassian, *Collationes XXIIII*, ed. Michael Petschenig, suppl. Gottfried Kreuz, CSEL 13 (Vienna: Verlag der Österreichischen Akademie der Wissenschaften, 1886; repr. 2004).

Cassiodorus, *Expositio Psalmorum*, Library of Latin Texts (based on CCSL 97–98, ed. Marcus Adriaen), Online (Turnhout: Brepols, 2010) http://clt.brepolis.net/llta/pages/Toc.aspx?ctx=1108218.

Chibnall, Marjorie, ed., *The Ecclesiastical History of Orderic Vitalis*, vols 1–6 (Oxford: Clarendon Press, 1969–80).

Clubb, Merrel Dare, ed., *Christ and Satan: an Old English Poem* (New Haven: Yale University Press, 1925).

Cockayne, Thomas Oswald, ed., *Leechdoms, Wortcunning, and Starcarft of Early England, Being a Collection of Documents, For the Most Part Never Before Printed, Illustrating the History of Science in this Country before the Norman Conquest*, Rolls Series 35 (London: Longman and Green, 1864–66).

Colgrave, Bertram, ed. and trans., *Two Lives of Saint Cuthbert: a Life by an Anonymous Monk of Lindisfarne and Bede's Prose Life* (Cambridge: Cambridge University Press, 1940).

_____, *The Earliest Life of Gregory the Great* (Lawrence: University of Kansas Press, 1968; repr., Cambridge: Cambridge University Press, 1985).

_____, *Felix's Life of St Guthlac* (Cambridge: Cambridge University Press, 1956).

Columbanus, *Regula Monachorum*, in *Sancti Columbani opera, Scriptores Latini Hiberniae II*, ed. and trans. George S. M. Walker (Dublin: Dublin Institute for Advanced Studies, 1957).

Crawford, Jane, 'Guthlac: an Edition of the Old English Prose Life together with the Poems in the Exeter Book', Unpub. D.Phil. (University of Oxford, 1967).

Cross, James E., ed., *Two Old English Apocrypha and their Manuscript Source: 'The Gospel of Nichodemus' and 'The Avenging of the Saviour'*, with contributions by Denis Brearley, Julia Crick, Thomas N. Hall, and Andy Orchard (Cambridge: Cambridge University Press, 1996).

DeVriend, Hubert Jan, ed., *The Old English Herbarium and Medicina de quadrupedibus* (London: Published for the Early English Text Society by Oxford University Press, 1984).

Doane, Alger Nicolaus, ed., *The Saxon Genesis: an Edition of the West Saxon Genesis B and the Old Saxon Vatican Genesis* (Madison: University of Wisconsin Press, 1991).

_____, *Genesis A: a New Edition, Revised* (Tempe: Arizona Center for Medieval and Renaissance Studies, 2013).

Dobbie, Elliot Van Kirk, ed., *The Anglo-Saxon Minor Poems*, Anglo-Saxon Poetic Records 6 (London: Routledge, 1942).

Evagrius, *Vita Antonii*, in Pascal H. E. Bertrand, 'Die Evagriusübersetzung der Vita Antonii. Rezeption, Überlieferung, Edition. Unter besonderer Berücksichtigung der Vitas Patrum-Tradition', Ph.D. diss. (University of Utrecht, 2006): http://igitur-archive.library.uu.nl/dissertations/2006-0221-200251/index.htm

_____, *On Thoughts*, in *Evagrius Ponticus*, ed. and trans. Augustine M. Casiday (London: Routledge, 2006).

_____, *Évagre le Pontique: Sur Les Pensées*, ed. Paul Géhin, Claire Guillamont and Antoine Guillamont (Paris: Les Éditions de Cerf, 1998).

Fortunatus, *De Virginitate*, ed. Friedrich Leo, Monumenta Germaniae Historica, Auctores Antiquissimi 4 (Berlin: Weidmann, 1881).

Fulk, Robert D., Robert E. Bjork, and John D. Niles, ed., *Klaeber's Beowulf and The Fight at Finnsburg*, 4th edn (Toronto and London: University of Toronto Press, 2008).

Godden, Malcolm, and Susan Irvine, ed., *The Old English Boethius: an Edition of the Old English Versions of Boethius's* De consolation philosophiae (Oxford: Oxford University Press, 2009).

Goossens, Louis, ed., *The Old English Glosses of MS. Brussels, Royal Library, 1650 (Aldhelm's* De laudibus virginitatis*)* (Brussels: Paleis der Academiën, 1974).

Gonser, Paul, ed., *Das angelsächsische Prosa-Leben des hl. Guthlac*, Anglistische Forschungen 27 (Heidelberg: Carl Winters Universitätsbuchhandlung, 1909).

Grattan, John H. G., and Charles Singer, ed., *Lacnunga*, in *Anglo-Saxon Magic and Medicine: Illustrated Specially from the Semi-Pagan Text*

'Lacnunga' (London: Wellcome Historical Medical Museum by Geoffrey Cumberlege, 1952).

Gregory the Great, *Grégoire le Grand: Dialogues*, ed. Adalbert de Vogüé, 3 vols, Sources chrétiennes 260 (Paris: Les Éditions du Cerf, 1978–1980).

_____, *Responsoria sive Antiphonae de Assumptione sanctae Mariae, Patrologia Latina* 38, ed. Jacques-Paul Migne (Paris: J-P. Garnier Frères, 1841; repr. 1993).

Heist, William W., ed., 'Vita Sancti Albei', in *Vitae Sanctorum Hiberniae: ex codice olim salamanticensi nunc Bruxellensi* (Brussels: Société des Bollandistes, 1965).

Holy Bible Douay-Rheims Version, with Challoner Revisions 1749–52 (Baltimore: John Murphy Company, 1899).

Holy Bible: New Revised Standard Version Catholic Edition (Oxford: Oxford University Press, 1999).

The Holy Bible: New International Version (London: Hodder and Stoughton, 2011).

Isidore, *Isidori Hispalensis Episcopi Etymologiarum sive Originum Libri XX*, ed. Wallace Martin Lindsay, 2 vols (Oxford: Clarendon Press, 1911; repr. 1957).

_____, *Isidorus episcopus Hispalensis: Expositio in Vetus Testamentum. Genesis*, ed. Michael M. Gorman and Martine Dulaey (Freiburg, Basel, and Vienna: Herder, 2009).

Jerome, *Epistulae*, ed. Isidorus Hilberg and Margit Kamptner, 4 vols, CSEL 54–56.2 (Vienna: Austrian Academy of Sciences, 1996).

_____, *Sancti Hieronymi Presbyteri. Commentariorum in Mathaeum libri iv*, ed. David Hurst and Marc Adrian, CCSL 77 (Turnhout: Brepols, 1969).

_____, *Sancti Hieronymi presbyteri Tractatus sive Homiliae in Psalmos, in Marci evangelium, aliaque Varia Argumenta*, ed. Germanus Morin, CCSL 78 (Turnhout: Brepols, 1958)

_____, *Jérôme: Trois vies de moines (Paul, Malchus, Hilarion)*, ed. Pierre Leclerc, Edgardo M. Morales, and Adalbert de Vogüé. Sources chrétiennes 508 (Paris: Les Éditions du Cerf, 2007).

Jonas, *Ionae Vitae Sanctorum Columbani, Vedastis, Iohannis*, ed. Bruno Krusch (Hanover and Leipzig: Impensis Bibliopolii Hahniani, 1905).

Juvencus, *Gai Vetti Aquilini Iuvenci Evangeliorum Libri Quattuor*, ed. Johann Huemer, CSEL 24 (Vienna: Tempsky, 1891).

King Alfred's West-Saxon Version of Gregory's Pastoral Care, ed. Henry Sweet, EETS original series 45, 2 vols (London: Kegan Paul, 1871; repr. Millwood, N.Y.: Kraus Reprint, 1978).

Klinck, Anne L., ed., *The Old English Elegies: A Critical Edition and Genre Study* (Montreal; London: McGill-Queen's University Press, 1992).

Lindelöf, Uno Lorenz, ed., *Der Lambeth-psalter: Eine Altenglische Interlinearversion Des Psalters in Der Hs. 427 Der Erzbischöflichen Lambeth Palace Library* (Helsinki: Druckerei der Finnischen litteraturgesellschaft, 1909–14).

Lucas, Peter J., ed., *Exodus* (London: Methuen, 1977).

Marsden, Richard, ed., *The Old English Heptateuch and Ælfric's* Libellus de Veteri Testamento et Novo, EETS (Oxford: Published for the Early English Text Society by Oxford University Press, 2008).

Miller, Thomas, ed., *The Old English Version of Bede's Ecclesiastical History of the English People*, EETS (London: Published for the Early English Text Society by N. Trübner, 1890–98).

Muir, Bernard J., ed., *The Exeter Anthology of Old English Poetry: an Edition of Exeter Dean and Chapter MS 3501*, second edition (Exeter: University of Exeter Press, 2000).

New English Bible (Oxford: Oxford University Press, 1970).

NKJV Study Bible (Nashville: Thomas Nelson, 2016).

North, Richard, and Michael Bintley, eds., *Andreas: an Edition* (Liverpool: Liverpool University Press, 2016).

Origen, *Commentarii in epistulam ad Romanos = Römerbriefkommentar*, ed. Theresia Heither (Freiburg: Herder, 1990–99).

Paulinus of Nola, *Opera, I. Epistulae*, ed. Guilelmus de Hartel, CSEL 29 (Vienna: F. Tempsky, 1894; repr., ed. with suppl. Margit Kamptner, 1999).

_____, *Opera, II. Carmina*, ed. Guilelmus de Hartel, CSEL 30 (Vienna: F. Tempsky, 1894).

Plummer, Charles, ed., *Vita Sanctorum Hiberniae partim hactenus ineditae ad fidem codicum manuscriptorum recognovit prolegomenis notis indicibus instruxit Carolus Plummer* (Oxford: Clarendon Press, 1968).

Pope Damasus, *Carmen XV. De Sancto Felice, Patrologia Latina* 13, ed. Jacques-Paul Migne (Paris: J-P. Garnier Frères, 1845; repr. 1992).

Roberts, Jane, ed., *The Guthlac Poems of the Exeter Book* (Oxford: Clarendon Press, 1979).

Scragg, Donald G., ed., *The Vercelli Homilies and Related Texts*, EETS (Oxford: Published for the Early English Text Society by Oxford University Press, 1992).

Severus, Sulpicius, *Vita di Martino, Vita di Ilarione, in memoria di Paola*, ed. Christina Mohrmann, Antonius A. R. Bastiaensen, Jan W. Smit, Luca Canali, and Claudio Moreschini, second edition (Milan: Fondazione Lorenzo Valla - A. Mondadori, 1983).

_____, *Dialogi, Patrologia Latina* 20, ed. Jacques-Paul Migne (Paris: J-P. Garnier Frères, 1845; repr. 1993).

Stryker, William G., 'The Latin-Old English Glossary in MS. Cotton Cleopatra A.III', Unpub. Ph.D. Dissertation (Stanford University, 1951).

Swanton, Michael, ed., *The Dream of the Rood* (Exeter: Exeter University Press, 1987; rev. repr. 1996).

Virgil, *P. Vergili Maronis Opera*, ed. Roger A. B. Mynors (Oxford: Oxford University Press, 1972).

Vita Sancti Joannis, Eleemosynarii, Patrologia Latina 73, ed. Jacques-Paul Migne (Paris: J-P. Garnier Frères, 1849).

Vita Sancti Onuphrii, *Patrologia Latina* 73, ed. Jacques-Paul Migne (Paris: J-P. Garnier Frères, 1849).

Yerkes, David, ed., *The Old English Life of Machutus* (Toronto: University of Toronto Press, 1984).

Zupitza, Julius, ed., *Ælfrics Grammatik und Glossar: Text und Varianten*, intro. Helmut Gneuss (Hildesheim: Weidmann, 2003).

Secondary Sources

Alexander, Dominic, *Saints and Animals in the Middle Ages* (Woodbridge: Boydell Press, 2008).

Andreopoulos, Andreas, *Metamorphosis: The Transfiguration in Byzantine Theology and Iconography* (Crestwood, N.Y.: St Vladimir's Seminary Press, 2005).

Anlezark, Daniel, 'The Anglo-Saxon World View', in *The Cambridge Companion to Old English Literature*, ed. Malcolm Godden and Michael Lapidge, 2nd edn (Cambridge: Cambridge University Press, 2013), pp. 66–81.

Appleton, Helen, 'Anglo-Saxon Landscapes: The Construction of the Environment in Old English and Related Texts', Unpub. Ph.D. Thesis (University of Sydney, 2012).

_____, 'The Old English *Durham* and the Cult of Cuthbert', *JEGP* 115 (2016), 346–69.

Barrar, Kathleen, 'A Spacious, Green and Hospitable Land: Paradise in Old English Poetry', *Bulletin of the John Rylands University Library of Manchester* 86 (2004), 105–25.

Berschin, Walter, '*Opus deliberatum ac perfectum*: Why Did the Venerable Bede Write a Second Prose Life of St Cuthbert?', in *St Cuthbert, His Cult and His Community to A.D. 1200*, ed. Gerald Bonner, David Rollason, and Clare Stancliffe (Woodbridge: Boydell Press, 1989), pp. 95–102.

Biggam, Carole Patricia, 'Sociolinguistic Aspects of Old English Colour Lexemes', *ASE* 24 (1995), 51–65.

Billett, Jesse D., *The Divine Office in Anglo-Saxon England, 597–c.1000* (Woodbridge and London: Boydell Press/Henry Bradshaw Society, 2014).

Bitel, Lisa M., *Isles of the Saints: Monastic Settlement and Christian Community in Early Ireland* (Ithaca, N.Y.: Cornell University Press, 1990).

Black, John R., 'Tradition and Transformation in the Cult of St. Guthlac in Early Medieval England', *The Heroic Age* 10 (2007), 1–22.

Bloomfield, Morton W.,'Patristics and Old English Literature: Notes on Some Poems', *Comparative Literature* 14 (1962), 36–43.

Bonner, Gerald, 'Bede and Medieval Civilization', *ASE* 2 (1973) 71–90.

Boyd, Nina, 'Doctrine and Criticism: A Revaluation of *Genesis A*', *Neuphilologische Mitteilungen* 83 (1982), 230–38.

Brady, Lindy, 'Echoes of Britons on a Fenland Frontier in the Old English *Andreas*', *RES* 61 (2010), 669–89.

Bridges, Margaret, 'The Heroic and Elegiac Contexts of Two Old English Laments of the Fallen Angels: Towards a Theory of Medieval Daemonization', *Swiss Papers in English Language and Literature* 4 (1988), 117–32.

Brooks, Britton, 'A New Source for the Anonymous *Vita S. Cuthberti*', *Notes and Queries* 62 (2015), 356–58.

_____, 'Felix's Construction of the English Fenlands: Literal Landscape, Authorizing Allusion, and Lexical Echo', in *Guthlac of Crowland: Celebrating 1300 Years*, ed. Jane Roberts and Alan Thacker (Paul Watkins Publishing, forthcoming 2019).

Brown, George Hardin, *A Companion to Bede* (Woodbridge: Boydell Press, 2009).

Bullough, Donald A., 'A Neglected Early-Ninth-Century Manuscript of the Lindisfarne *Vita S. Cuthberti*', *ASE* 27 (1998), 105–37.

Burt, Donald X., 'Courageous Optimism: Augustine on the Good of Creation', *Augustinian Studies* 21 (1990), 55–66.

Butler, Brian, 'The Whitby Life of Gregory the Great: Exegesis and Hagiography', Unpub. Ph.D. Thesis (University College Cork, 2005).

Carruthers, Mary, *The Book of Memory: A Study of Memory in Medieval Culture*, 2nd edn (Cambridge: Cambridge University Press, 2008).

Cavill, Paul, 'Some Dynamics of Story-Telling: Animals in the Early *Lives* of St Cuthbert', *Nottingham Medieval Studies* 43 (1999), 1–20.

Clanchy, Michael T., *From Memory to Written Record: England, 1066–1307*, 3rd edn (Oxford: Wiley-Blackwell, 2013).

Clark, Stephanie, 'A More Permanent Homeland: Land Tenure in *Guthlac A*', *ASE* 40 (2011), 75–102.

Clarke, Catherine A. M., *Literary Landscapes and the Idea of England, 700–1400* (Cambridge: D. S. Brewer, 2006).

_____, *Writing Power in Anglo-Saxon England: Texts, Hierarchies, Economies* (Cambridge: D. S. Brewer, 2012).

Classen, Albrecht, *Water in Medieval Literature: An Ecocritical Reading* (Lanham, Md: Lexington Books, 2017).

Clayton, Mary, 'Hermits and the Contemplative Life in Anglo-Saxon England', in *Holy Men and Holy Women: Old English Prose Saints' Lives and Their Contexts*, ed. Paul E. Szarmach (Albany, N.Y.: State University of New York Press, 1996), pp. 147–75.

Clemoes, Peter, *Interactions of Thought and Language in Old English Poetry* (Cambridge: Cambridge University Press, 1995).

Coates, Simon J., 'The Bishop as Pastor and Solitary: Bede and the Spiritual Authority of the Monk-Bishop', *Journal of Ecclesiastical History* 47 (1996), 601–19.

_____, 'The Construction of Episcopal Sanctity in Early Anglo-Saxon England: the Impact of Venantius Fortunatus', *Historical Research* 71 (1998), 1–13.

Cohen, Jeffrey Jerome, *Medieval Identity Machines* (Minneapolis and London: University of Minnesota Press, 2003).

Conner, Patrick W., 'Source Studies, the Old English *Guthlac A* and the English Benedictine Reformation', *Revue bénédictine* 103 (1993), 380–413.

Crane, Susan, *Animal Encounters: Contacts and Concepts in Medieval Britain* (Philadelphia: University of Pennsylvania Press, 2012).

Crépin, André, 'Bede and the Vernacular', in *Famulus Christi: Essays in Commemoration of the Thirteenth Centenary of the Birth of the Venerable Bede*, ed. Gerald Bonner (London: Society for Promoting Christian Knowledge, 1976), pp. 170–92.

Cubitt, Catherine, 'Memory and Narrative in the Cult of Early Anglo-Saxon Saints', in *The Uses of the Past in the Early Middle Ages*, ed. Yitzhak Hen and Matthew Innes (Cambridge: Cambridge University Press, 2000), pp. 29–66.

_____, 'Universal and Local Saints in Anglo-Saxon England', in *Local Saints and Local Churches in the Early Medieval West*, ed. Alan Thacker and Richard Sharpe (Oxford: Oxford University Press, 2002), pp. 423–53.

Curtius, Ernst Robert, *European Literature and the Latin Middle Ages*, trans. Willard R. Trask (Princeton, N.J.: Princeton University Press, 1990).

Dale, Corinne, *The Natural World in the Exeter Book Riddles* (Cambridge: D. S. Brewer, 2017).

Darby, Peter, *Bede and the End of Time* (Farnham: Ashgate, 2012).

DeGregorio, Scott, 'Affective Spirituality: Theory and Practice in Bede and Alfred the Great', *Essays in Medieval Studies* 22 (2005), 129–39.

_____, 'The Venerable Bede on Prayer and Contemplation', *Traditio* 54 (1999), 1–39.

Dendle, Peter, *Satan Unbound: The Devil in Old English Narrative Literature* (Toronto: University of Toronto Press, 2001).

Discenza, Nicole Guenther, *Inhabited Spaces: Anglo-Saxon Constructions of Place* (Toronto: University of Toronto Press, 2017).

Di Sciacca, Claudia, '*Concupita, quaesita, ac petita solitudinis secreta*: The Desert Ideal in Bede's *Vita S. Cuthberti* and Ælfric's Life of St Cuthbert', in *Hagiography in Anglo-Saxon England: Adopting and Adapting Saints' Lives into Old English Prose (c. 950–1150)*, ed. Loredana Lazzari, Patrizia Lendinara, and Claudia Di Sciacca (Barcelona: Fédération internationale des instituts d'études médiévales; Online edition by Brepols, 2014), pp. 121–81.

Downey, Sarah, 'Too much of too little: Guthlac and the Temptation of Excessive Fasting', *Traditio* 63 (2008), 89–127.

_____, 'Intertextuality in the Lives of St. Guthlac', Unpub. Ph.D. thesis (University of Toronto, 2004).

Downey, Sarah, Drout, Michael D. C., Kahn, Michael J., and LeBlanc, Mark D., '"Books Tell Us": Lexomic and Traditional Evidence for the Sources of *Guthlac A*', *Modern Philology* 110 (2012), 153–81.

Duncan, Sandra, '*Signa de Caelo* in the Lives of St Cuthbert: the Impact of Biblical Images and Exegesis on Early Medieval Hagiography', *The Heythrop Journal* 41 (2000), 399–412.

Edwards, Nancy, 'Celtic Saints and Early Medieval Archaeology', in *Local Saints and Local Churches in the Early Medieval West*, ed. Alan Thacker and Richard Sharpe (Oxford: Oxford University Press, 2002), pp. 225–65.

Estes, Heidi, *Anglo-Saxon Literary Landscapes: Ecotheory and the Environmental Imagination* (Amsterdam: Amsterdam University Press, 2017).

Evan, Peter, 'Word-Play as Evidence for the Date of *Durham*', *MÆ* 82 (2013), 314–17.

Foot, Sarah, *Monastic Life in Anglo-Saxon England: c. 600–900* (Cambridge: Cambridge University Press, 2006).

Fox, Michael, 'Ælfric on the Creation and Fall of the Angels', *ASE* 31 (2002), 175–200.

Foxhall Forbes, Helen, *Heaven and Earth in Anglo-Saxon England: Theology and Society in an Age of Faith* (Farnham: Ashgate, 2013).

Friesen, Bill, 'The *Opus Geminatum* and Anglo-Saxon Literature', *Neophilologus* 95 (2011), 123–44.

Fulk, Robert Dennis, *A History of Old English Meter* (Philadelphia: University of Pennsylvania Press, 1992).

Garde, Judith N. and Bernard J. Muir, 'Patristic Influence and the Poetic Intention in Old English Religious Verse', *Literature and Theology* 2 (1988), 49–68.

Gatch, Milton McC., 'Noah's Ravens in *Genesis A* and the Illustrated Old English Hexateuch', *Gesta* 14 (1975), 3–15.

Giliberto, Concetta, 'The *Descensus ad inferos* in the Old English Prose Life of St Guthlac and Vercelli Homily XXIII', in *Hagiography in Anglo-Saxon England: Adopting and Adapting Saints' Lives into Old English Prose (c. 950–1150)*, ed. Loredana Lazzari, Patrizia Lendinara, and Claudia Di Sciacca (Barcelona: Fédération internationale des instituts d'études médiévales; Online edition by Brepols, 2014), pp. 229–53.

Gittos, Helen, *Liturgy, Architecture, and Sacred Places in Anglo-Saxon England* (Oxford: Oxford University Press, 2013).

Gneuss, Helmut, and Michael Lapidge, eds, *Anglo-Saxon Manuscripts: a Bibliographical Handlist of Manuscripts and Manuscript Fragments Written or Owned in England up to 1100* (Toronto: University of Toronto Press, 2014).

Godman, Peter, 'The Anglo-Latin "Opus Geminatum": from Aldhelm to Alcuin', *MÆ* 50 (1981), 215–29.

Gretsch, Mechthild, *Ælfric and the Cult of Saints in Late Anglo-Saxon England* (Cambridge: Cambridge University Press, 2005, second ed. 2008).

Grossi, Joseph, 'Preserving the Future in the Old English *Durham*', *JEGP* 111 (2012), 42–73.

Haines, Dorothy, 'Vacancies in Heaven: The Doctrine of Replacement and *Genesis A*', *Notes and Queries* 44 (1997), 150–54.

Hall, Alaric, 'Constructing Anglo-Saxon Sanctity: Tradition, Innovation and Saint Guthlac', in *Images of Medieval Sanctity: Essays in Honour of Gary Dickson*, ed. Debra Higgs Strickland (Leiden: Brill, 2007), pp. 207–35.

Harris, Alexandra, *Weatherland: Writers and Artists under English Skies* (London: Thames and Hudson, 2015).

Harrison, Carol, *Beauty and Revelation in the Thought of Saint Augustine* (Oxford: Clarendon Press, 1992).

Hart-Hasler, Joan, 'Bede's Use of Patristic Sources: The Transfiguration', *Studia Patristica* 28 (1993), 197–204.

Harvey, Katherine, 'Episcopal Emotions: Tears in the Life of the Medieval Bishop', *Historical Research* 87 (2014), 591–610.

Hill, Thomas D., 'The Fall of Angels and Man in the Old English Genesis B', in *Anglo-Saxon Poetry: Essays in Appreciation, for John C. McGalliard*, ed. Lewis E. Nicholson and Dolores Warwick Frese (Notre Dame, Ind., and London: University of Notre Dame Press, 1975), pp. 279–90.

_____, 'The Middle Way: *Idel-Wuldor* and *Egesa* in the Old English *Guthlac A*', *RES* 30 (1979), 182–87.

Holder, Arthur G., 'Bede and the New Testament', in *The Cambridge Companion to Bede*, ed. Scott DeGregorio (Cambridge: Cambridge University Press, 2010), pp. 142–55.

Hooke, Della, *The Landscape of Anglo-Saxon England* (London: Leicester University Press, 1998).

Howe, Nicholas, *Across an Inland Sea: Writing in Place from Buffalo to Berlin* (Princeton, N.J., and Oxford: Princeton University Press, 2003).

_____, 'Looking for Home in Anglo-Saxon England', in *Home and Homelessness in the Medieval and Renaissance World*, ed. Nicholas Howe (Notre Dame, Ind.: University of Notre Dame Press, 2004), pp. 143–63.

_____, *Writing the Map of Anglo-Saxon England: Essays in Cultural Geography* (New Haven; London: Yale University Press, 2008).

Huppé, Bernard F., *Doctrine and Poetry: Augustine's Influence on Old English Poetry* (Albany, New York: State University of New York Press, 1959).

Jager, Eric, 'Speech and the Chest in Old English Poetry: Orality or Pectorality?', *Speculum* 65 (1990), 845–59.

_____, *The Tempter's Voice: Language and the Fall in Medieval Literature* (Ithaca, N.Y.: Cornell University Press, 1993).

Jayatilaka, Rohini, 'The *Regula Sancti Benedicti* in Late Anglo-Saxon England: The Manuscripts and Their Readers', Unpub. D.Phil. (Oxford University, 1996).

Jones, Christopher A., 'Ælfric and the Limits of "Benedictine Reform"', in *A Companion to Ælfric*, ed. Hugh Magennis and Mary Swan (Boston: Brill, 2009), pp. 67–108.

_____, 'Early Medieval Chaos', in *Verbal Encounters: Anglo-Saxon and Old Norse Studies for Roberta Frank*, ed. Antonina Harbus and Russell Poole (Toronto: University of Toronto Press, 2005), pp. 15–38.

_____, 'Envisioning the *Cenobium* in the Old English *Guthlac A*', *Mediaeval Studies* 57 (1995), 259–91.

Kabir, Ananya Jahanara, *Paradise, Death and Doomsday in Anglo-Saxon Literature* (Cambridge: Cambridge University Press, 2001).

Ker, Neil Ripley, *Catalogue of Manuscripts Containing Anglo-Saxon* (Oxford: Clarendon Press, 1957; repr. with suppl. 1990).

Kilpatrick, Kelly A., 'The Place-Names in Felix's *Vita sancti Guthlaci*', *Nottingham Medieval Studies* 58 (2014), 1–56.

Knibbs, Eric, 'Exegetical Hagiography: Bede's Prose *Vita Sancti Cuthberti*', *Revue bénédictine* 114 (2004), 233–52.

Kurtz, Benjamin P., 'From St. Antony to St. Guthlac: a Study in Biography', *University of California Publications in Modern Philology* 12 (1926), 103–46.

Ladner, Gerhart B., '*Homo Viator*: Medieval Ideas on Alienation and Order', *Speculum* 42 (1967), 233–59.

Lapidge, Michael, *Anglo-Latin Literature, 600–899* (London: Hambledon Press, 1996).

_____, 'Bede's "Metrical Vita S. Cuthberti"', in *St Cuthbert, His Cult and His Community to A.D. 1200*, ed. Gerald Bonner, David W. Rollason, and Clare Stancliffe (Woodbridge: Boydell Press, 1989), pp. 77–93.

_____, *The Anglo-Saxon Library* (Oxford: Oxford University Press, 2006).

_____, 'The Saintly Life in Anglo-Saxon England', in *The Cambridge Companion to Old English Literature*, 2nd edn, ed. Malcolm Godden and Michael Lapidge (Cambridge: Cambridge University Press, 2013), pp. 251–72.

_____, 'Versifying the Bible in the Middle Ages', in *The Text in the Community: Essays on Medieval Works, Manuscripts, Authors, and Readers*, ed. Jill Mann and Maura Nolan (Notre Dame, Ind.: University of Notre Dame Press, 2006), pp. 11–40.

Lee, Alvin A., *The Guest-Hall of Eden: Four Essays on the Design of Old English Poetry* (New Haven and London: Yale University Press, 1972).

Lee, Christina, 'Reluctant Appetites: Anglo-Saxon Attitudes towards Fasting', in *Saints and Scholars: New Perspectives on Anglo-Saxon Literature and Culture in Honour of Hugh Magennis*, ed. Stuart McWilliams (Cambridge: D. S. Brewer, 2012), pp. 164–86.

Leneghan, Francis, 'Making the Psalter Sing: the Old English *Metrical Psalms*, Rhythm and *Ruminatio*', in *The Psalms and Medieval English Literature: From the Conversion to the Reformation*, ed. Tamara Atkin and Francis Leneghan (Cambridge: D. S. Brewer, 2017).

_____, 'Preparing the Mind for Prayer: *The Wanderer*, *Hesychasm* and *Theosis*', *Neophilologus* 100 (2015), 121–42.

_____, 'Teaching the Teachers: The Vercelli Book and the Mixed Life', *English Studies* 94 (2013), 627–58.

Licence, Tom, *Hermits and Recluses in English Society 950–1200* (Oxford: Oxford University Press, 2011).

Lipp, Frances Randall, '*Guthlac A*: an Interpretation', *Mediaeval Studies* 33 (1971), 46–62.

Low, Soon-Ai, 'Approaches to the Old English Vocabulary for "Mind"', *Studia Neophilologica* 73 (2001), 11–22.

Loyn, Henry R., 'Kinship in Anglo-Saxon England', *ASE* 3 (1974), 197–209.

Lucas, Peter J., 'Easter, The Death of St Guthlac and the Liturgy for Holy Saturday in Felix's *Vita* and the Old English *Guthlac B*', *MÆ* 61 (1992), 1–16.

Lutterkort, Karl, 'Beda Hagiographicus: Meaning and Function of Miracle Stories in the *Vita Cuthberti* and the *Historia ecclesiastica*', in *Beda Venerabilis: Historian, Monk and Northumbrian*, ed. Luuk A. J. R. Houwen and Alasdair A. MacDonald (Groningen: Egbert Foresten, 1996), pp. 81–106.

Magennis, Hugh, 'Ælfric's Scholarship', in *A Companion to Ælfric*, ed. Hugh Magennis and Mary Swan (Boston: Brill, 2009), pp. 5–34.

_____, *Images of Community in Old English Poetry* (Cambridge: Cambridge University Press, 1996).

Marsden, Richard, 'Ælfric as Translator: The Old English Prose *Genesis*', *Anglia* 109 (1991), 319–58.

Martin, Lawrence T., 'The Influence of *Arator* in Anglo-Saxon England', *Proceedings of the PMR Conference* 7 (1982), 75–81.

Mathewes, Charles T., 'A Worldly Augustinianism: Augustine's Sacramental Vision of Creation', *Augustinian Studies* 41 (2010), 333–48.

_____, *Evil and the Augustinian Tradition* (Cambridge: Cambridge University Press, 2001).

Mathewson, David, *A New Heaven and a New Earth: the Meaning and Function of the Old Testament in Revelation 21.1–22.5* (London: Sheffield Academic Press, 2003).

McEntire, Sandra, 'The Doctrine of Compunction from Bede to Margery Kempe', in *The Medieval Mystical Tradition in England IV: The Exeter Symposium IV: Papers Read at Dartington Hall, July 1987*, ed. Marion Glasscoe (Cambridge: D. S. Brewer, 1987), pp. 77–90.

McKill, Larry N., 'Patterns of the Fall: Adam and Eve in the Old English *Genesis A*', *Florilegium* 14 (1995–96), 25–41.

Mckinnon, James W., 'The Origins of the Western Office', in *The Divine Office in the Latin Middle Ages: Methodology and Source Studies, Regional Developments, Hagiography*, ed. Margot Elsbeth Fassler and Rebecca Anne Baltzer (Oxford: Oxford University Press, 2000).

McMullen, A. Joseph, 'Rewriting the Ecclesiastical Landscape of Early Medieval Northumbria in the Lives of Cuthbert', *ASE* 43 (2014), 57–98.

Michelet, Fabienne, *Creation, Migration, and Conquest: Imaginary Geography and Sense of Space in Old English Literature* (Oxford: Oxford University Press, 2006).

Milfull, Inge B., *The Hymns of the Anglo-Saxon Church: A Study and Edition of the Durham Hymnal* (Cambridge: Cambridge University Press, 1996).

Mize, Britt, 'The Representation of the Mind as an Enclosure in Old English Poetry', *ASE* 35 (2006), 57–90.

Morton, Timothy, *Ecology Without Nature: Rethinking Environmental Aesthetics* (Cambridge, Mass. and London: Harvard University Press, 2007).

Neville, Jennifer, *Representations of the Natural World in Old English Poetry* (Cambridge: Cambridge University Press, 1999).

Newlands, Carole E., 'Bede and Images of Saint Cuthbert', *Traditio* 52 (1997), 73–109.

O'Brien O'Keeffe, Katherine, 'Guthlac's Crossings', *Quaestio: Selected Proceedings of the Cambridge Colloquium in Anglo-Saxon, Norse and Celtic* 2 (2001), 1–26.

Offler, H. S., 'The Date of Durham ("Carmen De Situ Dunelmi")', *JEGP* 61 (1962), 591–94.

O'Loughlin, Thomas, 'Living in the Ocean', in *Studies in the Cult of Saint Columba*, ed. Cormac Bourke (Dublin: Four Courts Press, 1997), pp. 11–23.

Orchard, Andy, 'Poetic Inspiration and Prosaic Translation: The Making of *Cædmon's Hymn*', in *Studies in English Language and Literature: 'Doubt Wisely', Papers in Honour of E.G. Stanley*, ed. Mary Jane Toswell and Elizabeth M. Tyler (London: Routledge, 1996), pp. 402–22.

_____, 'The Word Made Flesh: Christianity and Oral Culture in Anglo-Saxon Verse', *Oral Tradition* 24 (2009), 293–318.

O'Reilly, Jennifer, 'Reading the Scriptures in the Life of Columba', in *Studies in the Cult of Saint Columba*, ed. Cormac Bourke (Dublin: Four Courts Press, 1997), pp. 80–106.

O'Sullivan, Catherine Marie, *Hospitality in Medieval Ireland, 900–1500* (Dublin: Four Courts Press, 2004).

O'Sullivan, Tomás, 'The Miraculous Production of Water from Rock and the Impact of Exegesis on Early Irish Hagiography', *Eolas: The Journal of the American Society of Irish Medieval Studies* 3 (2009), 19–50.

Palmer, James M., '*Compunctio* and the Heart in the Old English Poem *The Wanderer*', *Neophilologus* 88 (2004), 447–60.

Pollmann, Karla, 'Human Sin and Natural Environment: Augustine's Two Positions on Genesis 3:18', *Augustinian Studies* 41 (2010), 69–85.

Pont, Jeanne-Marie, 'Homo angelorum decimus ordo', *Cahiers de civilisation médiévale* 31 (1988), 43–48.

Powell, Hilary, 'Sanctifying Landscapes: Topographical References in Anglo-Latin Saints' Lives', Unpub. D.Phil. (University of Oxford, 2007).

Reichardt, Paul F., 'Guthlac A and the Landscape of Spiritual Perfection', *Neophilologus* 58 (1974), 331–38.

Reinmuth, Oskar W., 'Vergil's Use of *Interea*, a Study of the Treatment of Contemporaneous Events in Roman Epic', *The American Journal of Philology* 54 (1933), 323–39.

Reuling, Hanneke, *After Eden: Church Fathers and Rabbis on Genesis 3:16–21* (Leiden and Boston: Brill, 2006).

Riedinger, Anita R., '*Home* in Old English Poetry', *Neuphilogische Mitteilungen* 96 (1995), 51–59.

Roberts, Jane, '*Guthlac A*: Sources and Source Hunting', in *Medieval English Studies Presented to George Kane*, ed. Edward Donald Kennedy, Ronald Waldron, and Joseph S. Wittig (Cambridge: D. S. Brewer, 1988), pp. 1–18.

_____, 'Hagiography and Literature: The Case of Guthlac of Crowland', in *Mercia: an Anglo-Saxon Kingdom in Europe*, ed. Michelle P. Brown and Carol A. Farr (London and New York: Leicester University Press, 2001), pp. 69–86.

_____, '*The Old English Prose Translation of Felix's* Vita Sancti Guthlaci', in *Studies in Earlier Old English Prose: Sixteen Original Contributions*, ed. Paul E. Szarmach (Albany, N.Y.: State University of New York Press, 1986), pp. 363–79.

Rosenthal, Joel T., 'Bede's Use of Miracles in "The Ecclesiastical History"', *Traditio* 31 (1975), 328–35.

Rudolf, Winfried, 'The Spiritual Islescape of the Anglo-Saxons', in *The Sea and Englishness in the Middle Ages: Maritime Narratives, Identity and Culture*, ed. Sebastian I. Sobecki (Cambridge: D. S. Brewer, 2011), pp. 31–57.

Schmidt, Gary, D., *The Iconography of the Mouth of Hell: Eighth-Century Britain to the Fifteenth Century* (Selinsgrove, Pa.: Susquehanna University Press, 1995).

Scragg, Donald G., 'The Corpus of Vernacular Homilies and Prose Saints' Lives before Ælfric', *ASE* 8 (1979) 223–77.

Semple, Sarah, 'A Fear of the Past: The Place of the Prehistoric Burial Mound in the Ideology of Middle and Later Anglo-Saxon England', *World Archaeology* 30 (1998), 109–26.

_____, 'Anglo-Saxon Attitudes to the Past: A Landscape Perspective: a Study of the Secondary Uses and Perceptions of Prehistoric Monuments in Anglo-Saxon Society', Unpub. D.Phil. (Oxford University, 2002).

Shockro, Sally, 'Bede and the Rewriting of Sanctity', *The Haskins Society Journal* 21 (2009), 1–19.

_____, 'Saints and Holy Beasts: Pious Animals in Early Medieval Insular Saints' *Vitae*', in *Animal Languages in the Middle Ages: Representations of Interspecies Communication*, ed. Alison Langdon (Basingstoke: Palgrave Macmillan, 2018), pp. 51–68.

Shook, Laurence K., 'The Burial Mound in "Guthlac A"', *Modern Philology* 58 (1960), 1–10.

Siewers, Alfred K., 'Landscapes of Conversion: Guthlac's Mound and Grendel's Mere as Expressions of Anglo-Saxon Nation-Building', *Viator* 34 (2003), 1–39.

Smith, Scott T., *Land and Book: Literature and Land Tenure in Anglo-Saxon England* (Toronto: University of Toronto Press, 2012).

Stancliffe, Clare, 'Cuthbert and the Polarity between Pastor and Solitary', in *St Cuthbert, His Cult and His Community to A.D. 1200*, ed. Gerald Bonner, David Rollason, and Clare Stancliffe (Woodbridge: Boydell Press, 1989), pp. 21–44.

_____, 'Red, White and Blue Martyrdom', in *Ireland in Early Mediaeval Europe: Studies in Memory of Kathleen Hughes*, ed. Dorothy Whitelock, Rosamond McKitterick, and David N. Dumville (Cambridge: Cambridge University Press, 1982), pp. 21–46.

Stapleton, Paul J., '*Kontrastimitation* and Typology in Alcuin's York Poem', *Viator* 43 (2012), 67–78.

Steen, Janie, *Verse and Virtuosity: the Adaptation of Latin Rhetoric in Old English Poetry* (Toronto: University of Toronto Press, 2008).

Stevenson, Jane Barbara, 'The Monastic Rules of Columbanus', in *Columbanus: Studies on the Latin Writings*, ed. Michael Lapidge (Woodbridge: Boydell Press, 1997), pp. 203–16.

Stewart, Columba, 'Benedictine Monasticism and Mysticism', in *The Wiley-Blackwell Companion to Christian Mysticism*, ed. Julia A. Lamm (Oxford: Wiley-Blackwell, 2013), pp. 216–34.

Stuart, Anthony J., and Lister, Adrian M., 'Extinction Chronology of the Cave Lion *Panthera spelaea*', *Quaternary Science Reviews* 30 (2011), 2329–40.

Sykes, Naomi and O'Connor, Terence Patrick, eds, *Extinctions and Invasions: A Social History of British Fauna* (Oxford: Windgather Press, 2010).

Szarmach, Paul E., 'The Vercelli Prose and Anglo-Saxon Literary History', in *New Readings in the Vercelli Book*, ed. Andy Orchard and Samantha Zacher (Toronto, London: University of Toronto Press, 2009), pp. 12–40.

Thacker, Alan, 'Bede and History', in *The Cambridge Companion to Bede*, ed. Scott DeGregorio (Cambridge: Cambridge University Press, 2010), pp. 170–89.

_____, 'Bede's Ideal of Reform', in *Ideal and Reality in Frankish and Anglo-Saxon Society: Studies Presented to J. M. Wallace-Hadrill*, ed. Patrick Wormald, with Donald A. Bullough and Roger Collins (Oxford: Blackwell, 1983), pp. 130–53.

_____, 'Lindisfarne and the Origins of the Cult of Saint Cuthbert', in *St Cuthbert, His Cult and His Community to A.D. 1200*, ed. Gerald Bonner, David Rollason, and Clare Stancliffe (Woodbridge: Boydell Press, 1989), pp. 103–22.

_____, 'The Making of a Local Saint', in *Local Saints and Local Churches in the Early Medieval West*, ed. Alan Thacker and Richard Sharpe (Oxford: Oxford University Press, 2002), pp. 45–73.

_____, 'The Social and Continental Background to Early Anglo-Saxon Hagiography', Unpub. D.Phil. (Oxford University, 1977).

Thornbury, Emily, *Becoming a Poet in Anglo-Saxon England* (Cambridge: Cambridge University Press, 2014).

Toswell, Mary Jane, *The Anglo-Saxon Psalter* (Turnhout: Brepols, 2014).

Van der Walt, Andries G. P., 'Reflections of the Benedictine Rule in Bede's Homiliary', *Journal of Ecclesiastical History* 37 (1986), 367–76.

Voss, Manfred, 'Strykers Edition des alphabetischen Cleopatraglossars: Corrigenda und Addenda', *Arbeiten aus Anglistik und Amerikanistik* 13 (1988), 123–38.

Ward, Benedicta, 'The Spirituality of St. Cuthbert', in *St Cuthbert, His Cult and His Community to A.D. 1200*, ed. Gerald Bonner, David Rollason, and Clare Stancliffe (Woodbridge: Boydell Press, 1989), pp. 65–76.

Waugh, Robin, 'The Blindness Curse and Nonmiracles in the Old English Prose Life of Saint Guthlac', *Modern Philology* 106 (2009), 399–426.

Webb, Diana, *Pilgrimage in Medieval England* (London: Hambledon, 2000).

Weber, Benjamin D., 'A Harmony of Contrasts: The Guthlac Poems of the Exeter Book', *JEGP* 114 (2015), 201–18.

Wehlau, Ruth, 'Rumination and Re-Creation: Poetic Instruction in *The Order of the World*', *Florilegium* 13 (1994), 65–77.

Whatley, E. Gordon, 'Lost in Translation: Omission of Episodes in some Old English Prose Saints' Legends', *ASE* 26 (1997), 187–208.

Whyte, Nicola, 'The After-Life of Barrows: Prehistoric Monuments in the Norfolk Landscape', *Landscape History* 25 (2003), 5–16.

Wickham-Crowley, Kelley M., 'Living on the *Ecg*: The Mutable Boundaries of Land and Water in Anglo-Saxon Contexts', in *A Place to Believe In: Locating Medieval Landscapes*, ed. Clare A. Lees and Gillian R. Overing (University Park, Pa.: Pennsylvania State University Press, 2006), pp. 85–110.

Wieland, Gernot, '*Geminus stilus*: Studies in Anglo-Latin Hagiography', in *Insular Latin Studies: Papers on Latin Texts and Manuscripts of the British Isles: 550–1066*, ed. Michael W. Herren (Toronto: Pontifical Institute of Mediaeval Studies, 1981), pp. 113–33.

Woolf, Rosemary E., 'The Devil in Old English Poetry', *RES* 4 (1953), 1–12.

Wright, Charles D., '*Genesis A ad litteram*', in *Old English Literature and the Old Testament*, ed. Michael Fox and Manish Sharma (Toronto and London: University of Toronto Press, 2012), pp. 121–71.

_____, *The Irish Tradition in Old English Literature* (Cambridge: Cambridge University Press, 1993).

Yalden, Derek William, with contributions by Priscilla Barrett, *The History of British Mammals* (London: Poyser, 1999).

Zanker, Graham, 'Enargeia in the Ancient Criticism of Poetry', *Rheinisches Museum für Philologie* 124 (1981), 291–311.

Dictionaries and Databases

An Anglo-Saxon Dictionary Based on the Manuscript Collections of the Late Joseph Bosworth, ed. by Thomas Northcote Toller (Oxford: Clarendon Press, 1898); *Supplement*, ed. by Thomas Northcote Toller (Oxford: Clarendon Press, 1921); *Revised and Enlarged Addenda*, ed. by Alistair Campbell (Oxford: Oxford University Press, 1972).

Dictionary of Medieval Latin from British Sources Online (Brepols): http://clt.brepolis.net/dmlbs/Default.aspx

Dictionary of Old English: A to I online, ed. by Angus F. Cameron, Ashley Crandell Amos, and Antonette diPaolo Healey (pub. online, 2018): http://tapor.library.utoronto.ca/doe/dict/index.html

Dictionary of Old English Web Corpus, compiled by Antonette diPaolo Healey with John Price Wilkin and Xin Xiang (Toronto: Dictionary of Old English Project, 2009): https://tapor.library.utoronto.ca/doecorpus/

Fontes Anglo-Saxonici Project, ed., *Fontes Anglo-Saxonici: World Wide Web Register*: http://fontes.english.ox.ac.uk/

Oxford Latin Dictionary, ed. Peter G. W. Glare (Oxford: Clarendon Press, 1982).

Perseus Digital Library, ed. Gregory R. Crane (Tufts University): http://www.perseus.tufts.edu

Revised Medieval Latin Word-List: from British and Irish Sources, ed. Ronald E. Latham (Oxford: Published for the British Academy by Oxford University Press, 1965).

The Electronic Sawyer, Peter Sawyer, *Anglo-Saxon Charters: an Annotated List and Bibliography*, Royal Historical Society Guides and Handbooks 8 (London, 1968): http://www.esawyer.org.uk/

Index